TABLE OF CONTENTS

LIST OF FIGURES vii
LIST OF TABLES xiii
LIST OF SYMBOLS xv
LIST OF ACRONYMS xix
GLOSSARY xxi
ACKNOWLEDGMENTS xxxvii

CHAPTER 1 - INTRODUCTION 1.1

1.1 INTRODUCTION 1.1

1.1.1 Background 1.1
1.1.2 Purpose 1.1
1.1.3 History of Bridge Hydraulics 1.2

1.2 HYDRAULIC ANALYSIS OVERVIEW 1.3
1.3 MANUAL ORGANIZATION 1.4

1.3.1 Chapter 1 - Introduction 1.4
1.3.2 Chapter 2 - Design Considerations and Regulatory Requirements ... 1.4
1.3.3 Chapter 3 - Governing Equations and Flow Classification 1.4
1.3.4 Chapter 4 - Hydraulic Analysis Considerations 1.5
1.3.5 Chapter 5 - One-Dimensional Bridge Hydraulic Analysis 1.5
1.3.6 Chapter 6 - Two-Dimensional Bridge Hydraulic Analysis 1.5
1.3.7 Chapter 7 - Unsteady Flow Analysis 1.5
1.3.8 Chapter 8 - Bridge Scour Considerations and Scour Countermeasure Hydraulic Analysis 1.5
1.3.9 Chapter 9 - Sediment Transport and Alluvial Channel Concepts 1.5
1.3.10 Chapter 10 - Other Considerations 1.6
1.3.11 Chapter 11 - Literature Cited 1.6

1.4 DUAL SYSTEM OF UNITS 1.6

CHAPTER 2 - DESIGN CONSIDERATIONS AND REGULATORY REQUIREMENTS 2.1

2.1 INTRODUCTION 2.1
2.2 BRIDGE OPENING AND ROAD GRADE DESIGN CONSIDERATIONS 2.1
2.3 FLOODPLAIN AND FLOODWAY REGULATIONS 2.2
2.4 SCOUR AND STREAM STABILITY CONSIDERATIONS AND GUIDANCE ... 2.4
2.5 NAVIGATION REQUIREMENTS 2.5
2.6 SECTION 404 REQUIREMENTS 2.5
2.7 AASHTO SPECIFICATIONS AND DESIGN CRITERIA 2.6

CHAPTER 3 - GOVERNING EQUATIONS AND FLOW CLASSIFICATION 3.1

3.1 INTRODUCTION 3.1

3.1.1 Streamlines and Streamtubes 3.1
3.1.2 Definitions 3.2
3.1.3 Classification of Open Channel Flow 3.3

3.2	THREE BASIC EQUATIONS OF OPEN CHANNEL FLOW	3.4
3.2.1	Conservation of Mass	3.5
3.2.2	Conservation of Energy	3.8
3.2.3	Conservation of Linear Momentum	3.10
3.3	FLOW RESISTANCE AND OTHER HYDRAULIC EQUATIONS	3.13
3.3.1	Flow Resistance	3.13
3.3.2	Drag Force	3.20
3.3.3	Weir Flow	3.23
3.3.4	Gate and Orifice Equations	3.28
3.4	FLOW CLASSIFICATION	3.30
3.4.1	Steady Versus Unsteady Flow	3.30
3.4.2	Subcritical Versus Supercritical Flow	3.32
3.4.3	Uniform, Gradually, and Rapidly Varied Flow	3.37
3.4.4	Profiles for Gradually Varied Flow	3.42

CHAPTER 4 - HYDRAULIC ANALYSIS CONSIDERATIONS **4.1**

4.1	INTRODUCTION	4.1
4.2	HYDRAULIC MODELING CRITERIA AND SELECTION	4.1
4.2.1	One-Dimensional Versus Two-Dimensional Modeling	4.1
4.2.2	Steady Versus Unsteady Flow Modeling	4.9
4.2.3	Three-Dimensional Modeling and Computational Fluid Dynamics	4.9
4.2.4	Physical Modeling	4.11
4.3	SELECTING UPSTREAM AND DOWNSTREAM MODEL EXTENT	4.12
4.4	IDENTIFYING AND SELECTING MODEL BOUNDARY CONDITIONS	4.13
4.4.1	Water Surface	4.14
4.4.2	Normal Depth and Energy Slope	4.14
4.4.3	Rating Curve	4.14
4.4.4	Critical Depth	4.15
4.5	RIVERINE HYDROLOGIC INFORMATION	4.15
4.6	NUMERICAL MODEL EVALUATION	4.15
4.7	DATA REQUIREMENTS AND SOURCES	4.16

CHAPTER 5 - ONE-DIMENSIONAL BRIDGE HYDRAULIC ANALYSIS **5.1**

5.1	INTRODUCTION	5.1
5.2	HDS 1 METHOD	5.1
5.3	WATER SURFACE PROFILE COMPUTATIONS	5.4
5.3.1	Standard-Step Methods	5.4
5.3.2	Other Water Surface Profile Methods	5.8
5.3.3	Mixed-Flow Regime	5.9

| 5.4 | CROSS-SECTION SUBDIVISION AND INEFFECTIVE FLOW | 5.10 |

| 5.4.1 | Cross Section Subdivision | 5.10 |
| 5.4.2 | Ineffective Flow | 5.13 |

| 5.5 | FLOW CONSTRICTIONS AND CROSS-SECTION PLACEMENT | 5.14 |

5.5.1	Effects of Highway Crossings	5.14
5.5.2	Cross Section Placement	5.15
5.5.3	Ineffective Flow Specifications at Bridges	5.17

| 5.6 | BRIDGE HYDRAULIC CONDITIONS | 5.17 |

5.6.1	Free-Surface Bridge Flow	5.18
5.6.2	Overtopping-Flow	5.18
5.6.3	Flow Submerging the Bridge Low Chord	5.19

| 5.7 | BRIDGE MODELING APPROACHES | 5.20 |

5.7.1	Modeling Approaches for Free-Surface Bridge Flow Conditions	5.20
5.7.2	Selection of Free-Surface Bridge Flow Modeling Approach	5.26
5.7.3	Modeling Approaches for Overtopping and Orifice Bridge Flow	5.27
5.7.4	Selection of the Overtopping and Submerged-Low-Chord Modeling Approach	5.31

| 5.8 | SPECIAL CASES IN ONE-DIMENSIONAL HYDRAULICS | 5.33 |

5.8.1	Skewed Crossing Alignment	5.33
5.8.2	Crossings with Parallel Bridges	5.34
5.8.3	Split Flow Conditions	5.36
5.8.4	Crossings with Multiple Openings in the Embankment	5.37
5.8.5	Lateral Weirs	5.39

| 5.9 | ONE-DIMENSIONAL MODELING ASSUMPTIONS AND LIMITATIONS | 5.41 |

5.9.1	Water Surface, Velocity, and Cross Section Orientation	5.42
5.9.2	Total Energy and Flow Distribution	5.42
5.9.3	Cross Section Spacing	5.43
5.9.4	Assumptions for Multiple Openings	5.44
5.9.5	Cross-Section, Bridge, and Pier Skew	5.44

CHAPTER 6 - TWO-DIMENSIONAL BRIDGE HYDRAULIC ANALYSIS 6.1

6.1	INTRODUCTION	6.1
6.2	GOVERNING EQUATIONS	6.2
6.3	TYPES OF TWO-DIMENSIONAL MODELS	6.3

| 6.3.1 | Finite Element Method | 6.3 |
| 6.3.2 | Finite Difference Method | 6.5 |

| 6.4 | GEOMETRIC REQUIREMENTS AND MESH QUALITY | 6.6 |

| 6.4.1 | Geometric Requirements | 6.6 |
| 6.4.2 | Mesh Quality | 6.9 |

6.5	BRIDGE MODELING APPROACHES	6.11
6.5.1	Pier Losses	6.11
6.5.2	Pressure Flow	6.13
6.5.3	Weir Flow and Road Overtopping	6.14
6.6	MULTIPLE OPENINGS IN AN EMBANKMENT	6.15
6.7	MULTIPLE OPENINGS IN SERIES	6.15
6.8	UPSTREAM FLOW DISTRIBUTION	6.16
6.9	SPECIAL CASES IN TWO-DIMENSIONAL MODELING	6.16
6.9.1	Split Flow	6.16
6.9.2	Lateral Weirs	6.17
6.9.3	Culverts	6.17
6.9.4	Debris	6.18
6.10	TWO-DIMENSIONAL MODELING ASSUMPTIONS AND REQUIREMENTS	6.18
6.10.1	Gradually Varied Flow	6.18
6.10.2	Flow Distribution and Water Surface at Boundaries	6.18
6.10.3	Model Step-Down and Convergence	6.18
6.10.4	Wetting and Drying	6.19

CHAPTER 7 - UNSTEADY FLOW ANALYSIS 7.1

7.1	INTRODUCTION	7.1
7.1.1	Unsteady Flow Equations - Saint-Venant Equations	7.1
7.1.2	Unsteady Continuity Equation	7.3
7.1.3	Dynamic Momentum Equation	7.5
7.2	MODEL UPSTREAM AND DOWNSTREAM EXTENTS	7.7
7.3	DIFFERENCES BETWEEN STEADY AND UNSTEADY FLOW ANALYSIS	7.8
7.4	SOLUTION SCHEMES FOR THE SAINT-VENANT EQUATIONS	7.10
7.5	CHANNEL AND FLOODPLAIN CROSS SECTION	7.12
7.6	STORAGE AREAS AND CONNECTIONS	7.14
7.7	HYDRAULIC PROPERTY TABLES	7.14
7.8	NUMERICAL STABILITY	7.18
7.9	TWO-DIMENSIONAL UNSTEADY FLOW MODELS	7.20
7.10	TIDAL WATERWAYS	7.21

CHAPTER 8 - BRIDGE SCOUR CONSIDERATIONS AND SCOUR COUNTERMEASURE HYDRAULIC ANALYSIS 8.1

8.1	INTRODUCTION	8.1
8.2	SCOUR CONCEPTS FOR BRIDGE DESIGN	8.2
8.3	TYPES OF SCOUR	8.3
8.3.1	Contraction Scour	8.3
8.3.2	Local Scour	8.7
8.3.3	Debris Scour	8.10
8.3.4	Channel Instability	8.12
8.3.5	Evaluating Channel Instability	8.14

| 8.4 | COMPUTING SCOUR | 8.14 |

| 8.4.1 | One-Dimensional Models | 8.15 |
| 8.4.2 | Two-Dimensional Models | 8.16 |

| 8.5 | HYDRAULIC ANALYSIS OF SCOUR COUNTERMEASURES | 8.17 |

8.5.1	Revetments and Vegetation	8.18
8.5.2	Guide Banks	8.18
8.5.3	Spurs and Bendway Weirs	8.20

CHAPTER 9 - SEDIMENT TRANSPORT AND ALLUVIAL CHANNEL CONCEPTS ... 9.1

9.1	INTRODUCTION	9.1
9.2	SEDIMENT CONTINUITY	9.2
9.3	SEDIMENT TRANSPORT CONCEPTS	9.4

9.3.1	Initiation of Motion	9.4
9.3.2	Modes of Sediment Transport	9.5
9.3.3	Bed-Forms	9.7

9.4	OVERVIEW OF SEDIMENT TRANSPORT EQUATIONS AND SELECTION	9.9
9.5	OVERVIEW OF SEDIMENT TRANSPORT MODELING	9.11
9.6	ALLUVIAL FANS	9.14

CHAPTER 10 - OTHER CONSIDERATIONS ... 10.1

| 10.1 | HYDRAULIC FORCES ON BRIDGE ELEMENTS | 10.1 |

10.1.1	General	10.1
10.1.2	Hydrostatic Force	10.1
10.1.3	Buoyancy Force	10.1
10.1.4	Stream Pressure and Lift	10.2
10.1.5	Wave Forces	10.5
10.1.6	Effects of Debris	10.6
10.1.7	Effects of Ice	10.7
10.1.8	Vessel Collision	10.8

| 10.2 | BACKWATER EFFECTS OF BRIDGE PIERS | 10.8 |
| 10.3 | COINCIDENT FLOWS AT CONFLUENCES | 10.9 |

| 10.3.1 | Significance of Coincident Flows at Confluences | 10.9 |
| 10.3.2 | Available Guidance | 10.10 |

| 10.4 | ADVANCED BRIDGE HYDRAULICS MODELING | 10.11 |

10.4.1	Background	10.11
10.4.2	Applications of Advanced Modeling	10.11
10.4.3	Example of Advanced Modeling Applied to Bridge Design Projects	10.14
10.4.4	Limitations of Advanced Modeling	10.15

10.5	BRIDGE DECK DRAINAGE DESIGN	10.16
10.5.1	Objectives of Bridge Deck Drainage Design	10.16
10.5.2	Bridge Deck Drainage Considerations	10.17
10.5.3	Design Rainfall Intensity	10.18
10.5.4	Practical Considerations in Design of Bridge Deck Inlets and Drainage Systems	10.18

CHAPTER 11 - LITERATURE CITED **11.1**

APPENDIX A - Metric System, Conversion Factors, and Water Properties A.1

LIST OF FIGURES

Figure 3.1. Flow in the X-Z plane and flow in terms of streamline and normal coordinates 3.2

Figure 3.2. Example of uniform flow ($Y_2 = Y_1$) 3.3

Figure 3.3. Example of nonuniform flow where the depth of flow $Y_2 \neq Y_1$ 3.4

Figure 3.4. Streamtube with fluid flowing from Section 1 to Section 2 .... 3.5

Figure 3.5. Net flow through a control volume 3.7

Figure 3.6. Surfaces forces acting on a fluid element in the X and Y directions for an inviscid fluid 3.8

Figure 3.7. Nonuniform velocity distribution 3.10

Figure 3.8. The control volume for conservation of linear momentum 3.11

Figure 3.9. Forces acting on a control volume for uniform flow conditions 3.13

Figure 3.10. Floodplain roughness example 3.19

Figure 3.11. Looking upstream from the right bank on Indian Fork, near New Cumberland, Ohio 3.20

Figure 3.12. An example of debris blockage on piers 3.22

Figure 3.13. An example of upstream bridge cross section with debris accumulation on a single pier 3.22

Figure 3.14. Weir flow over a sharp crested weir 3.23

Figure 3.15. Example of simplified weir flow for a sharp crested weir 3.24

Figure 3.16. Ogee spillway crest 3.25

Figure 3.17. Discharge coefficients for vertical-face ogee crest 3.26

Figure 3.18. Ratio of discharge coefficients caused by tailwater effects 3.26

Figure 3.19. Broad-crested weir 3.27

Figure 3.20. Discharge reduction factor versus percent of submergence 3.28

Figure 3.21. Sluice and tainter gates 3.29

Figure 3.22. Flow classification according to change in depth with respect to space and time 3.31

Figure 3.23. Propagation of a water wave in shallow water illustrating subcritical and supercritical flow 3.33

Figure 3.24. Specific energy description 3.33

Figure 3.25. Specific energy diagram for a constant discharge 3.35

Figure 3.26. Definition sketch of uniform flow 3.38

Figure 3.27. Definition sketch of a typical non-uniform water surface profile 3.38

Figure 3.28. Hydraulic jump 3.40

Figure 3.29. Control volume for the hydraulic jump 3.40

Figure 3.30. Water surface slopes versus channel bottom slope 3.43

Figure 3.31. Flow profile curves 3.44

Figure 3.32. Example flow profiles for gradually varied flow with a change in slope 3.46

Figure 4.1. Two-dimensional model velocities, US 1 crossing Pee Dee River 4.3

Figure 4.2. Backwater at a skewed crossing 4.4

Figure 4.3. Two-dimensional model velocities, I-35W over Mississippi River 4.5

Figure 4.4. Channel network, Altamaha Sound, Georgia 4.6

Figure 4.5. Two-dimensional flow within a bridge opening 4.7

Figure 4.6. Three-dimensional CFD modeling (a) flow prior to scour (b) flow at ultimate scour 4.10

Figure 4.7. Bed shear from LES prior to scour (a) mean bed shear (b) fluctuation of bed shear 4.11

Figure 4.8. Physical model of the I-90 Bridge over Schoharie Creek, New York 4.12

Figure 4.9. Flow profiles with downstream boundary uncertainty 4.13

Figure 5.1. Sketch illustrating positions of Cross Sections 1 through 4 in HDS 1 backwater method 5.2

Figure 5.2. Cross section layout for a standard-step floodplain model 5.4

Figure 5.3. Illustration of water surface profile between two cross sections 5.5

Figure 5.4. Illustration of the friction slope (the slope of the energy grade line) at each cross section in a stream segment 5.7

Figure 5.5. Example mixed-flow regime profile in HEC-RAS 5.11

Figure 5.6. Example of a properly subdivided cross section 5.12

Figure 5.7. Discontinuity of computed conveyance for a non-subdivided cross section .. 5.12

Figure 5.8. Example of the use of ineffective flow at a bridge 5.14

Figure 5.9. Illustration of flow transitions upstream and downstream of a bridge crossing 5.16

Figure 5.10. Illustration of free-surface bridge flow Classes A, B, and C 5.19

Figure 5.11. Plan view layout showing cross section identifiers referenced by the bridge hydraulics equations 5.21

Figure 5.12. Cross section view illustrating the area and $\overline{Y}$ variables in momentum equation 5.22

Figure 5.13. Profile view with definitions of variables in the Yarnell Equation 5.24

Figure 5.14. Profile view with definition of terms in the WSPRO Method 5.25

Figure 5.15. Illustration of flow overtopping a roadway embankment 5.28

Figure 5.16. Guidance on discharge coefficients for flow over roadway, embankments, from HDS 1 .... 5.29

Figure 5.17. Sketch of orifice bridge flow 5.29

Figure 5.18. Relationship between orifice bridge flow discharge coefficient and submergence of the low chord, from HDS 1 .... 5.30

Figure 5.19. Sketch of submerged-orifice bridge flow 5.31

Figure 5.20. Illustration of a skewed bridge crossing 5.34

Figure 5.21. Aerial image of Interstate 70 over the Colorado River 5.35

Figure 5.22. Backwater multiplication factor for parallel bridges 5.35

Figure 5.23. Illustration of a cross-section layout to model split flow conditions 5.36

Figure 5.24. Plan view sketch of a multiple-opening bridge crossing 5.38

Figure 5.25. HEC-RAS multiple-opening example 5.39

Figure 5.26. Illustration of a model incorporating lateral weir flow 5.40

Figure 5.27. One-dimensional model cross sections 5.42

Figure 5.28. Cross section location at abrupt vegetation transitions 5.44

Figure 6.1.	Three-dimensional coordinate system and two-dimensional hydraulic variables	6.1
Figure 6.2.	Terms in two-dimensional momentum transport equations, x direction	6.3
Figure 6.3.	Element types and shapes	6.4
Figure 6.4.	Example finite element network layout	6.5
Figure 6.5.	Example finite difference network	6.6
Figure 6.6.	Example surface of finite element network	6.7
Figure 6.7.	Topography and boundary of a finite element network	6.8
Figure 6.8.	Illustration of curved mesh boundary	6.8
Figure 6.9.	Finite element mesh quality considerations	6.10
Figure 6.10.	Disabled element approach for large piers	6.12
Figure 6.11.	Pier drag force	6.12
Figure 6.12.	Weir and culvert connections	6.14
Figure 6.13.	Two-dimensional model velocities with multiple bridge openings	6.15
Figure 6.14.	Illustration of split flow two-dimensional model	6.17
Figure 7.1.	Unsteady flow hydrographs	7.2
Figure 7.2.	Unsteady solution of discharge versus x and t	7.2
Figure 7.3.	x-t plane illustrating locations for known and computed values	7.3
Figure 7.4.	Definition sketch for continuity equation	7.4
Figure 7.5.	Channel cross section relating topwidth to area	7.4
Figure 7.6.	Approximations of the Momentum Equation	7.7
Figure 7.7.	Study limits upstream and downstream	7.8
Figure 7.8.	Looped rating curves showing the rising and recession limbs of a hydrograph	7.9
Figure 7.9.	Discrete x-t solution domain for any variable	7.11
Figure 7.10.	Channel and floodplain flows	7.13
Figure 7.11.	Routed hydrograph characteristics	7.15

Figure 7.12. Illustration of an off-line storage area using the HEC-RAS computed model 7.15

Figure 7.13. Cross section hydraulic table increments 7.16

Figure 7.14. Conveyance properties versus elevation for a single cross section 7.17

Figure 7.15. Channel cross section for an unsteady flow model 7.18

Figure 7.16. Conveyance properties versus elevation for a single cross section with right overbank subdivided 7.19

Figure 8.1. Velocity and streamlines at a bridge constriction 8.2

Figure 8.2. Live-bed contraction scour variables 8.5

Figure 8.3. Clear-water contraction scour variables 8.6

Figure 8.4. Vertical contraction scour 8.7

Figure 8.5. The main flow features forming the flow field at a cylindrical pier 8.8

Figure 8.6. Flow structure in floodplain and main channel at a bridge opening 8.9

Figure 8.7. View down at debris and scour hole at upstream end of pier 8.10

Figure 8.8. Idealized flow pattern and scour at pier with debris 8.11

Figure 8.9. Debris in HEC-RAS hydraulic model 8.11

Figure 8.10. Headcut downstream of a bridge 8.13

Figure 8.11. Meander migration on Wapsipinicon River near De Witt, Iowa 8.13

Figure 8.12. Channel widening and meander migration on Carson River near Weeks, Nevada 8.14

Figure 8.13. Cross section locations at bridge crossings in one-dimensional models 8.15

Figure 8.14. Flow distribution from one-dimensional models 8.16

Figure 8.15. Velocity and flow lines in two-dimensional models 8.17

Figure 8.16. Typical guide bank 8.18

Figure 8.17. Guide bank in a two-dimensional network 8.19

Figure 8.18. Flow field at a bridge opening with guide bank 8.20

Figure 8.19. Two-dimensional analysis of flow along spurs, (a) flow field without spurs, and (b) flow field with spurs 8.21

Figure 9.1.	Definition sketch of the sediment continuity concept	9.3
Figure 9.2.	Definitions of sediment load components	9.5
Figure 9.3.	Suspended sediment concentration profiles	9.6
Figure 9.4.	Bed forms in sand channels	9.8
Figure 9.5.	Relative resistance to flow in sand-bed channels	9.9
Figure 9.6.	Velocity and sediment concentration profiles	9.11
Figure 9.7.	Channel profiles from sediment routing model	9.13
Figure 9.8.	Contraction scour and water surface for fixed-bed and mobile-bed models	9.14
Figure 10.1.	CFD results plot showing velocity direction and magnitude from a model of a six-girder bridge	10.2
Figure 10.2.	Definition sketch for deck force variables	10.3
Figure 10.3.	Drag coefficient for 6-girder bridge	10.4
Figure 10.4.	Lift coefficient for 6-girder bridge	10.4
Figure 10.5.	Moment coefficient for 6-girder bridge	10.5
Figure 10.6.	Photograph of a bridge damaged by Hurricane Katrina from HEC-25	10.6
Figure 10.7.	Illustration of a confluence situation	10.10
Figure 10.8.	Velocity from physical modeling using Particle Image Velocimetry	10.12
Figure 10.9.	Velocity result from CFD (RANS) modeling	10.12
Figure 10.10.	Hybrid modeling of pier scour	10.13
Figure 10.11.	Illustration of CFD modeling of Woodrow Wilson Bridge pier and dolphins	10.14
Figure 10.12.	Photograph of the post-scour condition of a small-scale physical model test of a Woodrow Wilson Bridge pier	10.15
Figure 10.13.	Sketch illustrating spread width of bridge deck drainage	10.17
Figure 10.14.	Installed underdeck bridge drainage system	10.19

LIST OF TABLES

Table 1.1. Commonly Used Engineering Terms in U.S. Customary and SI Units 1.6

Table 3.1. Values for the Computation of the Manning Roughness Coefficient using Equation 3.43 .... 3.17

Table 3.2. Values of the Manning Roughness Coefficient for Natural Channels 3.18

Table 3.3. Typical Drag Coefficients for Different Pier Shapes 3.21

Table 3.4. Summary of the Flow Profiles 3.45

Table 4.1. Bridge Hydraulic Modeling Selection 4.8

Table 4.2. Data Used in Bridge Hydraulic Studies 4.17

Table 5.1. Ranges of Expansion Rates, ER 5.16

Table 5.2. Ranges of Contraction Rates, CR 5.17

Table 5.3. Yarnell Pier Shape Coefficients 5.24

(page intentionally left blank)

LIST OF SYMBOLS

a	=	acceleration, ft/s² (m/s²)
a	=	pier width, ft (m)
a_{proj}	=	projected pier width for skewed piers, ft (m)
A	=	area, ft² (m²)
A_E	=	area of an element, ft² (m²)
A_P	=	area of a pier projected in the direction of flow, ft² (m²)
A_n	=	area below normal stage in HDS 1 method, ft² (m²)
c	=	wave celerity, ft/s (m/s)
c	=	sediment concentration at height y above the bed
c_a	=	sediment concentration at height a above the bed
C	=	Chezy conveyance coefficient
C	=	weir or discharge coefficient
b	=	bridge length, ft (m)
b_{proj}	=	projected bridge length for skewed bridges, ft (m)
B	=	flow top width, ft (m)
C_c	=	contraction coefficient
C_d	=	discharge coefficient
C_D	=	drag coefficient
C_L	=	lift coefficient
C_m	=	moment coefficient
CR	=	flow contraction ratio
C_r	=	Courant condition
C_W	=	weir flow coefficient
D	=	pipe diameter, ft (m)
D_m	=	hydraulic depth (mean depth), ft (m)
D_s	=	sediment particle size, ft (m)
D_{50}	=	bed material median size, ft (m)
e	=	roughness height, ft (m)
E	=	specific energy (depth plus kinetic energy head), ft (m)
EGL	=	energy grade line, ft (m)
ER	=	flow expansion ratio
f	=	Darcy-Weisbach friction factor
F	=	force, lb (N)
F_D	=	drag force, lb (N)
F_f	=	friction force in momentum balance bridge method, ft³ (m³)
F_M	=	force in a model for dynamic similitude, lb (N)
F_P	=	force in a prototype for dynamic similitude, lb (N)
F_r	=	Froude number
g	=	acceleration due to gravity, ft/s² (m/s²)
h	=	height, ft (m)
h_b	=	height from channel bed to bridge low chord for a bridge under pressure flow, ft (m)
h_f	=	head loss between two locations, ft (m)
h_u	=	depth of flow upstream of bridge under pressure flow, ft (m)
h_1^*	=	total backwater from HDS 1 method, ft (m)
h_e	=	expansion loss, ft (m)
h_L	=	head loss between two sections, ft (m)
h^*	=	submergence ratio of a bridge deck under pressure flow
H	=	head, head difference, or total height, ft (m)
HGL	=	hydraulic grade line, ft (m)

K, K_i^j	=	any variable in a finite difference method
K	=	Conveyance, ft³/s (m³/s)
K^*	=	backwater coefficient for HDS 1 method
k_s	=	Shields parameter
K_u	=	units coefficient for U.S. Customary or SI
L	=	length, ft (m)
L	=	pier length, ft (m)
M, m	=	mass, slugs (kg)
M	=	bridge opening discharge ratio in HDS 1 method
M_{cg}	=	resultant moment from drag and lift of a bridge under pressure flow, ft-lb (N-m)
m	=	Cowan factor for increased Manning n due to channel meandering
n	=	Manning n roughness coefficient
n_c	=	Manning n of the channel
n_e	=	effective Manning n including pier drag
n_{ob}	=	Manning n of the overbank (floodplain)
$n_{0,1,2,3,4}$	=	Cowan factors for base Manning n and adjustments for channel properties
n	=	distance normal to a streamline, ft (m)
P, p	=	pressure, lb/ft² (Pa, N/m²)
q	=	unit discharge, ft²/s (m²/s)
q_b	=	unit bed material discharge, ft²/s (m²/s)
q_l	=	unit discharge of lateral inflow, ft²/s (m²/s)
Q	=	discharge, ft³/s (m³/s)
q_s	=	unit suspended load discharge, ft²/s (m²/s)
Q_s	=	sediment discharge, ft³/s (m³/s)
Q_w	=	weir discharge, ft³/s (m³/s)
r	=	radius of curvature, ft⁻¹ (m⁻¹)
R	=	hydraulic radius (flow area divided by top width), ft (m)
R_e	=	Reynolds number
s	=	distance along a streamline, ft (m)
s	=	deck thickness of a bridge under pressure flow, ft (m)
S	=	slope
S_0	=	slope of the channel bed
S_c	=	critical channel slope
S_e	=	energy grade slope
S_f	=	friction slope
S_w	=	water surface slope
t	=	boundary layer thickness for vertical contraction scour, ft (m)
t	=	time, s
T	=	top width of flow, ft (m)
U	=	average flow velocity in the x-direction, ft/s (m/s)
v	=	point flow velocity, ft/s (m/s)
V	=	average flow velocity, ft/s (m/s)
V	=	average flow velocity in the y-direction, ft/s (m/s)
$\forall$	=	volume, ft³ (m³)
V_c	=	critical velocity (Froude number = 1.0), ft/s (m/s)
V_c	=	critical velocity for initiation of motion of bed material, ft/s (m/s)
V_n	=	velocity for normal stage in HDS 1 method, ft/s (m/s)
v^*	=	shear velocity, ft/s (m/s)
W	=	flow width, ft (m)
W	=	deck width of a bridge under pressure flow, ft (m)
WP	=	wetted perimeter, ft (m)

WS	=	water surface elevation, ft (m)
WSEL	=	water surface elevation, ft (m)
W_x	=	water weight component in momentum balance bridge method, ft³ (m³)
X, x	=	Cartesian coordinate, ft (m)
Y, y	=	Cartesian coordinate, ft (m)
Y, y	=	flow depth, also average or hydraulic depth, ft (m)
y_0	=	average flow depth prior to scour, ft (m)
y_c	=	critical depth (Froude number = 1.0), ft (m)
y_n	=	normal depth (bed slope equals energy slope), ft (m)
y_s	=	scour depth, ft (m)
y_{s-a}	=	abutment scour depth, ft (m)
y_{s-c}	=	contraction scour depth, ft (m)
y_{s-p}	=	pier scour depth, ft (m)
y_{s-vc}	=	vertical contraction scour depth, ft (m)
$\overline{Y}$	=	Depth from water surface to the centroid of the total inundated area, ft (m)
$\overline{Y}_B$	=	Depth from water surface to the centroid of flow area of a bridge opening, ft (m)
$\overline{Y}_P$	=	Depth from water surface to the centroid of pier area, ft (m)
Z, z	=	Cartesian coordinate, ft (m)
z	=	Rouse number
α	=	energy correction coefficient
α_A	=	abutment scour amplification factor
β	=	momentum correction factor
β	=	momentum transfer coefficient due to turbulence
Δ	=	bed form height, ft (m)
Δ	=	submerged sediment weight ratio ($\rho_s/\rho - 1$)
η	=	bed material porosity (volume of voids/total volume)
ξ	=	shape factor in pier or abutment scour equations
γ	=	unit weight, lb/ft³ (N/m³)
γ_w	=	unit weight of water, lb/ft³ (N/m³)
γ_s	=	unit weight of sediment, lb/ft³ (N/m³)
κ	=	Von Karman's constant
σ	=	bed material gradation coefficient
ρ	=	density, slugs/ft³ (kg/m³)
ρ_s	=	sediment particle density, slugs/ft³ (kg/m³)
ν	=	fluid kinematic viscosity, ft²/s, (m²/s)
μ	=	fluid dynamic viscosity, slug/ft-s (kg/m-s)
τ	=	shear stress, lb/ft² (Pa)
τ_0	=	shear stress at the channel bed, lb/ft² (Pa)
τ_b	=	shear stress acting on the bed, lb/ft² (Pa)
τ_c	=	critical shear stress for bed material movement, lb/ft² (Pa)
τ_s	=	shear stress acting on the water surface, lb/ft² (Pa)
τ_{xx}, τ_{xy}	=	x-direction shear stress due to turbulence, lb/ft² (Pa)
τ_{yy}, τ_{yx}	=	y-direction shear stress due to turbulence, lb/ft² (Pa)
ω	=	particle fall velocity through quiescent fluid, ft/s (m/s)
Ω	=	Coriolis parameter, (1/s)
θ	=	slope of the channel bed
θ	=	time derivative weighting factor in finite difference method
θ	=	bridge skew to flow
ϕ	=	pier skew to flow

(page intentionally left blank)

LIST OF ACRONYMS

AASHTO	American Association of State Highway and Transportation Officials
AREMA	American Railroad Engineering and Maintenance-of-Way Association
ASCE	American Society of Civil Engineers
BEM	Boundary Element Method
CADD	Computer Aided Design and Drafting
CEM	Coastal Engineering Manual
CFD	Computational Fluid Dynamics
CFR	Code of Federal Regulations
CLOMR	Conditional Letter of Map Revision
CWA	Clean Water Act
DOT	Department of Transportation
EGL	Energy Grade Line
EPA	Environmental Protection Agency
FDM	Finite Difference Method
FEM	Finite Element Method
FEMA	Federal Emergency Management Agency
FESWMS	Finite Element Surface Water Modeling System
FHWA	Federal Highway Administration
FIS	Flood Insurance Study
FSA	Farm Service Agency
FST2DH	Flow and Sediment Transport - 2-Dimensional Horizontal plane
GIS	Geographical Information System
HDS	Hydraulic Design Series
HEC	Hydraulic Engineering Circular (FHWA)
HEC	Hydrologic Engineering Center (USACE)
HGL	Hydraulic Grade Line
LES	Large Eddy Simulation
LIDAR	Light Detection and Ranging
LRFD	Load Resistance Factor Design
NCHRP	National Cooperative Highway Research Program
NED	National Elevation Dataset
NEPA	National Environmental Policy Act
NOAA	National Oceanic and Atmospheric Administration
NOS	National Ocean Service
NRC	National Research Council
NRCS	Natural Resources Conservation Service
OHW	Ordinary High Water
PIV	Particle Image Velocimetry
RANS	Reynolds-Averaged Navier-Stokes
RAS	River Analysis System (HEC-RAS)
RMA2	Resource Management Associates 2-D hydraulic model
SBR	Set Back Ratio
SCS	Soil Conservation Service (now NRCS)
SI	System International (metric system of units)
SIAM	Sediment Impact Assessment Model (component in HEC-RAS)

SMS	Surface-water Modeling System
TAC	Transportation Agency of Canada
TRACC	Transportation Research and Analysis Computing Center
TRB	Transportation Research Board
UNET	Unsteady Network (model)
USACE	U.S. Army Corps of Engineers
USBR	U.S Bureau of Reclamation
USCG	U.S. Coast Guard
USDA	U.S Department of Agriculture
USDOT	U.S. Department of Transportation
USGS	U.S. Geologic Survey
WSEL	Water Surface Elevation
WSPRO	Water Surface Profile (model and bridge modeling approach in HEC-RAS)

GLOSSARY

adverse slope: The hydraulic condition where the bed slope in the direction of flow is negative and normal depth is undefined.

aggradation: General and progressive buildup of the longitudinal profile of a channel bed due to sediment deposition.

alluvial channel: Channel wholly in alluvium; no bedrock is exposed in channel at low flow or likely to be exposed by erosion.

alluvial fan: Fan shaped deposit of material at the place where a stream issues from a narrow valley of high slope onto a plain or broad valley of low slope. An alluvial cone is made up of the finer materials suspended in flow while a debris cone is a mixture of all sizes and kinds of materials.

alluvial stream: Stream which has formed its channel in cohesive or noncohesive materials that have been and can be transported by the stream.

alluvium: Unconsolidated material deposited by a stream in a channel, floodplain, alluvial fan, or delta.

annual flood: Maximum flow in 1 year (may be daily or instantaneous).

approach section: The cross section upstream of a bridge where flow is fully expanded in the floodplain and the discharge distribution is proportional to conveyance.

average velocity: The velocity at a given cross section determined by dividing discharge by cross sectional area.

avulsion: Sudden change in the channel course that usually occurs when a stream breaks through its banks; usually associated with a flood or a catastrophic event.

backwater: Increase in water surface elevation relative to elevation occurring under natural channel and floodplain conditions. It is induced by a bridge or other structure that obstructs or constricts the free flow of water in a channel.

bank: Sides of a channel between which the flow is normally confined.

bank, left or right: Sides of a channel as viewed in a downstream direction.

bank protection: Engineering works for the purpose of protecting streambanks from erosion.

bank revetment:	Erosion resistant materials placed directly on a streambank to protect the bank from erosion.
bankfull discharge:	Discharge that, on average, fills a channel to the point of overflowing.
base floodplain:	Floodplain associated with the flood with a 100-year recurrence interval.
bathymetry:	The below-water ground elevation at typical flow conditions.
bed:	Bottom of a channel bounded by banks.
bed layer:	Flow layer, several grain diameters thick (usually two) immediately above the bed.
bed load discharge (or bed load):	Quantity of bed load passing a cross section of a stream in a unit of time.
bed load:	Sediment that is transported in a stream by rolling, sliding, or skipping along the bed or very close to it; considered to be within the bed layer (contact load).
bed material:	Material found in and on the bed of a stream (may be transported as bed load or in suspension).
bedrock:	Solid rock exposed at the surface of the earth or overlain by soils and unconsolidated material.
bed shear (tractive force):	Force per unit area exerted by a fluid flowing past a boundary.
bed slope:	Inclination of the channel bottom.
boulder:	Rock fragment whose diameter is greater than 250 mm.
boundary condition:	A location along the model boundary where discharge and/or water surface are defined or set.
bridge opening:	Cross sectional area beneath a bridge that is available for conveyance of water.
bridge owner:	Any Federal, State, Local agency, or other entity responsible for a structure defined as a highway bridge by the National Bridge Inspection Standards (NBIS).
bridge section:	The cross section at a bridge. For scour calculations in HEC-RAS, typically the upstream adjacent section to the internal bridge sections.

bridge waterway:	Area of a bridge opening available for flow, as measured below a specified stage and normal to the principal direction of flow.
bulk density:	Density of the water sediment mixture (mass per unit volume), including both water and sediment.
causeway:	Rock or earth embankment carrying a roadway across water.
caving:	Collapse of a bank caused by undermining due to the action of flowing water.
channel:	Bed and banks that confine surface flow of a stream.
channel pattern:	Aspect of a stream channel in plan view, with particular reference to the degree of sinuosity, braiding, and anabranching.
channelization:	Straightening or deepening of a natural channel by artificial cutoffs, grading, flow control measures, or diversion of flow into an engineered channel.
check dam:	Low dam or weir across a channel used to control stage or degradation.
choking (of flow):	Excessive constriction of flow which may cause severe backwater effect.
clay (mineral):	Particle whose diameter is in the range of 0.00024 to 0.004 mm.
clear-water scour:	Scour at a pier or abutment (or contraction scour) when there is no movement of the bed material upstream of the bridge crossing at the flow causing bridge scour.
cobble:	Fragment of rock whose diameter is in the range of 64 to 250 mm.
Computational Fluid Dynamics (CFD):	A numerical hydraulic model that incorporates turbulence fluctuations.
confluence:	Junction of two or more streams.
constriction:	Natural or artificial control section, such as a bridge crossing, channel reach or dam, with limited flow capacity in which the upstream water surface elevation is related to discharge.
contact load:	Sediment particles that roll or slide along in almost continuous contact with the streambed (bed load).
contraction reach:	The river reach were flow is converging from being fully expanded in the floodplain into the bridge opening.

contraction scour:	Contraction scour, in a natural channel or at a bridge crossing, involves the removal of material from the bed and banks across all or most of the channel width. This component of scour results from a contraction of the flow area at the bridge which causes an increase in velocity and shear stress on the bed at the bridge. The contraction can be caused by the bridge or from a natural narrowing of the stream channel.
contraction:	Effect of channel or bridge constriction on flow streamlines.
control section:	A location where water surface is uniquely defined for a given discharge.
conveyance:	The capacity of the channel to accommodate flow.
Coriolis force:	Inertial force caused by the Earth's rotation that deflects a moving body to the right in the Northern Hemisphere.
countermeasure:	Measure intended to prevent, delay or reduce the severity of hydraulic problems.
critical depth:	In hydraulic analysis, the depth when flow has a Froude number of 1.0. Alternatively, in sediment transport analysis, the depth and velocity condition when a bed material particle size is at incipient motion.
critical shear stress:	Minimum amount of shear stress required to initiate soil particle motion.
critical slope:	The hydraulic condition where the bed slope is equal to critical slope and normal depth is equal to critical depth.
critical velocity:	In hydraulic analysis, the velocity when flow has a Froude number of 1.0. Alternatively, in sediment transport analysis, the velocity when a bed material particle size is at incipient motion.
cross section:	Section normal to the trend of a channel or flow.
crossing:	Relatively short and shallow reach of a stream between bends; also crossover or riffle.
current:	Water flowing through a channel.
cut bank:	Concave wall of a meandering stream.
cutoff:	(A) Direct channel, either natural or artificial, connecting two points on a stream, thereby shortening the original length of the channel and increasing its slope; (B) natural or artificial channel which develops across the neck of a meander loop (neck cutoff) or across a point bar (chute cutoff).

debris:	Floating or submerged material, such as logs, vegetation, or trash, transported by a stream.
degradation (bed):	General and progressive (long-term) lowering of the channel bed due to erosion, over a relatively long channel length.
depth of scour:	Vertical distance a streambed is lowered by scour below a reference elevation.
design flow (design flood):	Discharge that is selected as the basis for the design or evaluation of a hydraulic structure.
dike (groin, spur, jetty):	Structure extending from a bank into a channel that is designed to: (A) reduce the stream velocity as the current passes through the dike, thus encouraging sediment deposition along the bank (permeable dike); or (B) deflect erosive current away from the streambank (impermeable dike).
dike:	An impermeable linear structure for the control or containment of overbank flow. A dike-trending parallel with a streambank differs from a levee in that it extends for a much shorter distance along the bank, and it may be surrounded by water during floods.
discharge:	Volume of water passing through a channel during a given time.
drag force:	The force acting between the flow and an obstruction.
drift:	Alternate term for vegetative "debris."
eddy current:	Vortex type motion of a fluid flowing contrary to the main current, such as the circular water movement that occurs when the main flow becomes separated from the bank.
energy correction coefficient (α):	A correction coefficient the must be applied when the average velocity is used to compute kinetic energy because the total energy of a velocity distribution is not equal to the energy computed from the average velocity.
energy grade line (EGL):	The profile line that includes water surface elevation (hydraulic grade line) plus kinetic energy head.
ephemeral stream:	Stream or reach of stream that does not flow for parts of the year. As used here, the term includes intermittent streams with flow less than perennial.
equilibrium scour:	Scour depth in sand-bed stream with dune bed about which live bed pier scour level fluctuates due to variability in bed material transport in the approach flow.

erosion:	Displacement of soil particles due to water or wind action.
exit section:	The cross section downstream of a bridge where flow is fully expanded in the floodplain and the discharge distribution is proportional to conveyance.
expansion reach:	The river reach were flow is diverging from the bridge opening until it is fully expanded into the floodplain.
extent:	See model extent.
fall velocity:	Velocity at which a sediment particle falls through a column of still water.
fetch:	Area in which waves are generated by wind having a rather constant direction and speed; sometimes used synonymously with fetch length.
fetch length:	Horizontal distance (in the direction of the wind) over which wind generates waves and wind setup.
fill slope:	Side or end slope of an earth fill embankment. Where a fill slope forms the streamward face of a spill-through abutment, it is regarded as part of the abutment.
fine sediment load:	That part of the total sediment load that is composed of particle sizes finer than those represented in the bed (wash load). Normally, the fine sediment load is finer than 0.062 mm for sand bed channels. Silts, clays and sand could be considered wash load in coarse gravel and cobble-bed channels.
flashy stream:	Stream characterized by rapidly rising and falling stages, as indicated by a sharply peaked hydrograph. Typically associated with mountain streams or highly disturbed urbanized catchments. Most flashy streams are ephemeral, but some are perennial.
flood-frequency curve:	Graph indicating the probability that the annual flood discharge will exceed a given magnitude, or the recurrence interval corresponding to a given magnitude.
floodplain:	Nearly flat, alluvial lowland bordering a stream that is subject to inundation by floods.
flow hazard:	Flow characteristics (discharge, stage, velocity, or duration) that are associated with a hydraulic problem or that can reasonably be considered of sufficient magnitude to cause a hydraulic problem or to test the effectiveness of a countermeasure.
flow profile:	The longitudinal line of water surface elevation along a channel.

flow resistance:	The boundary impediment to flowing water depending on several factors, including boundary roughness, vegetation, irregularities, etc.
fluvial geomorphology:	Science dealing with morphology (form) and dynamics of streams and rivers.
fluvial system:	Natural river system consisting of (1) the drainage basin, watershed, or sediment source area; (2) tributary and mainstem river channels or sediment transfer zone; and (3) alluvial fans, valley fills and deltas, or the sediment deposition zone.
freeboard:	Vertical distance above a design stage that is allowed for waves, surges, drift, and other contingencies.
Froude Number:	Dimensionless number that represents the ratio of inertial to gravitational forces in open channel flow.
geomorphology/ morphology:	That science that deals with the form of the Earth, the general configuration of its surface, and the changes that take place due to erosion and deposition.
grade-control structure (sill, check dam):	Structure placed bank to bank across a stream channel (usually with its central axis perpendicular to flow) for the purpose of controlling bed slope and preventing scour or headcutting.
graded stream:	Geomorphic term used for streams that have apparently achieved a state of equilibrium between the rate of sediment transport and the rate of sediment supply throughout long reaches.
gradually varied flow:	Flow where streamlines are essentially parallel and vertical accelerations are negligible.
gravel:	Rock fragment whose diameter ranges from 2 to 64 mm.
guide bank:	Dike extending upstream from the approach embankment at either or both sides of the bridge opening to direct the flow through the opening. Some guide banks extend downstream from the bridge (also spur dike).
headcutting:	Channel degradation associated with abrupt changes in the bed elevation (headcut) that generally migrates in an upstream direction.
helical flow:	Three dimensional movement of water particles along a spiral path in the general direction of flow. These secondary type currents are of most significance as flow passes through a bend; their net effect is to remove soil particles from the cut bank and deposit this material on a point bar.

horizontal slope:	The hydraulic conditions where the bed slope is zero and normal depth is infinite.
hydraulic control:	See control section.
hydraulic grade line (HGL):	The profile line that is the water surface elevation.
hydraulic model:	Small-scale physical or mathematical representation of a flow situation.
hydraulic radius:	Cross sectional area of a stream divided by its wetted perimeter.
hydraulic structures:	Facilities used to impound, accommodate, convey or control the flow of water, such as dams, weirs, intakes, culverts, channels, and bridges.
hydraulics:	Applied science concerned with behavior and flow of liquids, especially in pipes, channels, structures, and the ground.
hydrograph:	The graph of stage or discharge against time.
hydrology:	Science concerned with the occurrence, distribution, and circulation of water on the earth.
hydrostatic pressure:	The pressure of as it varies with depth in still water. Also, the pressure of flowing water that is not affected by vertical accelerations other than gravity.
incipient overtopping:	The condition when flow is at the road crest, but not flowing over the road.
ineffective flow:	The portion of a cross section that is not actively conveying flow in the downstream direction.
initiation of motion:	The hydraulic condition when bed material, often the median grain size, begins to move and sediment transport of bed material occurs.
invert:	Lowest point in the channel cross section or at flow control devices such as weirs, culverts, or dams.
island:	A permanently vegetated area that divides the flow of a stream and is emergent at normal stage. Islands originate by establishment of vegetation on a bar, by channel avulsion, or at the junction of minor tributary with a larger stream.
lateral erosion:	Erosion in which the removal of material is extended horizontally as contrasted with degradation and scour in a vertical direction.

levee:	Embankment, generally landward of the top of bank, that confines flow during high-water periods, thus preventing overflow into lowlands.
live-bed scour:	Scour at a pier or abutment (or contraction scour) when the bed material in the channel upstream of the bridge is moving at the flow causing bridge scour.
load (or sediment load):	Amount of sediment being moved by a stream.
local scour:	Removal of material from around piers, abutments, spurs, and embankments caused by an acceleration of flow and resulting vortices induced by obstructions to the flow.
longitudinal profile:	Profile of a stream or channel drawn along the length of its centerline. In drawing the profile, elevations of the water surface or the thalweg are plotted against distance as measured from the mouth or from an arbitrary initial point.
lower bank:	That portion of a streambank having an elevation less than the mean water level of the stream.
mathematical model:	Numerical representation of a flow situation using mathematical equations (also computer model).
meander or full meander:	Meander in a river consists of two consecutive loops, one flowing clockwise and the other counter clockwise.
meandering stream:	Stream having a sinuosity greater than some arbitrary value. The term also implies a moderate degree of pattern symmetry, imparted by regularity of size and repetition of meander loops. The channel generally exhibits a characteristic process of bank erosion and point bar deposition associated with systematically shifting meanders.
median diameter:	Particle diameter of the 50th percentile point on a size distribution curve such that half of the particles (by weight, number, or volume) are larger and half are smaller (D_{50}).
migration:	Change in position of a channel by lateral erosion of one bank and simultaneous accretion of the opposite bank.
mild slope:	The hydraulic condition where the bed slope is less than critical slope and normal depth is greater than critical depth.
model extent:	The limits of a model domain including boundary conditions and fixed boundaries.

momentum correction coefficient (β):	A correction coefficient the must be applied when the average velocity is used to compute momentum because the total momentum of a velocity distribution is not equal to the momentum computed from the average velocity.
mud:	A soft, saturated mixture mainly of silt and clay.
multiple openings:	Road embankments that have two or more bridges (and/or culverts) located along the embankment.
natural levee:	Low ridge that slopes gently away from the channel banks that is formed along streambanks during floods by deposition.
non-uniform flow:	Flow of changing cross section or velocity through a reach of channel at a given time.
normal depth:	A condition when the water surface slope and energy grade slope are parallel to the bed slope. Also a boundary condition where the water surface is computed from a preset energy grade slope.
normal stage:	Water stage prevailing during the greater part of the year.
numerical model:	A computer representation of a prototype condition.
one-dimensional model:	A numerical hydraulic model that computes flow velocity in the downstream direction.
orifice flow:	Flow through a bridge where the upstream low-chord is submerged but the downstream low-chord is not.
overbank flow:	Water movement that overtops the bank either due to stream stage or to overland surface water runoff.
overtopping flow:	The bridge hydraulic condition when the approach embankment and/or bridge are being overtopped during a flood.
parallel bridges:	Bridges located in series along a channel where flow does not fully expand between the bridges.
perennial stream:	Stream or reach of a stream that flows continuously for all or most of the year.
phreatic line:	Upper boundary of the seepage water surface landward of a streambank.
physical model:	A laboratory hydraulic model of a prototype condition.

pile:	Elongated member, usually made of timber, concrete, or steel, that serves as a structural component of a river training structure or bridge.
probable maximum flood:	Very rare flood discharge value computed by hydro-meteorological methods, usually in connection with major hydraulic structures.
rapid drawdown:	Lowering of the water against a bank more quickly than the bank can drain without becoming unstable.
rapidly varied flow:	Flow where streamlines are not parallel or have significant curvature, pressure is not hydrostatic, and vertical accelerations cannot be ignored.
reach:	Segment of stream length that is arbitrarily bounded for purposes of study.
recurrence interval:	Reciprocal of the annual probability of exceedance of a hydrologic event (also return period, exceedance interval).
regime:	Condition of a stream or its channel with regard to stability. A stream is in regime if its channel has reached an equilibrium form as a result of its flow characteristics. Also, the general pattern of variation around a mean condition, as in flow regime, tidal regime, channel regime, sediment regime, etc. (used also to mean a set of physical characteristics of a river).
relief bridge:	An opening in an embankment on a floodplain to permit passage of overbank flow.
revetment:	Rigid or flexible armor placed to inhibit scour and lateral erosion. (See bank revetment).
riparian:	Pertaining to anything connected with or adjacent to the banks of a stream (corridor, vegetation, zone, etc.).
riprap:	Layer or facing of rock or broken concrete dumped or placed to protect a structure or embankment from erosion; also the rock or broken concrete suitable for such use. Riprap has also been applied to almost all kinds of armor, including wire enclosed riprap, grouted riprap, partially grouted riprap, sacked concrete, and concrete slabs.
river training:	Engineering works with or without the construction of embankment, built along a stream or reach of stream to direct or to lead the flow into a prescribed channel. Also, any structure configuration constructed in a stream or placed on, adjacent to, or in the vicinity of a streambank that is intended to deflect currents, induce sediment deposition, induce scour, or in some other way alter the flow and sediment regimes of the stream.

roughness coefficient:	Numerical measure of the frictional resistance to flow in a channel, as in the Manning or Chezy formulas.
runoff:	That part of precipitation which appears in surface streams of either perennial or intermittent form.
saltation load:	Sediment bounced along the streambed by energy and turbulence of flow, and by other moving particles.
sand:	Rock fragment whose diameter is in the range of 0.062 to 2.0 mm.
scour:	Erosion of streambed or bank material due to flowing water; often considered as being localized (see local scour, contraction scour, total scour).
sediment concentration:	Weight or volume of sediment relative to the quantity of transporting (or suspending) fluid.
sediment continuity:	An analysis that accounts for sediment inflow, sediment outflow, and erosion or storage of sediment along a river reach.
sediment discharge:	Quantity of sediment that is carried past any cross section of a stream in a unit of time. Discharge may be limited to certain sizes of sediment or to a specific part of the cross section.
sediment load:	Amount of sediment being moved by a stream.
sediment or fluvial sediment:	Fragmental material transported, suspended, or deposited by water.
sediment yield:	Total sediment outflow from a watershed or a drainage area at a point of reference and in a specified time period. This outflow is equal to the sediment discharge from the drainage area.
seepage:	Slow movement of water through small cracks and pores of the bank material.
shear stress:	See unit shear force.
silt:	Particle whose diameter is in the range of 0.004 to 0.062 mm.
similitude:	A relationship between full-scale flow and a laboratory flow involving smaller, but geometrically similar boundaries.
sinuosity:	Ratio between the thalweg length and the valley length of a stream.

skew:	The condition when a bridge opening is not perpendicular to flow or when a pier is not aligned with the flow.
slope (of channel or stream):	Fall per unit length along the channel centerline or thalweg.
slope-area method:	Method of estimating unmeasured flood discharges in a uniform channel reach using observed high water levels.
sloughing:	Sliding or collapse of overlying material; same ultimate effect as caving, but usually occurs when a bank or an underlying stratum is saturated.
slump:	Sudden slip or collapse of a bank, generally in the vertical direction and confined to a short distance, probably due to the substratum being washed out or having become unable to bear the weight above it.
specific energy:	Flow depth plus kinetic energy.
spill-through abutment:	Bridge abutment having a fill slope on the streamward side. The term originally referred to the "spill-through" of fill at an open abutment but is now applied to any abutment having such a slope.
spur dike:	See guide bank.
spur:	Permeable or impermeable linear structure that projects into a channel from the bank to alter flow direction, induce deposition, or reduce flow velocity along the bank.
stability:	Condition of a channel when, though it may change slightly at different times of the year as the result of varying conditions of flow and sediment charge, there is no appreciable change from year to year; that is, accretion balances erosion over the years.
stable channel:	Condition that exists when a stream has a bed slope and cross section which allows its channel to transport the water and sediment delivered from the upstream watershed without aggradation, degradation, or bank erosion (a graded stream).
stage:	Water surface elevation of a stream with respect to a reference elevation.
steady flow :	Flow of constant discharge, depth and velocity at a cross section through time.
steep slope:	The hydraulic condition where the bed slope is greater than critical slope and normal depth is less than critical depth.

stream:	Body of water that may range in size from a large river to a small rill flowing in a channel. By extension, the term is sometimes applied to a natural channel or drainage course formed by flowing water whether it is occupied by water or not.
streambank erosion:	Removal of soil particles or a mass of particles from a bank surface due primarily to water action. Other factors such as weathering, ice and debris abrasion, chemical reactions, and land use changes may also directly or indirectly lead to bank erosion.
streambank failure:	Sudden collapse of a bank due to an unstable condition such as removal of material at the toe of the bank by scour.
streambank protection:	Any technique used to prevent erosion or failure of a streambank.
streamline:	An imaginary line within the flow that is tangent everywhere to the velocity vector.
streamtube:	An element of fluid bounded by a pair of streamlines.
subcritical, supercritical flow:	Open channel flow conditions with Froude Number less than and greater than unity, respectively.
submerged orifice flow:	Flow through a bridge where the upstream and downstream low-chords are submerged.
suspended sediment discharge:	Quantity of sediment passing through a stream cross section above the bed layer in a unit of time suspended by the turbulence of flow (suspended load).
thalweg:	Line extending down a channel that follows the lowest elevation of the bed.
three-dimensional model:	A numerical hydraulic model that computes three components of velocity.
toe of bank:	That portion of a stream cross section where the lower bank terminates and the channel bottom or the opposite lower bank begins.
topography:	The above-water ground elevation at typical flow conditions.
total scour:	Sum of long-term degradation, general (contraction) scour, and local scour.
total sediment load:	Sum of suspended load and bed load or the sum of bed material load and wash load of a stream (total load).

tractive force:	Drag or shear on a streambed or bank caused by passing water which tends to move soil particles along with the streamflow.
turbulence:	Motion of fluids in which local velocities and pressures fluctuate irregularly in a random manner as opposed to laminar flow where all particles of the fluid move in distinct and separate lines.
two-dimensional model:	A numerical hydraulic model that computes two components of velocity, usually the horizontal velocity components.
ultimate scour:	Maximum depth of scour attained for a given flow condition. May require multiple flow events and in cemented or cohesive soils may be achieved over a long time period.
uniform flow:	Flow of constant cross section and velocity through a reach of channel at a given time. Both the energy slope and the water slope are equal to the bed slope under conditions of uniform flow.
unit discharge:	Discharge per unit width (may be average over a cross section, or local at a point).
unit shear force (shear stress):	Force or drag developed at the channel bed by flowing water. For uniform flow, this force is equal to a component of the gravity force acting in a direction parallel to the channel bed on a unit wetted area. Usually in units of stress, lb/ft^2 or (Pa or N/m^2).
unsteady flow:	Flow of variable discharge and velocity through a cross section with respect to time.
upper bank:	Portion of a streambank having an elevation greater than the average water level of the stream.
velocity:	Time rate of flow usually expressed in ft/sec (m/s). Average velocity is the velocity at a given cross section determined by dividing discharge by cross-sectional area.
vertical abutment:	An abutment, usually with wingwalls, that has no fill slope on its streamward side.
vortex:	Turbulent eddy in the flow generally caused by an obstruction such as a bridge pier or abutment (e.g., horseshoe vortex).
wash load:	Suspended material of very small size (generally clays and colloids) originating primarily from erosion on the land slopes of the drainage area and present to a negligible degree in the bed itself.
watershed:	See drainage basin.
waterway opening width (area):	Width (area) of bridge opening at (below) a specified stage, measured normal to the principal direction of flow.

(page intentionally left blank)

ACKNOWLEDGMENTS

This manual is a major revision and replacement of HDS 1 "Hydraulics of Bridge Waterways," which was written by Joseph N. Bradley in 1960 and revised in 1970 and 1978. The authors wish to acknowledge the contributions made by Mr. Bradley through his landmark document.

DISCLAIMER

Mention of a manufacturer, registered or trade name does not constitute a guarantee or warranty of the product by the U.S. Department of Transportation or the Federal Highway Administration and does not imply their approval and/or endorsement to the exclusion of other products and/or manufacturers that may also be suitable.

(page intentionally left blank)

CHAPTER 1

INTRODUCTION – HYDRAULIC DESIGN OF SAFE BRIDGES

1.1. INTRODUCTION

1.1.1 Background

The Federal Highway Administration provides oversight of the Nation's bridges through the National Bridge Inspection Standards (NBIS) and other regulatory policies and programs. Bridge failures resulting from both natural and human factors led the U.S. Congress to express its concern about the safety, approaches, and oversight of the Nation's bridges.

Within the Conference Report for the Departments of Transportation and Housing and Urban Development, and Related Agencies Appropriations Act, 2010 (H.R. Rep. No. 111-366), the Congress recommended that the "... (FHWA) use a more risk-based, data-driven approach to its bridge oversight" to improve bridge safety. Congress stated its intention to monitor the progress that FHWA makes in identifying new approaches to bridge oversight, completing initiatives, and achieving results from its efforts. Congress directed that FHWA apply funds to focus and achieve these activities.

To address the conference report, FHWA undertook a combination of activities that contribute to four primary outcomes: more rigorous oversight of bridge safety; full NBIS compliance by all States; improved information for safety oversight and condition monitoring; and qualified and equipped bridge inspection personnel.

As hydraulic issues remain a leading factor in bridge failures, FHWA recognized that these activities need to include efforts to better collect, understand and deploy more recent and robust guidance and techniques to the accepted state of hydraulic and waterway related practice. This document is one of the products of these efforts.

1.1.2 Purpose

The purpose of HDS 7, Hydraulic Design of Safe Bridges, is to provide technical information and guidance on the hydraulic design of bridges. HDS 7 replaces the HDS 1 manual "Hydraulics of Bridge Waterways" (FHWA 1978) for guidance of bridge hydraulic analyses. Bridges should be designed as safely as possible while optimizing costs and limiting impacts to property and the environment. Many significant aspects of bridge hydraulic design are discussed. These include regulatory topics, specific approaches for bridge hydraulic modeling, hydraulic model selection, bridge design impacts on scour and stream instability, and sediment transport.

The impacts of bridge design and construction on the economics of highway design, safety to the traveling public, and the natural environment can be significant. An economically viable and safe bridge is one that is properly sized, designed, constructed, and maintained. In general, although longer bridges are more expensive to design and build than shorter bridges, they cause less backwater, experience less scour, and can reduce impacts to the environment. Increased scour from too short a bridge can require deeper foundations and necessitate countermeasures to resist these effects. A properly designed bridge is one that balances the cost of the bridge with concerns of safety to the traveling public, impacts to the environment, and regulatory requirements to not cause harm to those that live or work in the floodplain upstream and downstream of the bridge.

1.1.3 History of Bridge Hydraulics

Determining the hydraulic capacity of bridges and culverts is a field that has been evolving in the United States since the mid 1800s. The earliest methods of sizing hydraulic openings were largely based on experience and historic performance. However, as the railroads expanded westward many crossings were encountered where there was no flood history or other up or downstream structures to use as the basis for determining bridge or culvert size. Therefore, tabular and empirical methods were developed that related waterway opening to size of drainage area and other coefficients that accounted for drainage basin and stream characteristics. The American Railroad Engineering and Maintenance-of-Way Association (AREMA) published a report in 1911 that presented six formulas for waterway area and 21 formulas for design discharge. A report by V.T. Chow in 1962 listed 12 formulas for waterway area and 62 formulas for design discharge (McEnroe 2007).

The earliest methods for determining waterway openings for bridges and culverts did not consider bridge or culvert configuration. Furthermore, the concept of a "design" discharge or recurrence interval of expected floods to use when determining structure size was not considered. Even though design discharges were not considered an early textbook on highway design and construction by Byrne (1893) suggested that the factors to be considered when determining the capacity of a hydraulic culvert depended on; (1) the rate of rainfall, (2) the kind and condition of the soil, (3) the character and inclination of the surface, (4) the condition of inclination of the bed of the stream, (5) the shape of the area to be drained, and the branches of the stream, (6) the form of the mouth and the inclination of the bed of the culvert, and (7) whether it is permissible to back the water up above the culvert, thereby causing it to discharge under a head. These same concepts were applied to the hydraulic sizing of bridges. As techniques for estimating discharge developed throughout the 1900s these same factors translated into many of the parameters found in methods used today to estimate recurrence intervals, peak discharges, and hydrographs.

At the same time these methods were being developed in the United States a formula developed by Robert Manning (Manning 1889) was becoming popular. Originally developed in SI units with a coefficient of 1.0, the form of the equation in U.S. Customary units is presented as:

$$V = \frac{1.486}{n} R^{2/3} S^{1/2} \qquad (1.1)$$

where:

V = Velocity, ft/s
n = Roughness Coefficient
R = Hydraulic Radius, ft
S = Slope

There were two things that Robert Manning did not like about his equation, (1) that it was dimensionally incorrect, and (2) it was difficult (at the time) to determine the cubed root of a number and then square it to arrive at a number to the $2/3^{rd}$ power. King's handbook (King 1918) presented a table of numbers from 0.01 to 10 to the $2/3^{rd}$ power which eliminated the problem of determining a number to the $2/3^{rd}$ power and is considered to be a leading reason in the early acceptance and of use of the Manning Equation.

As methods were being developed to estimate discharge, it was realized that one could make an estimate of the roughness coefficient based on known values from similar channels and floodplains, determine the slope of the channel, and then use an iterative solution to determine the "normal" depth at a cross-section or hydraulic opening. Through the 1950s this remained a popular method of determining the depth and velocity of flow at a cross-section or through a hydraulic opening.

The problem with using normal depth as the estimate of flow depth (and velocity) for determining the size of hydraulic opening is that it does not consider the effects of backwater. Backwater is the additional depth to accelerate flow through the bridge opening and overcome a variety of resistance and drag forces. These forces depend on a number of factors including bridge type, degree of contraction, embankment skew, pier number and type, debris blockage, etc.

To account for backwater, research was completed and methods were developed that examined the components of backwater (Liu et al. 1957). In HDS 1, the computed backwater was added to the "normal" depth at a location upstream of the bridge to evaluate the overall impacts of a bridge (FHWA 1978).

Another significant development that contributed to the development of the current state of bridge hydraulics was the publication of a textbook about open channel flow by V.T. Chow (Chow 1959). The publication presents and applies concepts of energy, momentum, and continuity to the flow of water in open channels. It is also one of the places where the direct and standard step methods for computing water surface profiles were first presented. The direct step method is applicable to prismatic channels and the standard step method to natural channels. The standard step method uses concepts of conservation of energy and flow, and is widely used for water surface profile calculations.

At the same time the physics of open channel flow and water surface profiles were being developed, mainframe computer and programming languages were developing. The application of computer programs made it possible to rapidly complete trial and error solutions required for computing water surface profiles. One of the first computer programs that was developed to compute water surface profiles in natural channels was HEC-2 (USACE 1992) with development dating back to at least 1964. The HEC-2 program has undergone continual development and was ported to the PC in 1984. HEC-2 has evolved into the HEC-RAS (River Analysis System) model (USACE 2010a, b, c). HEC-RAS performs steady non-uniform flow hydraulic calculations similar to HEC-2, but incorporates enhanced visualization, more complete bridge and culvert hydraulic computations, unsteady flow, and sediment transport. There were many other computer programs developed to compute water surface profiles. The USGS developed E431 (USGS 1976) and the Federal Highway Administration developed WSPRO (FHWA 1998) that had components specifically formulated for the analysis of flow through bridge openings. HEC-RAS has incorporated features from these programs including the WSPRO bridge routine.

More recent developments in the field of bridge hydraulics include the development of two-dimensional hydraulic and hydrodynamic models to compute the entire flow field. These models include FST2DH (FHWA 2003) and RMA-2 (USACE 2009).

1.2 HYDRAULIC ANALYSIS OVERVIEW

The hydraulic analysis of a bridge opening is a complicated undertaking. Decisions must be made regarding the type of model computational methods, model extent, and amount of topographic data that needs to be collected. An assessment of flow resistance caused by channel and floodplain conditions needs to be made and the impacts on flow due to different seasonal conditions also needs to be evaluated. An understanding of flow type, historic flow conditions, and flooding at the site also provides valuable insight into the approaches that need to be employed.

Once the preliminary data has been collected and an understanding of the flow complexity at the bridge opening is obtained, a decision must be made regarding the type of hydraulic model that should be used at the hydraulic crossing. Some situations call for a one-dimensional gradually-varied steady-state flow model while others require the use of unsteady flow models, or two-dimensional steady or unsteady flow models to more fully understand the flow conditions at the hydraulic crossing. Some situations call for a more sophisticated modeling approach because of other factors. These can include the need for a more complete understanding of the flow conditions because of bridge scour or bank stabilization.

There are also regulatory requirements that must be adhered to. The FHWA, USACE, FEMA, EPA, state and local agencies, and others have requirements that must be considered when determining the best overall approach for evaluating the flow through a bridge opening and its impact on adjacent land owners, the environment, and economic concerns. These types of issues must be considered when developing the best approach for analyzing the flow through a bridge opening or reach of river.

1.3 MANUAL ORGANIZATION

This manual is intended to be a general resource for bridge hydraulic design. The concepts are valid for a range of one- and two-dimensional numeric models, not just for the specific models that are mentioned.

1.3.1 Chapter 1 – Introduction

The purpose of HDS 7 is to provide FHWA guidance on hydraulic analysis and design of safe bridges. Significant aspects of bridge hydraulic design are discussed, which include regulatory topics, specific approaches for bridge hydraulic modeling, model selection, scour and stream instability, and sediment transport.

1.3.2 Chapter 2 – Design Considerations and Regulatory Requirements

Chapter 2 provides information and discussion on the range of design considerations, environmental considerations, and regulatory requirements that may be encountered during bridge design and construction. Topics include FHWA guidance, AASHTO Specifications, freeboard, setback and road grade requirements, design considerations, scour and channel instability concerns, Federal regulations, navigation permits, and environmental permits.

1.3.3 Chapter 3 – Governing Equations and Flow Classification

Chapter 3 provides background on fundamental open channel flow concepts. Although this is not a hydraulic engineering textbook, there is sufficient information to serve as a reference source on the equations used in open channel and bridge hydraulics.

1.3.4 Chapter 4 – Hydraulic Analysis Considerations

Chapter 4 builds on the background from Chapter 3 to discuss hydraulic modeling. One- and two-dimensional modeling are compared and contrasted as well as steady versus unsteady flow modeling. Criteria for selecting a modeling approach are identified. Topics that are relevant to the range of hydraulic modeling approaches are also included in this chapter. These include model extent, boundary conditions, hydrology, and model calibration.

1.3.5 Chapter 5 – One-Dimensional Bridge Hydraulic Analysis

Chapter 5 provides information and guidance on the use of one-dimensional models for bridge hydraulic analysis. Information focuses on the use of HEC-RAS, but the guidance is applicable to a wide range of one-dimensional models. The chapter covers standard applications as well as special cases. The chapter also provides a discussion of the assumptions that the one dimensional approach requires and how violating these assumptions leads to error and uncertainty in the modeling results.

1.3.6 Chapter 6 – Two-Dimensional Bridge Hydraulic Analysis

Chapter 6 provides information and guidance on the use of two-dimensional models for bridge hydraulic analysis. Information focuses on the use of FST2DH, but the guidance is also applicable to other finite element and finite difference models. The chapter covers standard applications as well as special cases, although finite element models are primarily used for complex cases.

1.3.7 Chapter 7 – Unsteady Flow Analysis

Chapter 7 discusses modeling unsteady flow with one- and two-dimensional models. Topics discussed in this chapter include the basic equations that define unsteady flow, upstream and downstream model extents, floodplain storage and connections, and boundary conditions. River and tidal applications are included.

1.3.8 Chapter 8 – Bridge Scour Considerations and Scour Countermeasure Hydraulic Analysis

Chapter 8 discusses an extremely important aspect of bridge safety. Scour during floods is a significant part of bridge design and is a primary contribution of the hydraulic engineer to the bridge structural design. This topic is covered in detail in HEC-20 and HEC-18 (FHWA 2012a,b). A general discussion of the types of scour and information on obtaining appropriate hydraulic variables from one- and two-dimensional models are the focus of this chapter. The importance of considering future channel change (width adjustments, changes in channel alignment and channel migration, and long-term aggradation and degradation) are addressed.

1.3.9 Chapter 9 – Sediment Transport and Alluvial Channel Concepts

Chapter 9 provides an overview of sediment transport concepts, which are covered thoroughly in HDS 6, River Engineering for Highway Encroachments (FHWA 2001).

1.3.10 Chapter 10 – Other Considerations

Chapter 10 is a resource for hydraulic engineers to identify additional factors that may impact bridge design and structure safety. These topics include bridge deck drainage, hydraulic forces on bridge decks, piers and pile groups, coincident flows at confluences of rivers, physical modeling, and computational fluid dynamics.

1.3.11 Chapter 11 – Literature Cited

Chapter 11 provides all the references for the document. For references that are produced by government agencies (FHWA, USGS, NCHRP etc.) the agency is indicated as the author. This format was selected to group all agency documents together in the reference list. The authors are listed within the reference.

1.4 DUAL SYSTEM OF UNITS

HDS 7 uses dual units (U.S. Customary and SI metric). **In Appendix A, the metric (SI) unit of measurement is explained. The conversion factors, physical properties of water in the U.S. Customary and SI systems of units, sediment particle size grade scale, and some common equivalent hydraulic units are also given.** This edition uses for the unit of length the foot (ft) or meter (m); of mass the slug or kilogram (kg); of weight/force the pound (lb) or newton (N); of pressure the lb/ft^2 or Pascal (Pa, N/m^2); and of temperature degrees Fahrenheit (°F) or Centigrade (°C). The unit of time is the same in U.S. Customary as in SI system (seconds, s). Sediment particle size is given in millimeters (mm), but in calculations the decimal equivalent of millimeters in meters is used (1 mm = 0.001 m) or for the U.S. Customary system in feet (ft). The value of some hydraulic engineering terms used in the text in U.S. Customary units and the equivalent SI units are given in Table 1.1.

Table 1.1. Commonly Used Engineering Terms in U.S. Customary and SI Units.

Term	U.S. Customary Units	SI Units
Length	3.28 ft	1 m
Volume	35.31 ft^3	1 m^3
Discharge	35.31 ft^3/s	1 m^3/s
Acceleration of Gravity	32.2 ft/s^2	9.81 m/s^2
Unit Weight of Water	62.4 lb/ft^3	9800 N/m^3
Density of Water	1.94 $slugs/ft^3$	1000 kg/m^3
Density of Quartz	5.14 $slugs/ft^3$	2647 kg/m^3
Specific Gravity of Quartz	2.65	2.65
Specific Gravity of Water	1	1
Temperature	°F	°C = 5/9 (°F - 32)

CHAPTER 2

DESIGN CONSIDERATIONS AND REGULATORY REQUIREMENTS

2.1 INTRODUCTION

Hydraulic engineers and designers are faced with a wide variety of choices when determining the capacity or location of a new bridge or an existing bridge that is to be replaced. In addition to the choices regarding hydrologic and hydraulic components of a bridge hydraulic analysis there are many other factors and requirements to consider.

One early consideration is the level of service the bridge is expected to provide. If the bridge is remote and carries a low volume of traffic, it can be designed with a lower hydraulic capacity resulting in a smaller and less expensive bridge. This means that the bridge and/or approach roadways will be overtopped more frequently and the bridge owner can expect the bridge and approach roadways to require more frequent maintenance and repair. On the other hand, if the bridge is on an important route such that significant hardships or economic impacts would be encountered if the bridge were out of service, then it should be designed with a higher hydraulic capacity resulting in a larger and more expensive bridge and higher approach embankments. These bridges and/or approach roadways would be rarely overtopped and would need less frequent maintenance or repair. A smaller bridge may be less expensive from a capital (initial) cost perspective, but this does not necessarily always hold true from a life-cycle cost perspective. Most states or local jurisdictions have policies and criteria that govern the level of service expected from their roadways and bridges.

There are also a significant number of permits that may be required when designing or replacing a bridge. Federal, state, and local agencies have diverse and important interests regarding the design and construction of bridges. A good hydraulic analysis conducted early in the design process and a thorough understanding of the permitting and approval process helps avoid costly redesigns or delays, and problems with permitting.

2.2 BRIDGE OPENING AND ROAD GRADE DESIGN CONSIDERATIONS

In general, given a particular design discharge at a given crossing, the shorter a bridge the more backwater it will create. This same smaller bridge will also have higher velocities through the bridge opening and an increased potential for scour at the bridge foundation. A longer bridge at this same crossing will generate a smaller amount of backwater and will have lower velocities and potential for scour. Policy considerations and economics require an understanding of the impacts that the bridge could have on the flow of water in the floodplain and impacts it might have on adjacent properties.

The bridge waterway width is directly associated with the bridge length, from abutment to abutment. Hydraulic capacity should be a primary consideration in setting the bridge length. The bridge must provide enough capacity to:

- Avoid excessive backwater in order to prevent adverse floodplain impacts
- Prevent excessive velocity and shear stress within the bridge waterway

Freeboard refers to the vertical distance from the water surface upstream of the bridge to the low chord of the bridge. The freeboard requirement is associated with a particular design recurrence-interval event, which is usually the 50- or 100-year event. Rural, low-traffic routes often allow a lower recurrence interval for establishing hydraulic capacity and freeboard.

The road profile can have a significant effect on bridge crossing hydraulics. Even if a bridge is designed to provide freeboard above a 100-year flood, the approach roadways may be overtopped by that same flood. When the overtopping occurs over a long segment of roadway, the associated weir flow is an important component of the overall hydraulic capacity of the crossing. In such a case, raising the road profile will have the potential to increase backwater unless additional capacity is provided in the bridge waterway to compensate for the lost roadway overtopping flow capacity.

The design of the piers and abutments has an effect on the bridge hydraulic capacity. Although this effect is small compared to the bridge length and road profile, it can still be important. For example, a bridge that crosses a regulatory floodway must be shown to cause no increase in backwater over existing conditions. In such a case the energy losses that are affected by the number of piers and their geometry can be significant. Spill-through abutments, set well back from the tops of the main channel banks, are advisable when bridge hydraulic capacity must be optimized.

Frequently the bridge waterway design includes subtle changes to the channel cross section under the bridge and for a short distance upstream and downstream of the bridge. These changes are intended to enhance channel stability and, in some cases, to improve hydraulic efficiency. Channel stability can be enhanced, for instance, by grading the channel banks to side slopes of 2H:1V or flatter, and by providing channel bank revetment. Capacity can be improved by a moderate widening of the channel bottom in the immediate vicinity of the bridge, with appropriate width transitions upstream and downstream.

There are several potential bridge opening and road grade considerations that impact hydraulic capacity and upstream flood risk, especially when a road is upgraded and the bridge is replaced. These include bridge length, deck width, abutment configuration (spill through or vertical wall), number and size of piers, low chord elevation, freeboard, and road grade. If a crossing with a 25-year level of service is improved to a 50-year level of service, the road elevation may need to be increased. To avoid increased flood risk, the replacement bridge may need to be considerably longer and higher than the existing bridge. If there is inadequate freeboard, debris may collect along the deck and reduce flow conveyance.

2.3   FLOODPLAIN AND FLOODWAY REGULATIONS

A number of federal regulations affect the design and construction of bridges across the nation's waterways. Executive Order (EO) 11988, which became law in 1977, is the source from which the federal floodplain regulations are derived. The Federal Emergency Management Agency (FEMA 2010) provides the following information regarding EO 11988.

> Executive Order 11988 requires federal agencies to avoid to the extent possible the long- and short-term adverse impacts associated with the occupancy and modification of flood plains and to avoid direct and indirect support of floodplain development wherever there is a practicable alternative. In accomplishing this objective, "each agency shall provide leadership and shall take action to reduce the risk of flood loss, to minimize the impact of floods on human safety, health, and welfare, and to restore and preserve the natural and beneficial values served by flood plains in carrying out its responsibilities" for the following actions:
>
> - Acquiring, managing, and disposing of federal lands and facilities

- Providing federally-undertaken, financed, or assisted construction and improvements
- Conducting federal activities and programs affecting land use, including but not limited to water and related land resources planning, regulation, and licensing activities

The guidelines address an eight-step process that agencies should carry out as part of their decision-making on projects that have potential impacts to or within the floodplain. The eight steps, which are summarized below, reflect the decision-making process required in Section 2(a) of the Order.

1. Determine if a proposed action is in the base floodplain (that area which has a one percent or greater chance of flooding in any given year).
2. Conduct early public review, including public notice.
3. Identify and evaluate practicable alternatives to locating in the base floodplain, including alterative sites outside of the floodplain.
4. Identify impacts of the proposed action.
5. If impacts cannot be avoided, develop measures to minimize the impacts and restore and preserve the floodplain, as appropriate.
6. Reevaluate alternatives.
7. Present the findings and a public explanation.
8. Implement the action.

Among a number of things, the Interagency Task Force on Floodplain Management clarified the EO with respect to development in floodplains, emphasizing the requirement for agencies to select alternative sites for projects outside the floodplains, if practicable, and to develop measures to mitigate unavoidable impacts.

FHWA regulations regarding the implementation of EO 11988 can be found in Title 23, Section 650, Subpart A - Location and Hydraulic Design of Encroachments on Flood Plains of the Code of Federal Regulations (23 CFR 650A). An FHWA policy statement referred to as Non-Regulatory Supplement Attachment 2 provides additional guidance on complying with the floodplain provisions of 23CFR650A. FEMA procedures for implementing this EO are found in Title 44 Part 9 of the Code of Federal Regulations (44 CFR 9).

Floodplain regulations create constraints on the allowable backwater for the design of a new bridge. The stringency of the constraint depends upon the status of the particular floodplain being crossed. When a bridge project is to cross a FEMA floodplain featuring an established regulatory floodway, the hydraulic engineer for the project must demonstrate that the fill and/or bridge elements to be constructed within the floodway will not cause any increase in the 100-year flood water surface elevation compared to existing conditions. This constraint is often termed a no-rise requirement. If meeting the no-rise requirement is not practicable, the bridge owner must coordinate with the local community floodplain administrator and with FEMA to revise the floodplain mapping and floodway boundaries as appropriate. In such a case the local community could be sanctioned or penalized by FEMA under the National Flood Insurance Program unless a Conditional Letter of Map Revision (CLOMR) request is submitted to and approved by FEMA prior to the beginning of project construction.

When crossing a FEMA floodplain without a regulatory floodway, federal regulations are less stringent. In such a case the federal regulations allow the project, combined with existing development that has occurred since the floodplain map became effective, and with other future developments that might reasonably be anticipated, to cause up a 1.0 foot increase in the 100-year flood water surface elevation. In many locations, however, state regulations or local ordinances may impose a more stringent constraint than the federal regulations.

Some bridge projects involve replacing an existing floodplain crossing that causes significant backwater over pre-bridge conditions. If the bridge to be replaced is known to cause more than 1 foot of backwater, it is advisable to design the replacement bridge to avoid any additional backwater, even if the floodplain regulations might allow a moderate increase. In such a case the new bridge should result in some reduction of the backwater.

Even though floodplain regulations are derived from federal laws, floodplain management is a function of state or local government. The project-specific enforcement of floodplain regulations, therefore, typically takes the form of floodplain permits from state or local agencies. It is the responsibility of the bridge owner to assure that any potential designs meet the criteria outlined in these regulations and assure that all required floodplain permits are applied for and received before the construction of a new or replacement bridge takes place.

2.4 SCOUR AND STREAM STABILITY CONSIDERATIONS AND GUIDANCE

Another critical component of the design and/or evaluation of a bridge opening is to design the bridge to be stable from scour at the piers, abutments, and across the contracted opening. From a hydraulic perspective, the magnitude of local scour at a pier is a function of depth and velocity of flow, alignment of the pier with flow, and pier type and location. Depending of foundation costs and complexity it will be necessary to balance the number and size of piers, length and height, and anticipated total scour depth against increased costs of the superstructure associated with longer spans (girder type and allowable span) and foundation required to resist scour.

The magnitude of local scour at an abutment is a function of depth and velocity of flow, the skew of the embankment to the floodplain, as well as the amount of flow from the overbank that passes through the bridge opening. It is also a function of where the abutment is located in relation to the main channel. It is recommended that an abutment not be located in or close to the main channel if at all possible.

The amount of contraction scour that occurs at a bridge crossing is a function of the degree that a bridge contracts floodplain flow. In general, bridges with higher degrees of contraction can be expected to have higher flow velocities and larger scour depths. If the depths of contraction scour are too large it may be necessary to increase the bridge length to reduce scour across the bridge opening.

Bridges should be designed to withstand scour from large floods and from stream instabilities expected over the life of a bridge. Recommended procedures for evaluating and designing bridges to resist scour can be found in FHWA publications HEC-20 (FHWA 2012a) and HEC-18 (FHWA 2012b).

2.5 NAVIGATION REQUIREMENTS

The FHWA regulation 23 CFR 650 Subpart H (Navigational Clearances for Bridges) establishes policy and sets forth coordination procedures for Federal-aid highway bridges which require navigational clearances. This regulation involves a bridge owner applying for and obtaining a bridge permit from the U.S. Coast Guard (USCG). However, the regulation also involves ensuring NEPA coordination and compliance with other guidelines and specifications to ensure that bridges do not interfere with navigational requirements in the waters of the United States. Specifically, the policy of FHWA is (a) to provide clearances which meet the reasonable needs of navigation and provide for cost-effective highway operations; (b) to provide fixed bridges wherever practicable, and; (c) to consider appropriate pier protection and vehicular protection and warning systems on bridges subject to ship collisions.

The American Association of State Highway and Transportation Officials (AASHTO) acknowledges these requirements within respective (i.e., Standard (AASHTO 2005) and LRFD (AASHTO 2010)) bridge design specifications. These specifications include articles on vertical and horizontal clearances. The U.S. Army Corps of Engineers (USACE) conducted a research investigation that helped determine the basis for setting these clearances (FHWA 1984a). The FHWA and AASHTO collaborated to produce the articles on vessel collision contained within the AASHTO specifications.

Not all bridges will require U.S. Coast Guard (USCG) permits, however only FHWA (in coordination with USCG) may make such a determination. Therefore, safe and prudent bridge design should seek to investigate and resolve such issues during the environmental documentation phase of a project.

2.6 SECTION 404 REQUIREMENTS

The USACE Regulatory Program regulates the discharge of fill placed within waters of the United States, through Section 404 of the Federal Clean Water Act (CWA). The Regulatory Program authorizes two types of permits: Standard Permit and General Permit. A Standard Permit authorizes impacts which have more than minimal impact to the waters of the United States. These permits are typically needed for the larger impact projects and require a more thorough review by the Regulatory Program. A General Permit authorizes minimal adverse impacts to waters of the United States. There are two types of General Permits: Regional General Permits and Nationwide Permits. Regional General Permits are issued for projects which are similar in nature and are typically issued for a specific geographic region by a local U.S. Army Corps of Engineers District. Nationwide Permits authorize only minimal adverse environmental impacts. As of January 2012 there are 50 Nationwide Permits, which authorize activities from utility line installation to minor fill discharge. The three Nationwide Permits which are often utilized by DOT's are Nationwide Permits 3 (Maintenance of currently serviceable structures), 13 (Bank Stabilization), and 14 (Linear Transportation Projects).

If a project involves work within a water of the United States, the bridge owner must coordinate with the local USACE regulatory office to determine whether a permit is required. Some activities for scour protection projects may be exempt but some may require a permit. The permitting process can take a few months to more than a year, so it is imperative to address the issue early in the design process. This is particularly true if the project requires a standard permit rather than a general permit.

2.7 AASHTO SPECIFICATIONS AND DESIGN CRITERIA

The FHWA regulation 23 CFR 625 (Design Standards for Highways) requires the use of AASHTO Specifications and Design Standards for bridge design. Specifically, there are three AASHTO documents applicable to bridge owners. The earliest specification is typically referred to as the "Standard Specifications" (AASHTO 2005). The newer specification uses Load and Resistance Factor Design (LRFD) principles and is known as the "LRFD Specifications" (AASHTO 2010). As of 2007, the FHWA requires all new bridges to be designed using the LRFD specifications. The FHWA also requires that a bridge owner follow guidance outlined in their state hydraulic or drainage manual or, if not available, the "AASHTO Highway Drainage Guidelines" (AASHTO 2007).

CHAPTER 3

GOVERNING EQUATIONS AND FLOW CLASSIFICATION

3.1 INTRODUCTION

This chapter provides background on the fundamentals of rigid boundary open channel flow. Although this is not a hydraulic engineering textbook, there is sufficient information to act as a source reference on the equations used in open channel and bridge hydraulics. In open channel flow the upper surface of the water is in contact with the atmosphere, therefore, the surface configuration, flow pattern and pressure distribution depend primarily on gravity. Because the flow involves a free surface, it has more degrees of freedom than flow in a closed conduit flowing full. The types of flow include:

- One-, two-, or three-dimensional
- Real or ideal fluid (viscid or inviscid)
- Incompressible or compressible
- Steady or unsteady
- Pressure or gravity
- Uniform or nonuniform (varied)
- Gradually or rapidly varied
- Laminar or turbulent
- Subcritical or supercritical (tranquil or rapid)

The following sections will emphasize open channel flow as being: (1) uniform or nonuniform (varied) flow; (2) steady or unsteady flow; (3) laminar or turbulent flow; and (4) subcritical or supercritical.

3.1.1 Streamlines and Streamtubes

The motion of each fluid particle is described in terms of its velocity vector, V, which is defined as the time rate of change of the position of the particle. The particle's velocity is a vector quantity with a magnitude (the speed, $V = |V|$) and direction. As the particle moves, it follows a particular path, which is governed by the velocity of the particle. The location of the particle along the path is a function of where the particle started at the initial time and its velocity along the path. If the flow is steady (i.e., nothing changes with time at a given location in the flow field), each successive particle that passes through a given point such as point (1) in Figure 3.1, will follow the same path. For such cases the path is a fixed line in the X-Z plane. Neighboring particles that pass either side of point (1) follow their own paths, which may be of different shape but do not cross the one passing through (1). The entire X-Z plane is filled with such paths.

For steady flow each particle progresses along its path and its velocity vector is everywhere tangent to the path. The lines that are tangent to the velocity vectors throughout the flow field are called streamlines. For many situations it is easiest to describe the flow in terms of the "streamline" coordinates based on the streamlines as shown in Figure 3.1. The particle motion is described in terms of its distance along the streamline. The distance along the streamline is related to the particle speed by $V = ds/dt$, and the radius of curvature is related to the shape of the streamline. In addition to the equal potential coordinates along the streamlines, the coordinate normal to the streamline, n, will be of use in the applications of open channel flow.

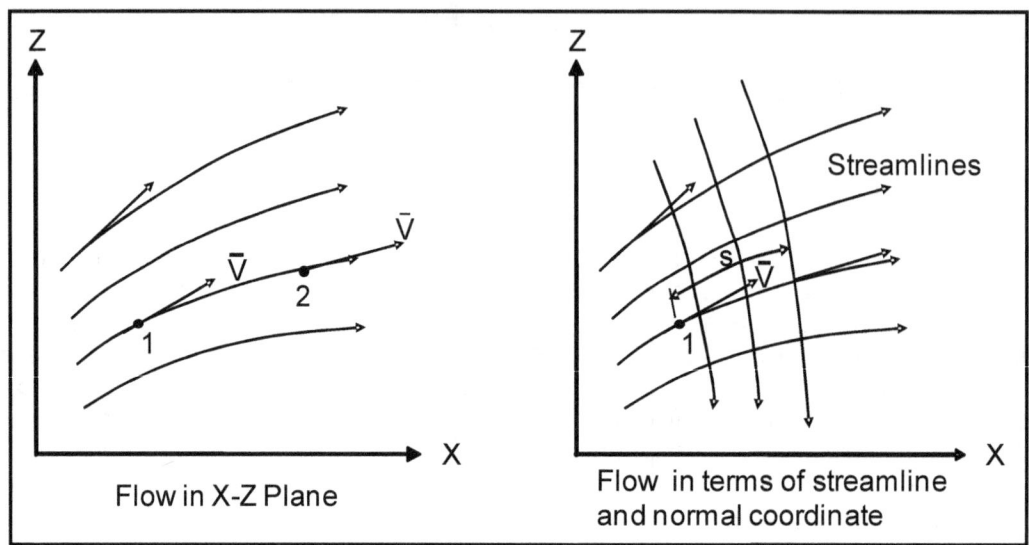

Figure 3.1. Flow in the X-Z plane and flow in terms of streamline and normal coordinates.

3.1.2 Definitions

Velocity: The velocity of a fluid particle is the rate of displacement of the particle from one point to another and is a vector quantity having both magnitude and direction. The mathematical formulation of velocity magnitude is given in Equation 3.1.

$$V = \frac{ds}{dt} \qquad (3.1)$$

Streamline: A streamline is an imaginary line within the flow that is tangent everywhere to the velocity vector, see Figure 3.1. Since the flow is tangent to the streamline, there cannot be any net movement of fluid across the streamline in any direction.

Streamtube: A streamtube is an element of fluid bounded by a pair of streamlines that enclose or confine the flow. Since there can be no net movement of fluid across a streamline, it follows that there can be no net movement of fluid in or out of the streamtube, except at the ends. This fact will be utilized in the development of the continuity equation.

Acceleration: Acceleration is the time rate of change in magnitude or direction of the velocity vector. Acceleration can be expressed by the total derivative of the velocity vector as follows:

$$a = \frac{dv}{dt} \qquad (3.2)$$

The vector acceleration, a, has components both tangential and normal to the streamline, the tangential component representing the change in magnitude of the velocity, and the normal component reflecting the change in direction:

$$a_s = \frac{dv_s}{dt} = \frac{\partial v_s}{\partial t} + 1/2 \frac{\partial (v_s^2)}{\partial s} \qquad (3.3)$$

$$a_n = \frac{dv_n}{dt} = \frac{\partial v_n}{\partial t} + \frac{(v^2)}{r} \tag{3.4}$$

The first terms in Equations 3.3 and 3.4 represent the change in velocity, both magnitude and direction, with time at a given point. This is called local acceleration. The second term in each equation is the change in velocity, both magnitude and direction, with distance. This is called convective acceleration.

3.1.3 Classification of Open Channel Flow

Uniform flow in open channels depends upon there being no change with distance in either the magnitude or the direction of the velocity along a streamline; that is both $\partial v/\partial s = 0$, and $\partial v/\partial n = 0$. Nonuniform flow in open channels occurs when either $\partial v/\partial s \neq 0$ or $\partial v/\partial n \neq 0$. The particular type of nonuniform flow that occurs when $\partial v/\partial s \neq 0$ in open channels is usually called varied flow. Figure 3.2 illustrates uniform flow in a straight channel having a constant depth of flow, a constant slope, and a constant cross section throughout. Obviously, this condition seldom exists in nature. Examples of nonuniform flow, where $\partial v/\partial n \neq 0$, are bends or curving sides of the channel. When $\partial v/\partial s \neq 0$, the flow is varied and occurs when there is a change in depth of flow due either to a change in slope, a barrier or drop, or a change in side or bottom, so that the velocity increases or decreases in the direction of flow, Figure 3.3.

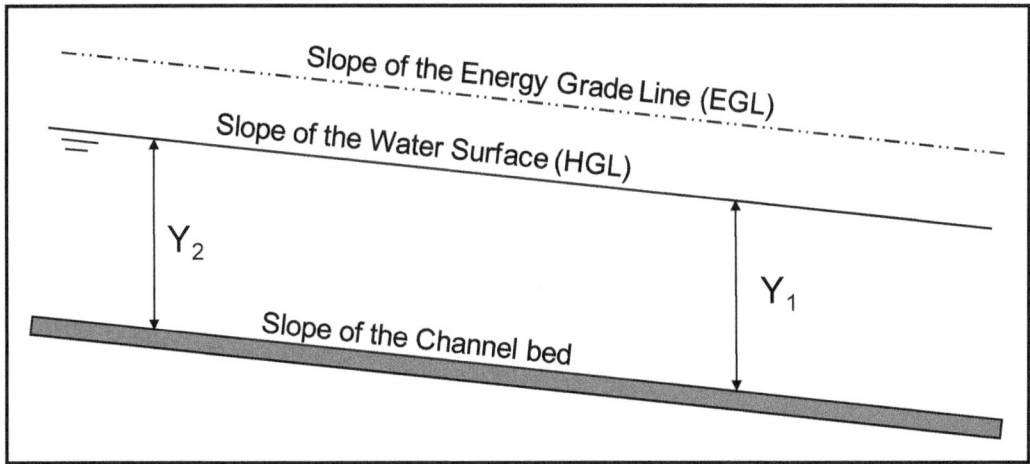

Figure 3.2. Example of uniform flow ($Y_2 = Y_1$).

There are three slopes of interest in Figures 3.2 and 3.3. The channel bed slope, S_o, water surface slope, S_w, often referred to as the hydraulic grade line (HGL), and slope of the energy grade line (EGL), S_f, often referred to as the friction slope.

Steady flow occurs when the velocity at a point does not change with time, that is $\partial v/\partial t = 0$. When the flow is unsteady, $\partial v/\partial t \neq 0$. Examples of unsteady flow are traveling surges and flood waves in an open channel.

Whether laminar or turbulent flow exists in an open channel depends upon the Reynolds Number, R_e, of the flow. Laminar flow occurs when viscous forces are predominant compared with the inertial forces, and turbulent flow occurs when the inertial forces are great compared with forces of viscosity. Laminar flow in open channels occurs very infrequently, except with special liquids such as oils or extreme concentration sediment mixtures.

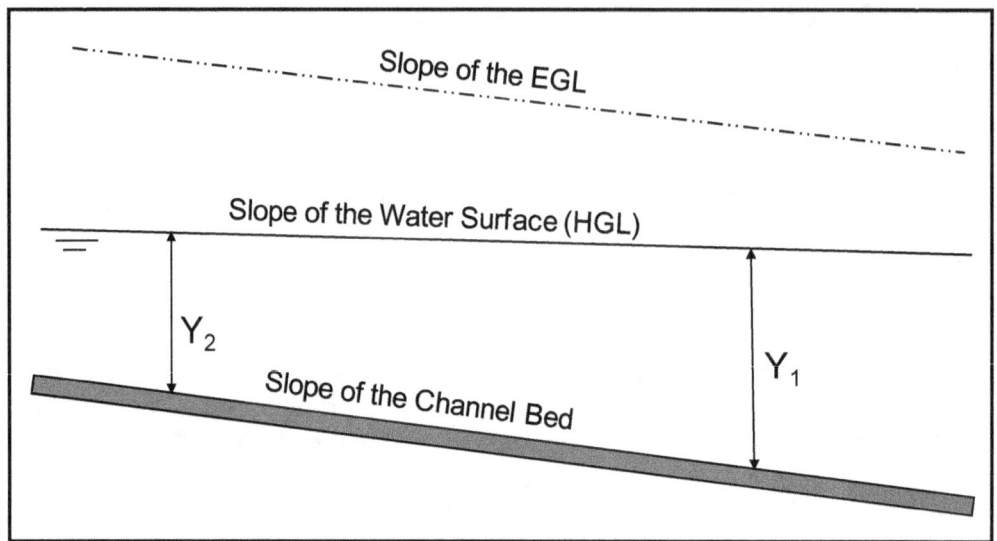

Figure 3.3. Example of nonuniform flow where the depth of flow $Y_2 \neq Y_1$.

Unlike laminar and turbulent flow, subcritical and supercritical flows exist only with a free surface or interface. The criterion for subcritical (tranquil) and supercritical (rapid) flow is the Froude Number, $F_R = \dfrac{V}{\sqrt{gy}}$, which like the Reynolds Number, $R_e = \dfrac{VD\rho}{\mu}$, is the ratio of two types of forces.

The Froude number is a ratio of the forces of inertia to the forces of gravity and is discussed in detail later in this chapter. It will suffice at this point to indicate that when the $F_R = 1$ the flow is critical, when $F_R \leq 1$ the flow is tranquil, and when the $F_R \geq 1$ the flow is rapid.

In the above discussion, there are four classifications needed to describe the type of flow in an open channel.

1. Uniform or nonuniform
2. Steady or unsteady
3. Laminar or turbulent
4. Subcritical or supercritical

One from each of these four types must exist. Because the classifying characteristics are independent, sixteen different types of flow can occur. These terms, uniform or nonuniform, steady or unsteady, laminar or turbulent, subcritical or supercritical, and the two dimensionless numbers (the Froude number and the Reynolds number) are more fully explained in the following sections.

3.2 THREE BASIC EQUATIONS OF OPEN CHANNEL FLOW

The basic equations of flow in open channels are derived from the three conservation laws. These are: (1) the conservation of mass, (2) the conservation of energy, and (3) the conservation of linear momentum. The conservation of mass is another way of stating that (except for mass-energy interchange) matter can neither be created nor destroyed. The conservation of energy is an empirical law of physics that in a closed system the energy remains constant over time. Similar to the conservation of mass, energy can neither be created nor destroyed, although it can be transformed from one state to another (i.e., kinetic energy to potential energy). The principle of conservation of linear momentum is based on Newton's second law of motion which states that a mass (of fluid) accelerates in the direction of and in proportion to the applied forces on the mass.

In the analysis of flow problems, much simplification can result if there is no acceleration of flow or if the acceleration is primarily in one direction and the accelerations in other directions are negligible. However, a very inaccurate analysis may occur if it is assumed that accelerations are small or zero when in fact they are not. The concepts in this chapter assume one-dimensional flow and the derivations of the equations utilize a control volume concept. A control volume (Figure 3.4) is a volume which is fixed in space or moving with the fluid and through whose boundary matter, mass, momentum, energy can flow. The volume is called a control volume and its boundary is a control surface.

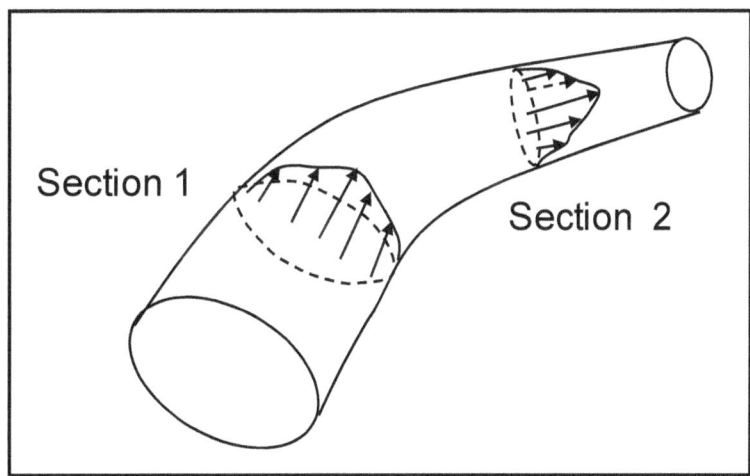

Figure 3.4. Streamtube with fluid flowing from Section 1 to Section 2.

3.2.1 Conservation of Mass

The application of the principle of conservation of mass to a steady flow in a streamtube results in the equation of continuity, which describes the continuity of the flow from section to section of the streamtube. The derivation uses a control volume and the fluid system that just fills the volume at a particular time, t.

<u>Integral Form of the One-Dimensional Continuity Equation</u>. The conservation of mass for the control volume in Figure 3.4 can be stated as:

$$\left| \begin{array}{c} \text{Mass flux out of} \\ \text{the control volume} \end{array} \right| - \left| \begin{array}{c} \text{Mass flux into the} \\ \text{control volume} \end{array} \right| + \left| \begin{array}{c} \text{Time rate of change in mass} \\ \text{in the control volume} \end{array} \right| = 0$$

Mass can enter or leave the control volume through any or all of the control volume surfaces. Rainfall would contribute mass through the surface of the control volume and seepage passes through the interface between the water and the banks and bed. In the absence of any lateral mass fluxes, the mass enters the control volume at section 1 and leaves at section 2, or

$$\frac{Dm}{Dt} = 0 \tag{3.5}$$

The fluid mass can be represented by the density times the volume, ρdxdydz, and Equation 3.5 can be written as:

$$\frac{d}{dt}\int_{CVol} \rho dxdydz + \int_{CS} \rho dxdydz = 0 \tag{3.6}$$

Orienting the axes for sections one and two to be perpendicular to dx, then

dx/dt = Velocity in the x direction (3.7)

In the absence of any lateral mass fluxes, the mass enters the control volume at section 1 and leaves at section 2. Further assuming steady flow, Equation 3.6 can be written as:

$$\int_{A_1}\left(\rho\frac{dx}{dt}dydz\right)_1 = \int_{A_2}\left(\rho\frac{dx}{dt}dydz\right)_2 \tag{3.8}$$

Substituting Equation 3.7 into 3.8 the continuity equation reduces to inflow equal outflow.

$$\int_{A_1}\rho_1(v_1 \bullet n_1)dA_1 = \int_{A_2}\rho_2(v_2 \bullet n_2)dA_2 \tag{3.9}$$

Where n is the outward normal unit vector, and A is the area of the control surface that the flow is passing through. The dot product of the velocity vector with the unit outward normal (V x n) determines the component of the velocity perpendicular to the surface since only that component can carry mass through the surface. For most open channel flow situations flow is virtually incompressible and the equation reduces to:

$$\int_{A_1}(v_1 \bullet n_1)dA_1 = \int_{A_2}(v_2 \bullet n_2)dA_2 \tag{3.10}$$

It is often convenient to work with the average conditions at a cross section, so an average velocity V is defined such that

$$V = \frac{1}{A}\int_A vdA \tag{3.11}$$

The symbol v represents the local velocity whereas V is the average velocity at the cross section. Therefore, for steady incompressible flow the continuity equation can be reduced to

$$A_1V_1 = A_2V_2 = Q = AV \tag{3.12}$$

where Q is the volume flow rate or the discharge. Equation 3.12 is the familiar form of the conservation of mass equation for steady flow in open channels. It is applicable when the fluid density is constant, the flow is steady and there is no significant lateral inflow or seepage. The velocity will generally vary in both direction and magnitude over the cross section and the summation (integral) of the area over the cross section must be at right angles to the velocity component.

Two-Dimensional Form of the Continuity Equation.

Considering flow through a small control volume as shown in Figure 3.5, assume a general flow, $V(x,y,z,t)$. Flow through surfaces 1 and 2 are perpendicular to the Y-Z plane. Note that the efflux rate through area 1 is $-\rho V_x$ per unit area and the flow through the area is given as $\rho V_x + \{\partial(\rho V_x)/\partial x\}dx$. Therefore, the net efflux through the surface would be $\{\partial(\rho V_x)/\partial x\}dx$. Performing similar computations for the other sides and adding the results the total net efflux rate is:

$$\text{Net Efflux Rate} = \left\{\frac{\partial(\rho V_x)}{\partial x} + \frac{\partial(\rho V_y)}{\partial y} + \frac{\partial(\rho V_z)}{\partial z}\right\} dx\, dy\, dz$$

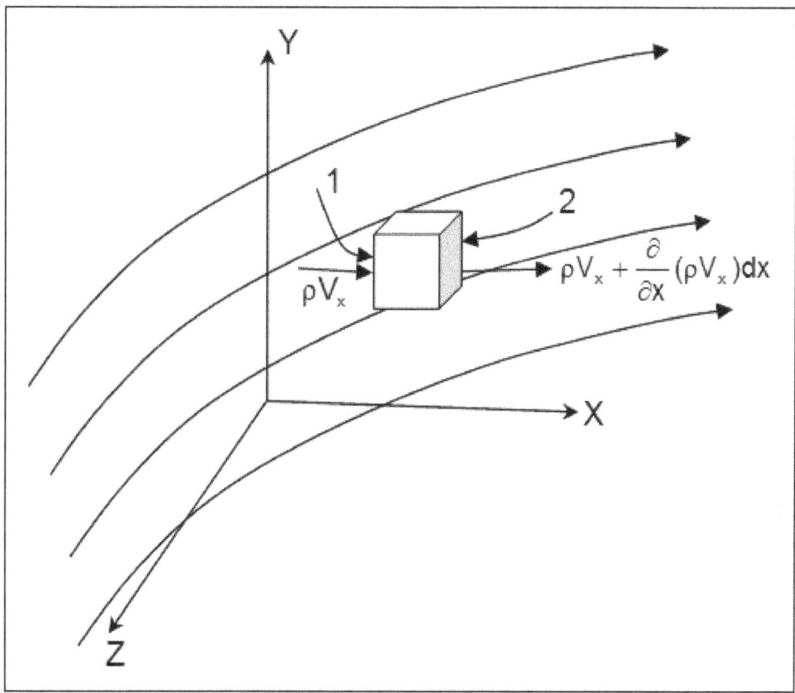

Figure 3.5. Net flow through a control volume.

The rate of decrease of mass inside the control volume equals $-\partial\rho/\partial t\,(dx\,dy\,dz)$. Dividing both sides of the equation by $dx\,dy\,dz$ yields the following relationship:

$$\left\{\frac{\partial(\rho V_x)}{\partial x} + \frac{\partial(\rho V_y)}{\partial y} + \frac{\partial(\rho V_z)}{\partial z}\right\} = \frac{-\partial\rho}{\partial t}$$

And for steady incompressible flow ρ = constant and $\partial/\partial t = 0$, which gives the simplified differential form of the continuity equation.

$$\left\{\frac{\partial(V_x)}{\partial x} + \frac{\partial(V_y)}{\partial y} + \frac{\partial(V_z)}{\partial z}\right\} = 0$$

The equation states that for steady flow the rate of flow into the control volume must be equal to the rate of flow out.

3.2.2 Conservation of Energy

The first law of thermodynamics is the application of the conservation of energy to heat and thermodynamic processes. The first law of thermodynamics stipulates that energy must at all times be conserved. The first law accounts for energy entering, leaving, and accumulating in either a system or a control volume. Significant insight into the basic laws of fluid flow can be obtained from a study of flow of a hypothetical ideal fluid. An ideal fluid is a fluid assumed to be inviscid (having no viscosity). In such a fluid there are no frictional effects between moving fluid layers or between these layers and the boundary walls, and therefore, no energy dissipation due to friction. The assumption that a fluid is ideal allows it to be treated as an aggregation of small particles that will support pressure forces normal to their surfaces but will slide over one another without resistance. The unbalanced forces can be solved with Newton's second law.

Under the assumption of frictionless motion, equations are considerably simplified and more easily understood. This section introduces the Bernoulli equation that can be used to relate and predict pressures and velocities in a flow field. Euler first applied Newton's second law to the motion of fluid particles. Consider a streamline and select a fluid particle shown in Figure 3.6.

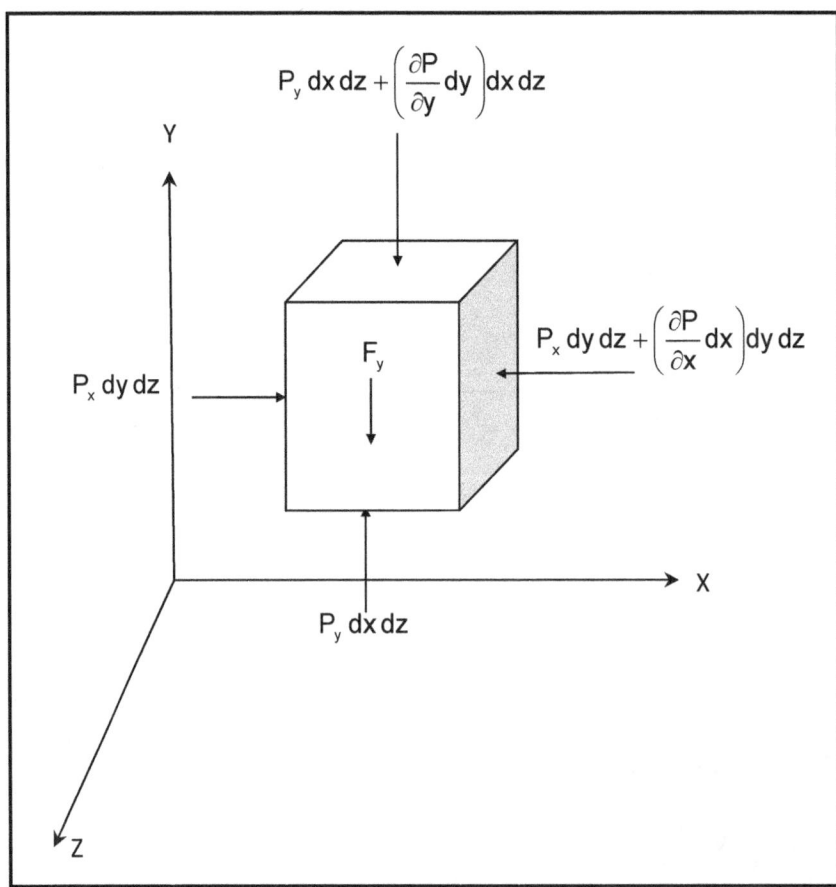

Figure 3.6. Surfaces forces acting on a fluid element in the X and Y directions for an inviscid fluid.

The forces tending to accelerate the particle are: pressure forces on the ends of the system, pdA - (p+dp)dA = dpdA (the pressure on the sides of the system have no effect on its acceleration), and the component of weight in the direction of motion, $-\rho g_n$ dsdA (dz / ds = $-\rho g_n$dAdz. The differential mass being accelerated by the action of these differential forces is dM = ρdsdA. Applying Newton's second law, dF = (dM)a, along a streamline and using the one dimensional expression for acceleration for steady flow, $a_s = v\dfrac{dv}{ds}$ gives:

$$-dpdA - \rho g_n dAdz = (\rho dsdA)v\frac{dv}{ds} \tag{3.13}$$

Dividing by pdA produces the one-dimensional Euler equation

$$\frac{dp}{\rho} + vdv + g_n dz = 0 \tag{3.14}$$

For incompressible flow the equation can be divided by g_n and written as

$$\frac{dp}{\gamma} + d\left(\frac{v^2}{2g_n}\right) + dz = 0 \tag{3.15}$$

For uniform density, Equation 3.15 can be rewritten as

$$d\left(\frac{p}{\gamma} + \frac{v^2}{2g_n} + z\right) = 0 \tag{3.16}$$

For incompressible flow of a uniform density fluid, the one-dimensional Euler equation can be integrated between any two points (because γ and g_n are both constant) along a streamline to obtain the Bernoulli equation

$$\frac{p_1}{\gamma} + \frac{V_1^2}{2g_n} + z_1 = \frac{p_2}{\gamma} + \frac{V_2^2}{2g_n} + z_2 \tag{3.17}$$

Equation 3.17 applies to all points on the streamline and thus provides a useful relationship between pressure, the magnitude of the velocity, and the height above the datum. An empirical relationship can be added to account for the losses in the system. These losses include friction and transition (expansion and contraction) losses that are described in Chapter 5.

Velocity Distribution. As shown in Figure 3.7 the requirement of zero velocity at a boundary for either laminar and turbulent flow produces velocity distributions that are not uniform (nonuniform). The term $v^2/2g$ is the kinetic energy per unit weight at a particular point. If the velocity distribution varies across the section of the flow, the total kinetic energy of the section will be greater than the kinetic energy computed from the average velocity (e.g., the average value of the sum of incremental velocity squared is greater than the average velocity squared).

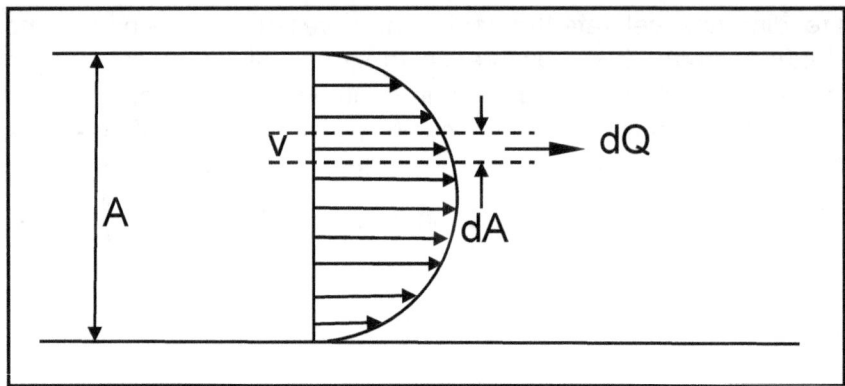

Figure 3.7. Nonuniform velocity distribution.

As indicated by Equations 3.11 and 3.12, the average velocity can be represented by:

$$V = \frac{1}{A}\int^A v\, dA = \frac{Q}{A}$$

The total kinetic energy per unit time that passes the section is determined by integrating the product of the kinetic energy per unit weight and the weight of the fluid passing per unit time from streamline to streamline across the section.

$$\text{Energy flux} = \int^A \frac{v^2}{2g}(\gamma\, dQ) = \gamma \int^A \frac{v^2}{2g} v\, dA = \frac{\gamma}{2g}\int^A v^3\, dA \qquad (3.18)$$

The energy flux using the average velocity would then require a correction coefficient α.

$$\text{Energy flux} = \alpha \frac{\gamma}{2g} A V^3 \qquad (3.19)$$

Solving for the energy correction coefficient α yields the following relationship:

$$\alpha = \frac{1}{A}\int^A \left(\frac{v^3}{V^3}\right) dA \qquad (3.20)$$

The energy correction coefficient is used in one-dimensional models (Chapter 5) to account for horizontal velocity variation across a cross section (channel versus floodplains) and is used by some two-dimensional models (Chapter 6) to account for vertical velocity variation at a point.

3.2.3 Conservation of Linear Momentum

The flow shown in Figure 3.4 is complex to analyze in terms of Newton's Second Law because of the curvature in the flow. Therefore, as a starting point, the differential length of reach dx is isolated as a control volume. For this control volume, shown in Figure 3.8, the pressure terms P_1 and P_2 are directed toward the control volume in a direction normal to the sections 1 and 2. Shear stress τ_o is exerted along the interface between the water and wetted perimeter and acts in a direction opposite to the flow. The following derivation was taken from FHWA Hydraulic Design Series (HDS) 6 (FHWA 2001). The statement of conservation of linear momentum is:

| Flux of momentum out of the control volume | - | Flux of momentum into the control volume | + | Time rate of change of momentum in the control volume | = | Sum of the forces acting on the fluid in the control volume |

The terms in the statement are vectors so direction as well as magnitude must be considered. Consider the conservation of momentum in the direction of flow (the x-direction in Figure 3.8). At the outflow section (section 2), the flux of momentum out of the control volume through the differential area dA_2 is:

$$\rho_2 v_2 dA_2 v_2 \tag{3.21}$$

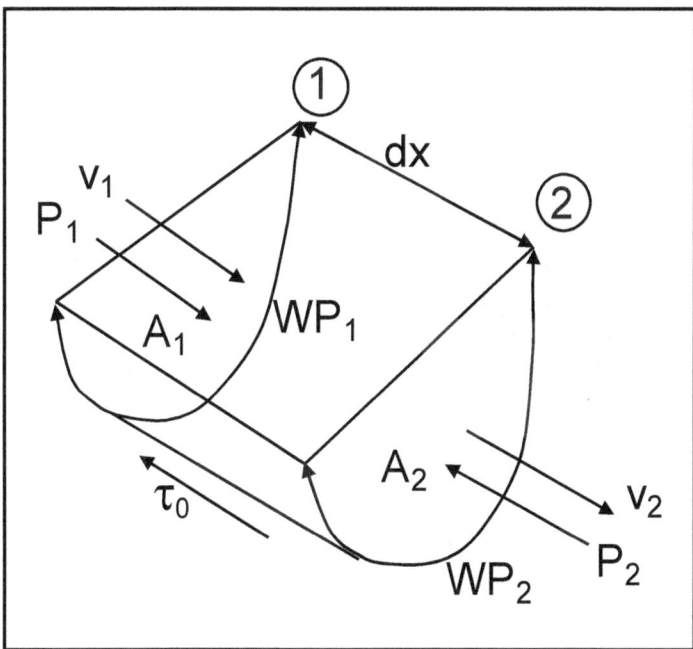

Figure 3.8. The control volume for conservation of linear momentum.

Here $\rho_2 v_2 dA_2$ is the mass flux (mass per unit of time) and $\rho_2 v_2 dA_2 v_2$ is the momentum flux through the area at section 2. The flux of momentum out of the control volume is then:

$$\int_{A_2} \rho_2 v_2 dA_2 v_2 \tag{3.22}$$

Similarly, at the inflow section (section 1), the flux of momentum into the control volume is:

$$\int_{A_1} \rho_1 v_1 dA_1 v_1 \tag{3.23}$$

The amount of momentum in the control volume is $\int_{Vol} \rho v d(Vol)$ and therefore the time rate of change of momentum in the control volume is given by:

$$\frac{\partial}{\partial t}\{\int_{Vol} \rho v d(Vol)\} \qquad (3.24)$$

At the upstream section, the force acting on the differential area dA_1 of the control volume is $p_1 dA_1$, where p_1 is the pressure from the upstream fluid on the differential area. The total force in the x-direction at section 1 is $\int_{A_1} p_1 dA_1$ and similarly at section 2 the total force is $\int_{A_2} p_2 dA_2$. There is also a fluid shear stress τ_o acting along the interface between the water and the bed and banks (over the wetted perimeter). The shear on the control volume is in the direction opposite to the direction of flow and results in a force $-\tau_o WPdx$ where τ_o is the average shear stress on the interface area, and WP is the average wetted perimeter and dx is the length of the control volume. The WPdx is the interface area (the area that the water is in contact with).

The forces affecting the body fall into two classes, surface forces (as were just identified) and body forces. Another surface force not included in this derivation is a wind stress acting on the water surface. Body forces are forces with a long range of influence which act on all the material particles in the body and which, as a rule, have their source in fields of force. The most important example of a body force is the earth's gravity field. The body force component in the x-direction is denoted by F_b and will be discussed in a subsequent section. The statement of conservation of momentum in the x-direction for the control volume is:

$$\int_{A_2} \rho_2 v_2^2 dA_2 - \int_{A_1} \rho_1 v_1^2 dA_1 + \frac{\partial}{\partial t}\int_{Vol} \rho v d(Vol) = \int_{A_1} p_1 dA_1 - \int_{A_2} p_2 dA_2 - \int_L \tau_o WPdx + F_b \qquad (3.25)$$

As with the conservation of mass equation, it is convenient to use average velocities instead of point velocities. The momentum coefficient β is defined so that when average velocities are used instead of point velocities, the correct momentum flux is considered.

$$\beta = \frac{1}{V^2 A}\int_A v^2 dA \qquad (3.26)$$

For steady incompressible flow, Equation 3.25 is combined with Equation 3.26 to give:

$$\rho \beta_2 V_2^2 A_2 - \rho \beta_1 V_1^2 A_1 = \int_{A_1} p_1 dA_1 - \int_{A_2} p_2 dA_2 - \int_L \tau_o WPdx + F_b \qquad (3.27)$$

The pressure force and shear force terms on the right-hand side of Equation 3.27 are usually abbreviated as ΣF_x so:

$$\Sigma F_x = \int_{A_1} p_1 dA_1 - \int_{A_2} p_2 dA_2 - \int_L \tau_o WPdx + F_b \qquad (3.28)$$

The conservation of linear momentum equation becomes:

$$\rho \beta_2 V_2^2 A_2 - \rho \beta_1 V_1^2 A_1 = \sum F_x \qquad (3.29)$$

For steady flow with constant density, combining Equation 3.12 with 3.29, the steady flow conservation of linear momentum equation takes the familiar form:

$$\rho Q (\beta_2 V_2 - \beta_1 V_1) = \sum F_x \qquad (3.30)$$

Depending on the situation, other external forces need to be applied, such as a surface force due to wind blowing over the control volume. This force plays a significant role when analyzing currents during a hurricane storm surge.

3.3 FLOW RESISTANCE AND OTHER HYDRAULIC EQUATIONS

3.3.1 Flow Resistance

This section provides basic information to determine the friction loss coefficients for steady flow in natural channels. The two most commonly used equations for the computation of steady flow in natural channels are the Chezy and Manning equations.

<u>Development of the Manning Formula</u>. In 1889 Manning developed an equation to estimate flow in an open channel as an alternative to measuring flow that passes over or through a hydraulic structure such as a weir or flume. He assumed that the flow was uniform. For this condition, the forces acting on a control volume (Figure 3.9) can be quantified and used to develop flow resistance equations.

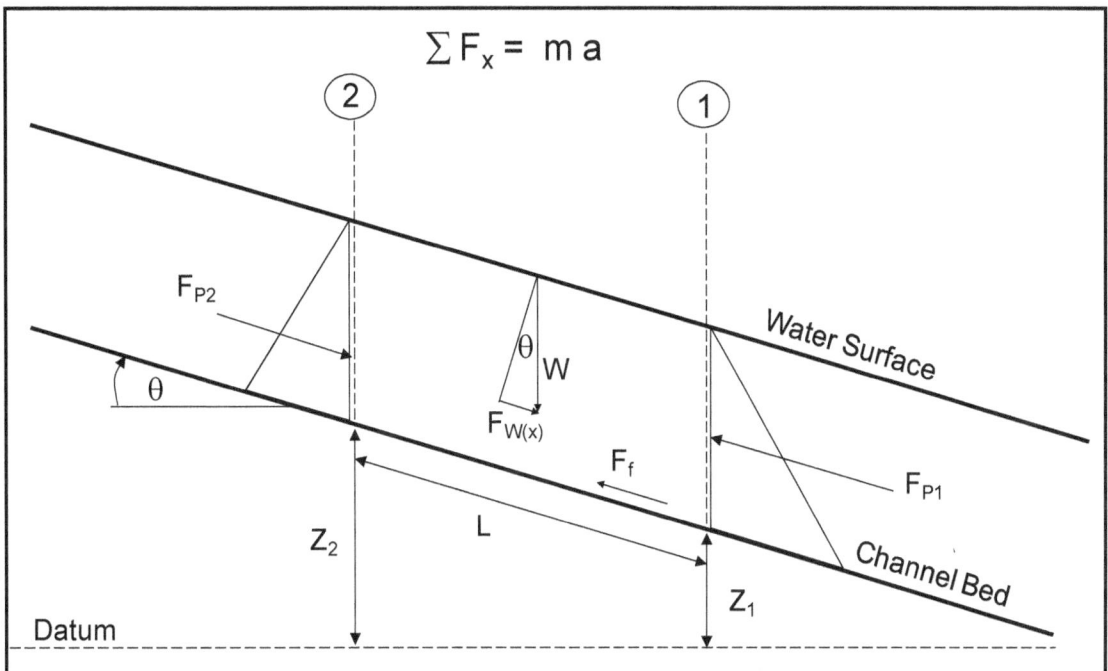

Figure 3.9. Forces acting on a control volume for uniform flow conditions.

Summing the forces acting on the control volume in the x direction gives the following:

$$\sum F_x = ma \qquad (3.31)$$

Noting that the mass times acceleration term is zero for uniform flow conditions (i.e., the velocity at section 1 and is the same as the velocity at section 2, therefore there is no acceleration of the fluid) Equation 3.31 becomes:

$$F_{p1} - F_{p2} + F_{w(x)} - F_f = 0 \qquad (3.32)$$

where:

F_{p1} and F_{p2} = Forces due to hydrostatic pressure at sections 1 and 2
$F_{w(x)}$ = Force due to weight of water in the control volume
F_f = Force opposing the flow due to friction
θ = Slope angle of channel bottom
L = Distance between section 2 and section 1
Z = Distance from channel invert to the datum elevation

Since the depth of flow is the same at sections 1 and 2 and the cross sections are the same, the hydrostatic forces at sections 1 and 2 are equal and opposite in sign.

$$F_{x1} = F_{x2} \qquad (3.33)$$

Therefore Equation 3.33 can be written as:

$$F_w - F_f = 0 \qquad (3.34)$$

Noting that $F_{w(x)} = \frac{\gamma(A_1 + A_2)}{2} LS_o$, weight of water multiplied by bottom of the bed slope, and $F_f = \tau(WP)L$. Substituting these expressions for F_f and F_w, yields:

$$\tau(WP)L = \frac{\gamma(A_1 + A_2)}{2} L \qquad (3.35)$$

Solving for the shear stress, τ, yields:

$$\tau = \gamma \frac{\overline{A}}{WP} S_o \qquad (3.36)$$

where:

$\overline{A}$ is the average area between sections 1 and 2, also note that $A_1 = A_2$ uniform flow. Flow velocity in the channel depends on its cross-sectional shape (among other factors), and the resistance to the flow depends upon the shear stress acting over the channel boundary, the wetted perimeter. The hydraulic radius is defined as the ratio of the channel cross-sectional area to the channel wetted perimeter (the portion in contact with the flow $R = A/WP$ and is defined as the hydraulic radius. Rewriting Equation 3.36 yields.

$$\tau = \gamma R S_o \tag{3.37}$$

For the remainder of this section the subscript on slope will be dropped and only use the symbol S, remembering for uniform flow, $S_f = S_w = S_o$. Therefore, Equation 3.37 can be expressed as:

$$\tau = \gamma R S \tag{3.38}$$

From fundamental fluid mechanics for pipe flow it is assumed that the shear stress $\tau = \dfrac{f \rho V^2}{8}$. Darcy, Weisbach and others developed an expression for the head loss in a long straight cylindrical pipe that is a function of the friction factor, f, of the pipe boundary, the diameter of the pipe, D, the length of the between points of interest (i.e., distance between 1 and 2), and the velocity head, $V^2/2g$ and proposed an equation of the form:

$$h_f = f \frac{L}{D} \frac{V^2}{2g} = SL \tag{3.39}$$

where:

h_f = Head loss between sections 1 and 2
f = Darcy-Weisbach friction coefficient
L = Distance between cross sections 1 and 2
D = Diameter of a pipe

Solving for the velocity in Equation 3.39 gives the following:

$$V = \sqrt{\frac{2g}{f} DS} \tag{3.40}$$

Noting that D = 4 R for a cylindrical pipe flowing full and substituting 4R for the diameter of the pipe gives:

$$V = \sqrt{\frac{8g}{f}} \sqrt{RS} \tag{3.41}$$

In 1775 Chezy established his relationship that identified the Chezy Coefficient C to equal $\sqrt{\dfrac{8g}{f}}$ and published his equation for uniform flow as:

$$V = C\sqrt{RS} \tag{3.42}$$

The Chezy Equation is used extensively in Europe. Manning performed experiments in a laboratory to develop a relationship for C and from his work and others he found that the C coefficient varied with the hydraulic radius according to:

$$C = \frac{1}{n} R^{1/6} \tag{3.43}$$

The Manning equation then becomes

$$V = \frac{1}{n} R^{2/3} S^{1/2} \tag{3.44}$$

Manning developed his formula in the metric system with the unit of length being the meter. To convert Equation 3.44 to U.S. Customary units and maintain the value of n in both systems, a factor of 1.486 needs to be included in the equation. Therefore, the Manning equation in U.S. Customary units becomes:

$$V = \frac{1.486}{n} R^{2/3} S^{1/2} \tag{3.45}$$

The 1.486 is due to the dimensional relationship of the equation ($1/n = t^{-1} L^{1/3}$) and converting meters to feet ($3.2808^{1/3} = 1.486$).

By applying the continuity equation, the Manning equation can be written in terms of discharge as:

$$Q = \frac{1.486}{n} A R^{2/3} S^{1/2} \tag{3.46}$$

It is important to estimate correctly the Manning resistance coefficient "n" in order to determine either the flow given the depth or determine the depth given the flow in a natural channel. There are several different methods to estimate the Manning resistance coefficient presented in the following sections.

Cowan (USGS 1956) published the following relationship for estimating the Manning resistance coefficient.

$$n = (n_o + n_1 + n_2 + n_3 + n_4) m_5 \tag{3.47}$$

Where n_o is a base n value for a straight, uniform, smooth channel, n_1 is the degree of surface irregularities of the channel, n_2 is the variation of the channel cross section, n_3 is the relative effect of obstructions, n_4 is due to the effect of vegetation and flow conditions, and m_5 relates to the degree of meandering. Table 3.1 is a reproduction of Cowan's summary table taken from Chow (1959). Chow also presented an excellent table listing typical n values for a range of conditions. The minimum, normal, and maximum values of n are shown in the table. A more complete discussion can be found in Chow (1959, pp. 108-113). Table 3.2 shows a portion of the table for Natural Streams taken from Chow's Open-Channel Hydraulics book.

Table 3.1. Values for the Computation of the Manning Roughness Coefficient Using Equation 3.43 (after Chow 1959).

Channel Conditions			Values
Material involved	Earth	n_0	0.020
	Rock cut		0.025
	Fine gravel		0.024
	Coarse Gravel		0.028
Degree of irregularity	Smooth	n_1	0.000
	Minor		0.005
	Moderate		0.010
	Severe		0.020
Variation of channel cross section	Gradual	n_2	0.000
	Alternating occasionally		0.005
	Alternating frequently		0.010-0.015
Relative effect of obstructions	Negligible	n_3	0.000
	Minor		0.010-0.015
	Appreciable		0.020-0.030
	Severe		0.040-0.060
Vegetation	Low	n_4	0.005-0.010
	Medium		0.010-0.025
	High		0.025-0.050
	Very High		0.050-0.100
Degree of meandering	Minor	m_5	1.000
	Appreciable		1.150
	Severe		1.300

Table 3.2. Values of the Manning Roughness Coefficient for Natural Channels.

Type of channel and description	Minimum	Normal	Maximum
D. Natural Streams			
D-1 Minor stream (top width at flood stage < 100 ft)			
a. Stream on plain			
1. Clean, straight, full stage, no rifts or deep pools	0.025	0.030	0.033
2. Same as above, but more stones and weeds	0.030	0.035	0.040
3. Clean, winding, some pools and shoals	0.033	0.040	0.045
4. Same as above, but some weeds and stones	0.035	0.045	0.050
5. Same as above, lower stages, more ineffective slopes and sections	0.040	0.048	0.055
6. Same as above, but more stones	0.045	0.050	0.060
7. Sluggish reaches, weedy, deep pools	0.050	0.070	0.080
8. Very weedy reaches, deep pools, or floodways with heavy stand of timber and underbrush	0.075	0.100	0.150
b. Mountain streams, no vegetation in channel, banks usually steep, trees and brush along banks submerged at high stages			
1. Bottom: gravels, cobbles, and few boulders	0.03	0.04	0.05
2. Bottom: cobbles with large boulders	0.04	0.05	0.07
D-2 Flood Plains			
a. Pasture, no brush			
1. Short grass	0.025	0.030	0.035
2. High grass	0.030	0.035	0.050
b. Cultivated areas			
1. No crop	0.020	0.030	0.040
2. Mature row crops	0.025	0.035	0.045
3. Mature field crops	0.030	0.040	0.050
c. Brush			
1. Scattered brush, heavy weeds	0.035	0.05	0.07
2. Light brush and trees, in winter	0.035	0.05	0.06
3. Light brush and trees, in summer	0.040	0.06	0.08
4. Medium to dense brush, in winter	0.045	0.07	0.11
5. Medium to dense brush, in summer	0.070	0.10	0.16
d. Trees			
1. Dense willows, summer, straight	0.11	0.15	0.20
2. Cleared land with tree stumps, no sprouts	0.03	0.04	0.05
3. Same as above, but with heavy growth of sprouts	0.05	0.06	0.08
4. Heavy stand of timber, a few down trees, little undergrowth, flood stage below branches	0.08	0.10	0.12
5. Same as above, but with flood stage reaching branches	0.10	0.12	0.16
D-3 Major streams (top width at flood stage > 100 ft). The n value is less than that for minor streams of similar description, because banks offer less effective resistance.			
a. Regular section with no boulders or brush	0.025	up to	0.06
b. Irregular and rough section	0.035	up to	0.10

Other sources for determining the Manning Roughness factors include pictures of selected streams and over-bank floodplains to use as a guide for selecting n. The publication "Guide for Selecting Manning's Roughness Coefficients for Natural Channels and Flood Plains" (FHWA 1984b) is an excellent resource on how to estimate Manning n values for both channel and out-of-bank flows. The U.S. Geological Water Supply Paper 1849, "Roughness Characteristics of Natural Channels," (USGS 1967) is another example of color photographs of natural channels exhibiting various values of n. These publications can be found on the internet by searching the document titles. Other publications that have photographs of calibrated streams are: Arcement and Schneider (USGS 1989), Chow (1959), Hicks and Mason (1991) and NRCS (1963).

Figure 3.10 shows a floodplain with a computed roughness coefficient: Manning n of 0.20. The vegetation of the floodplain is a mixture of small and large trees, including oak, gum, and ironwood. The base is firm soil and has minor surface irregularities. Obstructions are minor. Ground cover is medium, and there is a large amount of undergrowth that includes vines and palmettos (FHWA 1984b). Similarly, Figure 3.11 is taken from the USGS (1967) Water Supply Paper and shows a channel with a computed Manning n of 0.026.

Figure 3.10. Floodplain roughness example (FHWA 1984b).

Figure 3.11. Looking upstream from the right bank on Indian Fork, near New Cumberland, Ohio (USGS 1967).

3.3.2 Drag Force

As stated earlier, resistance to flow can be divided into shear resistance and resistance due the difference in pressure from the upstream side to the downstream side of an object (form drag). When flow passes over an object a resistance to flow is created that depends upon the shape or form of the boundary of the object. The shape of the boundary (i.e., a bridge pier) causes a deflection of the streamlines and local acceleration of the fluid. Consequently a change in pressure takes place from the upstream to the downstream side of the boundary, which is also referred to as a normal stress. The summation of the forces over the surface results is a drag force on the boundary and a pressure resistance against the fluid.

<u>Derivation of the Drag Equation</u>. The determination of the drag of a flowing fluid past a boundary can be accomplished using dimensional analysis to determine the significant variables and through experimental data to determine the numerical relationships between the parameters. For incompressible flow the drag force can be written as a function of the following parameters that represent the geometry of the object (area, A), the flow (velocity of flow, V), the roughness of the boundary (roughness height, e) and the fluid (density and viscosity of the fluid, ρ and μ):

$$F_D = \text{func}\,(A, V, \rho, \mu, e) \tag{3.48}$$

The Buckingham π method of dimensional analysis shows that three non-dimensional groups can be formed. Choosing A, V, and ρ as the repeating independent variables yields the following relationship.

$$F_D = \text{func}\left(A\rho V^2, \frac{A^{1/2}\rho V}{\mu}, \frac{e}{L}\right) \tag{3.49}$$

Equation 3.48 can be rearranged where $F_D/\rho AV^2$ is defined as the coefficient of drag C_D.

$$\frac{F_D}{A\rho V^2} = C_D = \text{func}\left(R_e, \frac{e}{L}\right) \tag{3.50}$$

Since stagnation pressure can be represented as $\rho V^2/2$, Equation 3.49 can be rearranged as follows:

$$\frac{F_D/A}{\rho \frac{V^2}{2}} = C_D = \text{func}\left(R_e, \frac{e}{L}\right) \tag{3.51}$$

The force of drag then can be written as:

$$F_D = C_D \frac{\rho AV^2}{2} = \frac{1}{2} C_D \rho AV^2 \tag{3.52}$$

The force of drag then is a function of the area of the object, the velocity of the flow, and the density of the fluid. The drag coefficient C_D depends upon the Reynolds Number R_e which in turn depends upon the inertial effects of the flow relative to the viscous effects. The coefficient of drag for many objects has been determined experimentally.

Application of Drag to Piers. The placement of bridge piers in the channel or the floodplain of natural rivers will cause an additional backwater due to the pier obstruction to the flow. The drag force created by the pier can be computed using Equation 3.52 and knowing the coefficient of drag C_D. Lindsey (1938) provided drag coefficients for various objects as a function of Reynolds Number. The drag coefficient is dependent on the ratio of pier area to the total area of the bridge opening, the type of piers, pier shape, and the orientation of flow. Table 3.3 includes typical drag coefficient of piers as given in the HEC-RAS Reference Manual (USACE 2010c).

Table 3.3. Typical Drag Coefficients for Different Pier Shapes.

Pier Shape	Drag Coefficient C_D
Circular pier	1.20
Elongated piers with semi-circular ends	1.33
Elliptical piers with 2:1 length to width	0.60
Elliptical piers with 4:1 length to width	0.32
Elliptical piers with 8:1 length to width	0.29
Square nose piers	2.00
Triangular nose with 30 degree angle	1.00
Triangular nose with 60 degree angle	1.39
Triangular nose with 90 degree angle	1.60
Triangular nose with 120 degree angle	1.72

Drag Force due to the Addition of Debris. The accumulation of debris on bridge piers as illustrated in Figure 3.12 and on the superstructure can create significant forces on the structure. The hydraulic effects of debris can be analyzed using a one-dimensional model. However, depending upon the complexity of the hydraulics and the risk of failure of the structure, two- or three-dimensional models as well as physical model studies can be performed. The reader is referred to NCHRP Report 445, Debris Forces on Highway Bridges (NCHRP 2000), and Hydraulic Engineering Circular No. 9 (FHWA 2005), Debris Control Structures Evaluation and Countermeasures, Chapter 4 Analyzing and Modeling Debris Impacts to Structures.

Figure 3.12. An example of debris blockage on piers.

Modeling debris on a single pier is illustrated in Figure 3.13 where the pier is modeled by increasing the width of the pier by the area of the blockage. If the debris obstruction is large then it might also be necessary to model some of the flow area as ineffective.

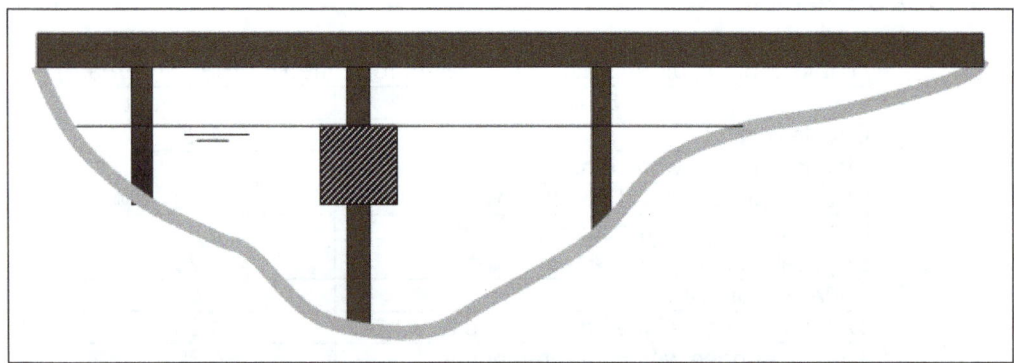

Figure 3.13. An example of upstream bridge cross section with debris accumulation on a single pier.

Effects of Submerged Structures.

If the high-water level reaches the bottom of the superstructure, the bridge may act like a short culvert. For bridges that are designed for submergence, it is advisable to analyze the structure for uplift and the additional drag that the flow will create on the bridge. Chapter 10 provides guidance on computing drag and lift forces for submerged bridge decks and the resulting moment from the combination of these forces.

3.3.3 Weir Flow

There is a wide application for utilizing weirs for measuring flow in a laboratory or in the field. A weir may be described as any regular obstruction over which water flows. The edge or surface is called the crest and the water flowing over the crest is called the nappe (Figure 3.14).

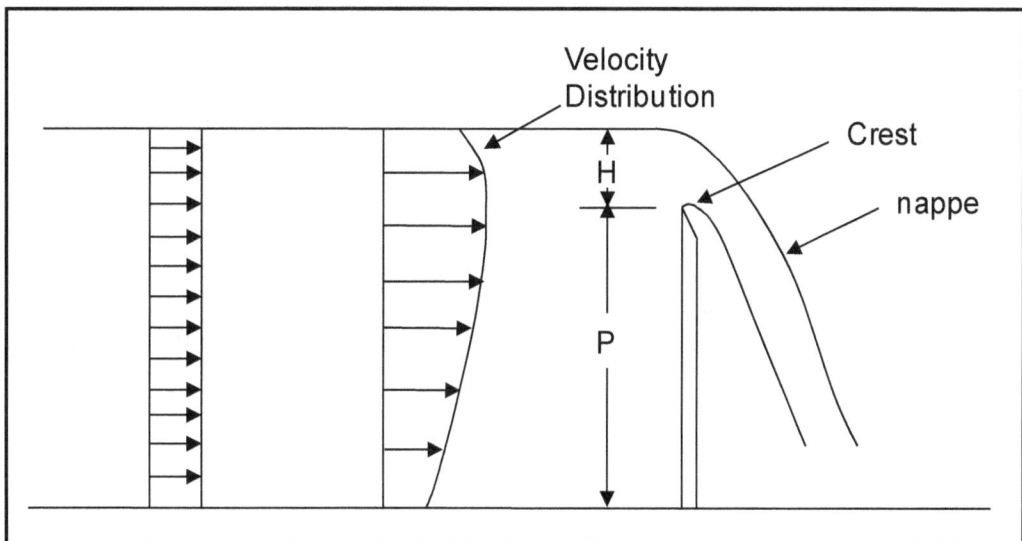

Figure 3.14. Weir flow over a sharp crested weir.

The two types of weir applications that relate to bridge hydraulics are a broad-crested weir and an ogee spillway structure. Both are discussed in the following sections. For a more complete coverage of weir flow the reader is referred to the following references (Chow 1959, Henderson 1966, USBR 1987).

When flow overtops a culvert, bridge and/or the roadway approaches, the flow is calculated using the standard weir equation. Since weir flow is a very complex two dimensional flow problem, the derivation is simplified by assuming a sharp crested weir and the equation can be generalized for other weir types. Figure 3.15 shows a sharp crested weir for a rectangular channel. The weir consists of a flat plate with the upper edges sharpened. The rectangular weir has a straight horizontal crest that extends over the entire length of the channel. The flow pattern produced by the weir is two dimensional.

For the simplified case the problem leads to an approximate result. The flow pattern depicted in Figure 3.15 results by assuming that the velocity distribution upstream of the weir is uniform, all fluid particles pass horizontally over the weir crest, the pressure in the nappe is atmospheric, and the influences of viscosity, turbulence, secondary flows, and surface tension are negligible.

Writing Bernoulli's equation between section 1, which is in the approach channel where the velocity profile is uniform and section 2, which is slightly downstream of the weir crest along the streamline AB, results in the following relationship:

$$H + \frac{V_1^2}{2g} = (H - h) + \frac{V_2^2}{2g} \tag{3.53}$$

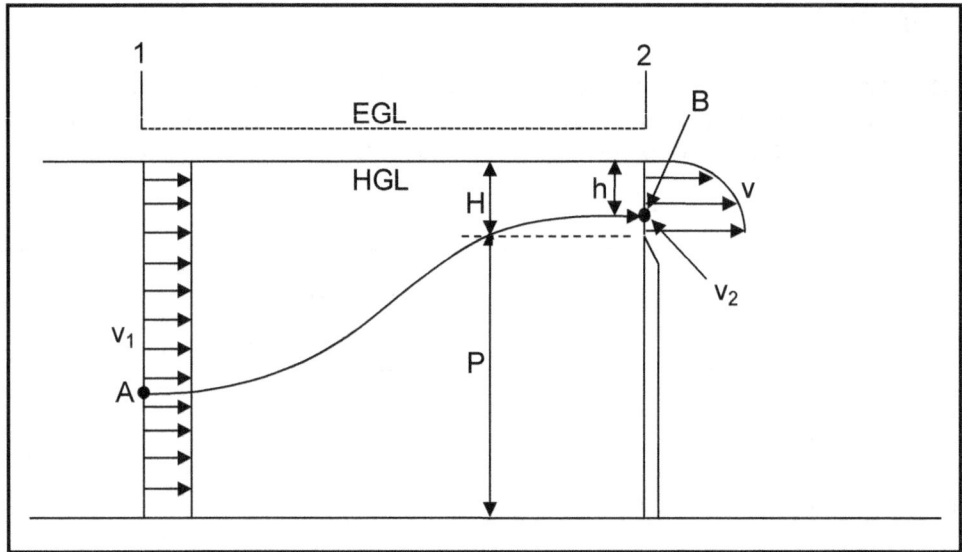

Figure 3.15. Example of simplified weir flow for a sharp crested weir.

Solving for the velocity at section 2 gives:

$$V_2 = \sqrt{2g\left(h + \frac{V_1^2}{2g}\right)} \tag{3.54}$$

To solve for the flow rate per unit width $q = V_2 H$ leads to the following equation:

$$q = \int_0^H V_2 \, dh = \sqrt{2g} \int_0^H \left(h + \frac{V_1^2}{2g}\right)^{1/2} dh \tag{3.55}$$

Integrating Equation 3.53 yields:

$$q = \frac{2}{3}\sqrt{2g}\left[\left(H + \frac{V_1^2}{2g}\right) - \left(\frac{V_1^2}{2g}\right)\right]^{3/2} \tag{3.56}$$

Typically Equation 3.56 can be simplified by assuming that V_1 is small and therefore $V_1^2/2g$ can be neglected. The kinetic energy term can be neglected for approach velocities of 2 ft/s (0.6 m/s) or less and for most practical problems there is more uncertainty in determining the appropriate coefficient. Therefore, Equation 3.56 can be written to solve for unit discharge, q:

$$q = \frac{2}{3}\sqrt{2g}\,H^{3/2} \tag{3.57}$$

Equation 3.57 is the basic equation for the typical rectangular weir. Because of the assumptions made in developing the simplified form of the weir equation a coefficient needs to be introduced to account for the properties that were neglected and therefore, Equation 3.57 can be expressed as:

$$q = C_w \frac{2}{3}\sqrt{2g}\,H^{3/2} \quad (3.58)$$

If C_w is considered to be a constant, then $q \propto H^{3/2}$. The coefficient is then used to transform the simplified equation into the actual weir equation with C_w being determined experimentally (King and Brater 1963). To write the continuity equation for the total flow $Q = qL$ and combining $C_w \frac{2}{3}\sqrt{2g}$ into a single coefficient C_w then the common form of the weir equation can be written as:

$$Q = C_w L H^{3/2} \quad (3.59)$$

There has been a significant amount of research to determine the coefficient for the various types of weirs that have been developed. Note that C_w in Equation 3.59 includes gravitational acceleration so the value depends on the system of units being used.

<u>Ogee Spillway Crest</u>. Diversion structures are used to divert water from an existing natural watercourse into an off-channel conveyance system. Very often the shape is that of an ogee weir (Figure 3.16). The ogee spillway is designed for a specific flow, Q_0, and a specific head, H_0, and therefore there is a unique coefficient C_0 for this flow. Any other flow and head will have a different coefficient. The United State Bureau of Reclamation in their Design of Small Dams (USBR 1987) has design curves for these conditions (Figure 3.17).

When flow overtops a weir and there is significant tailwater the discharge will be reduced due to the submergence of the weir. For an ogee shape weir, the USBR (1987) has developed a relationship that is presented in Figure 3.18 to account for the reduction in the discharge coefficient, which in turn reduces the flow over the weir.

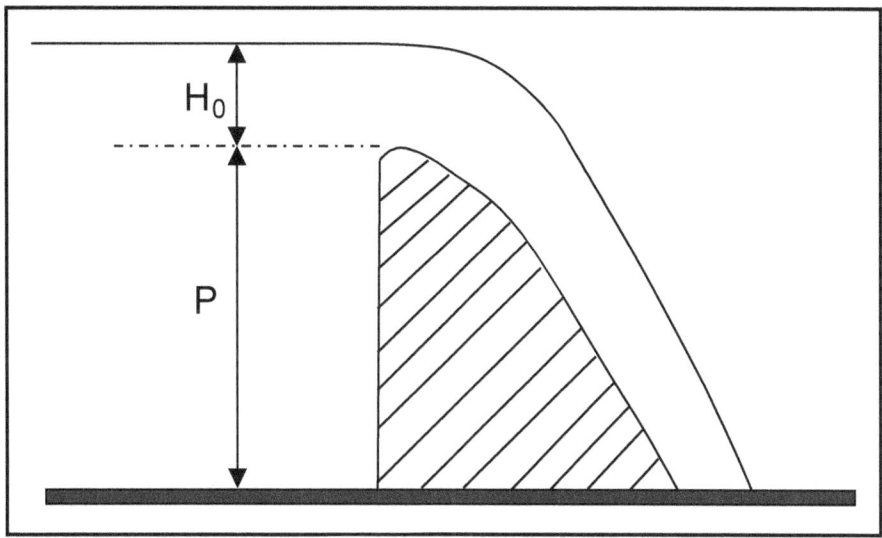

Figure 3.16. Ogee spillway crest (USBR 1987).

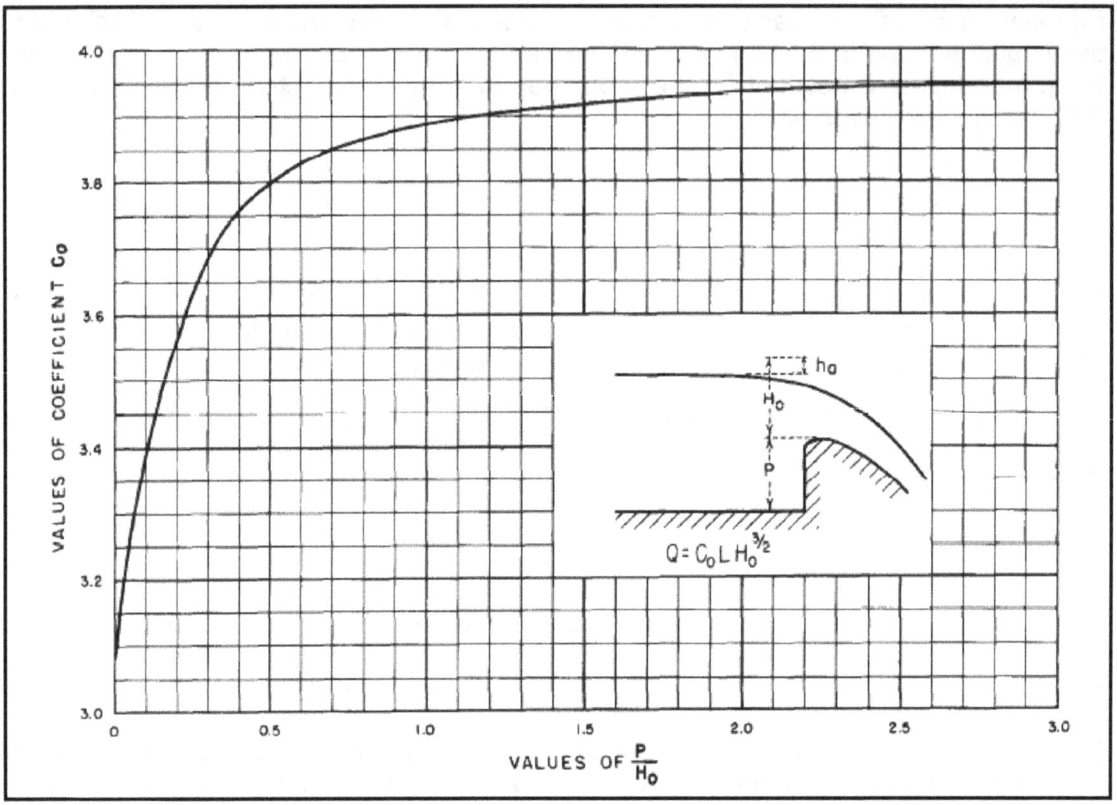

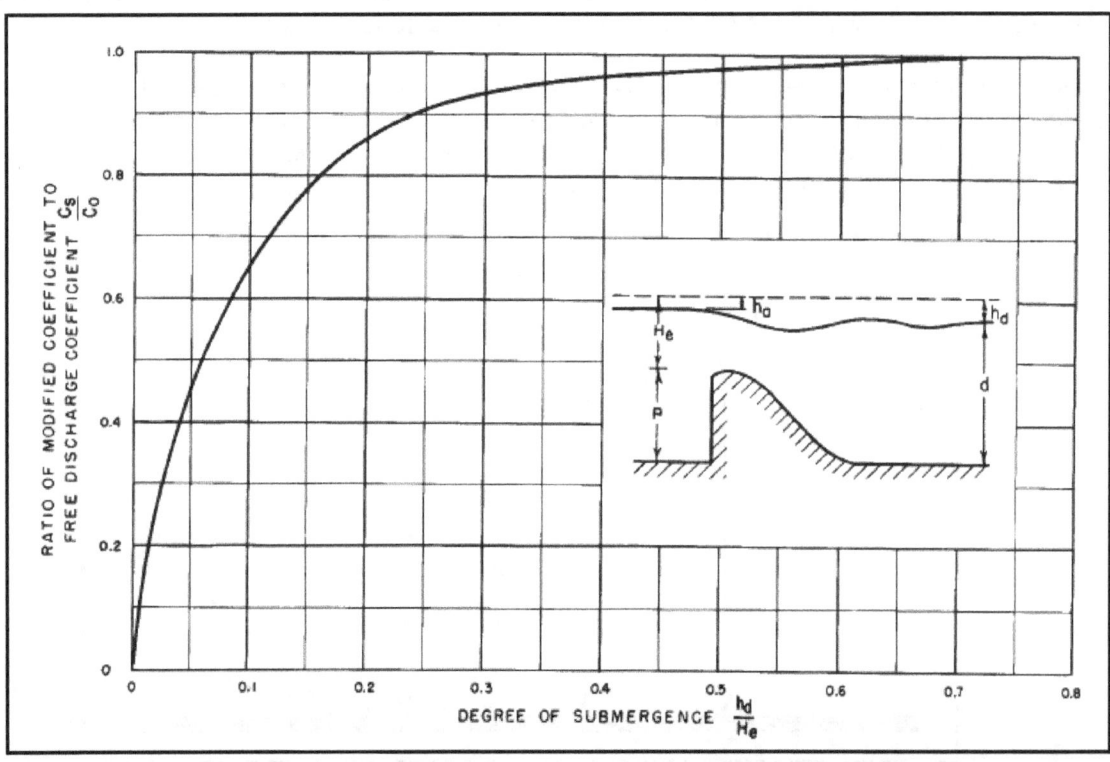

Figure 3.17. Discharge coefficients for vertical-face ogee crest (USBR 1987).

Figure 3.18. Ratio of discharge coefficients caused by tailwater effects (USBR 1987).

Broad-Crested Weir. The flow over a broad-crested weir will occur at critical depth for an ideal fluid flow (Figure 3.19). The flow over a broad-crested weir can be computed using the continuity equation and assuming that there is a hydrostatic pressure distribution where critical depth y_c occurs. For a rectangular channel the Froude Number at minimum energy (critical depth) is equal to one.

$$F_R = 1 = \frac{V_c}{\sqrt{gy_c}}$$

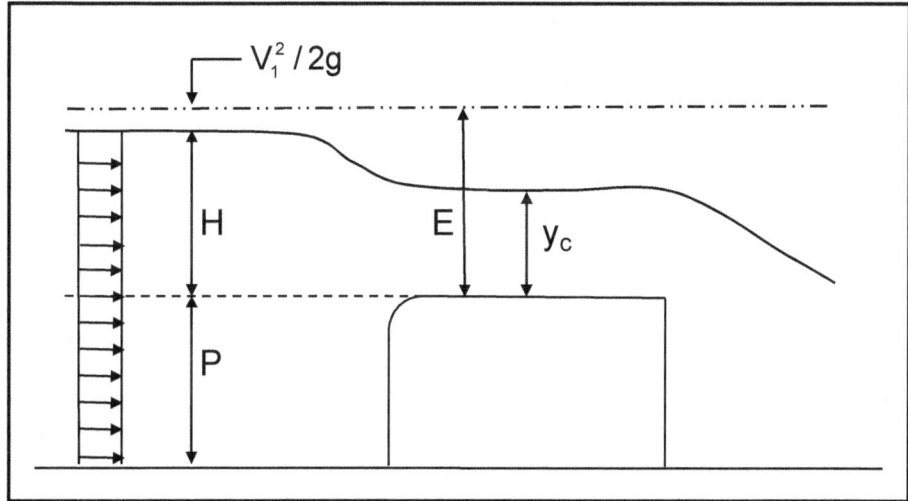

Figure 3.19. Broad-crested weir.

Noting that $Q = V_c\, y_c\, L$ the following equation is obtained:

$$Q = L\, y_c \sqrt{gy_c}$$

Also noting that for minimum energy occurring in a rectangular channel the critical depth is equal to 2/3 of the head H. Substituting yields:

$$Q = L\sqrt{g}\left(\frac{2}{3}H\right)^{3/2} = 3.089 L H^{3/2} \tag{3.60}$$

The coefficient for SI units is 1.705. In many situations the critical depth section may be in a area of strong streamline curvature and the boundary friction along the crest may reduce the true specific energy from the assumed value of H by the time the flow reaches the critical section. Therefore, the more general form of the equation for a broad-crested weir is:

$$Q = C\, L\, H^{3/2} \tag{3.61}$$

The coefficient is a function of P/H, L/H, R_e, shape of the weir, and roughness. Discharge coefficients for a broad-crested weir usually range from about 2.6 to 3.05 (1.44 to 1.68 for SI). Flow over a bridge and the roadway approaches (embankments) is usually calculated using the general broad-crested weir equation.

Weir Coefficient. The discharge coefficient for a broad-crested weir as stated above ranges between 2.6 and 3.05 (1.44 to 1.68 for SI). Flow overtopping a bridge deck is not an ideal broad-crested weir and it is generally recommended that the lower value be used for the discharge coefficient where increased resistance to flow caused by obstructions such as bridge railings, curbs, and debris would cause a decrease in the value of C. From King's Handbook (King and Brater 1963), weir coefficients are given with respect to the head on the weir and to the width of weir. The Hydraulics of Bridge Waterways manual (FHWA 1978) provides a curve of C versus head on the roadway. "Hydraulic Performance of Bridge Rails" (Charbeneau et al. 2008) shows the impact of bridge rails on the computation for the overtopping flow.

Similar to the Ogee Weir, when flow overtops a broad-crested weir and there is significant tailwater the discharge will be reduced due to the submergence of the weir. For broad-crested weirs FHWA (1978) has developed a relationship to account for weir submergence (Figure 3.20).

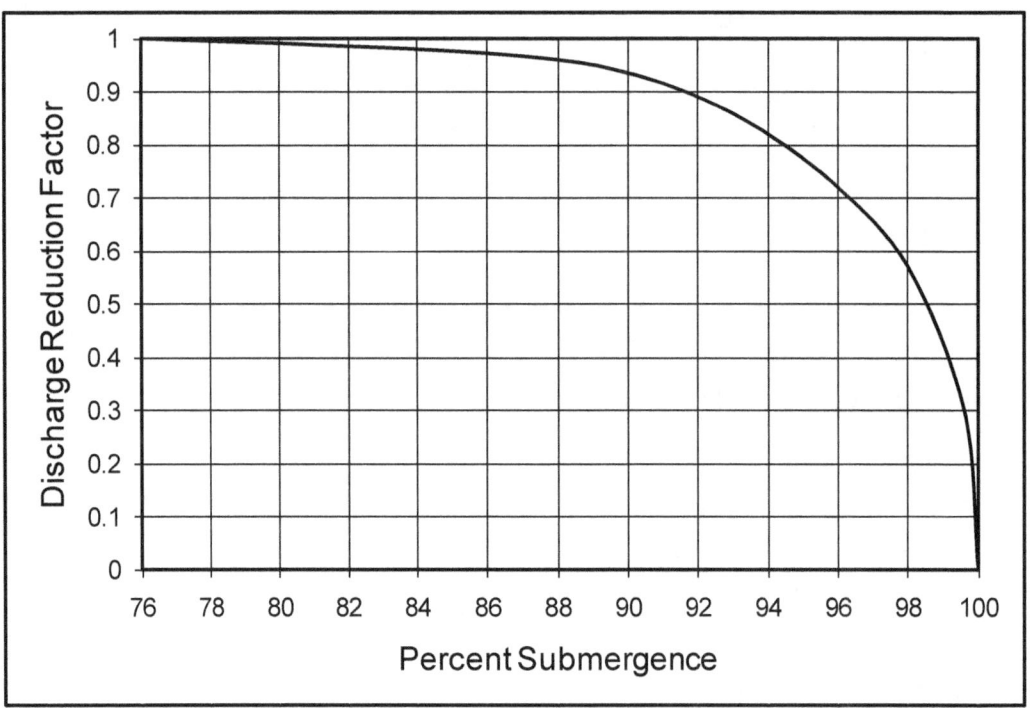

Figure 3.20. Discharge reduction factor versus percent of submergence.

3.3.4 Gate and Orifice Equations

This section deals with inline structures that are commonly found in open channel flow. Common examples of inline structures are the orifice and Tainter gates (radial gates). Other inline structures such as bridges can often be modeled with an orifice type equation.

Sluice and Tainter Gates. The two most important design features for the sluice (vertical lift) and Tainter gates (Figure 3.21) are the head (elevation) versus the discharge relationship and the pressure distribution over the gate surfaces. The structural design of the orifice involves consideration of the hydrostatic force on the gate, the hoisting force, the weight of the gate, and the roller friction (the friction on the gate is reduced by rollers that are typically attached to the gate). The Tainter gate has a circular segment for its face which rotates about the center of the curvature. Since the hydrostatic pressures are radial, passing through the trunnion bearing, the thrust on the gate is substituted for the roller friction of the orifice. The pin friction is usually much less than the roller friction, so that the Tainter gate is comparatively light and easy to operate.

For the sluice and Tainter gates shown in Figure 3.21, the Bernoulli equation for one-dimensional flow can be used to solve for the discharge.

$$\frac{V_2^2}{2g} + y_2 = \frac{V_1^2}{2g} + y_1 \tag{3.62}$$

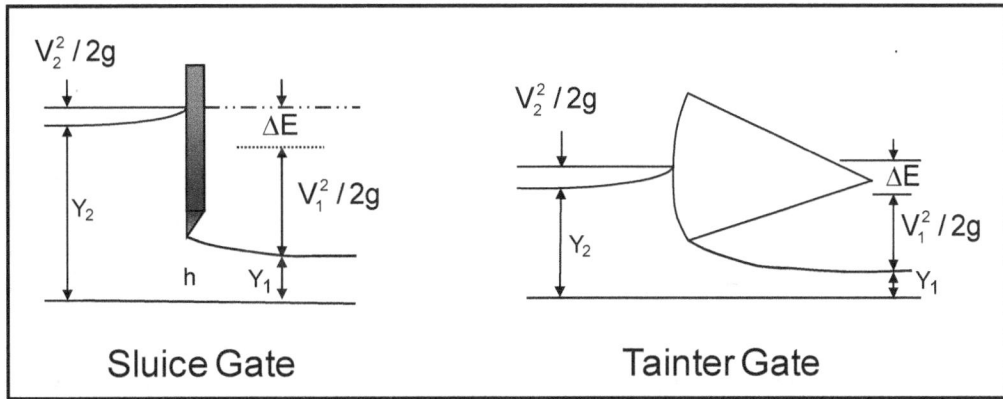

Figure 3.21. Sluice and tainter gates.

Noting that $Q = V_1 A_1 = V_1 C_c hL$ and substituting the discharge relationship for V_1 into the Bernoulli equation gives the following:

$$\frac{V_2^2}{2g} + y_2 = \frac{Q^2}{2g(C_c hL)^2} + y_1 \tag{3.63}$$

Then solving for the discharge gives the relation for flow passing under the gate as:

$$Q = C_c hL \sqrt{2g(y_2 - y_1) + \frac{V_2^2}{2g}} \tag{3.64}$$

Although this derivation ignored the losses due to the boundary development along the gate and floor, these are usually insignificant due to the short distances involved. The coefficient of contraction C_c is determined through experimental measurements or two-dimensional analysis of the curvilinear zone. For the simplifying assumption where the approach velocity is small and therefore, $V_2^2/2g = 0$, the equation can be further simplified to:

$$Q = C_d hL \sqrt{2g(y_2 - y_1)} \tag{3.65}$$

The discharge coefficient C_d is a function of the upstream and downstream depths, the gate opening, and the gate geometry. The form of the equation as stated above is for both free and submerged conditions.

3.4 FLOW CLASSIFICATION

Open channel flow can be classified in many ways. Flow can be classified as either steady or unsteady. Flow can also be classified as uniform or non-uniform (varied). Non-uniform flow can be further classified as gradually varied or rapidly varied. The flow can also be subcritical or supercritical, and depending upon the turbulence of the flow field, the flow can be classified as laminar (low R_e) or turbulent (high R_e). Since nearly all open channel flow situations are turbulent, laminar flow will not be discussed in the following sections. For further clarification of laminar and turbulent flow in open channels the reader is directed to Henderson (1966), Chow (1959), and Fundamentals of Fluid Mechanics by Munson et al. (2010).

3.4.1 Steady Versus Unsteady Flow

When the time derivatives of a flow field become insignificant, the flow is considered to be a steady flow. Steady-state flow refers to the condition where the fluid properties at a point in the system do not change over time. Otherwise, flow is called unsteady. Whether a particular flow can be treated as steady or unsteady can depend on the frame of reference. The simplest steady flow is uniform flow, in which no flow variable changes with distance. In a uniform steady flow, every flow variable such as depth and velocity is constant with respect to distance and time (i.e., $\partial/\partial x = 0$ and $\partial/\partial t = 0$). If the flow is not uniform, then it is classified as non-uniform and can be further divided into gradually varied and rapidly varied flow. In gradually varied flow, the flow variables may change with distance but are constant with respect to time. The changes with distance are assumed to be gradual so that the vertical accelerations are small. This allows use of the Manning uniform flow equation for computations of depth, velocity, slope, and discharge. In rapidly varied flow, substantial variations are present in the vertical and/or transverse flow. A good example is a hydraulic jump. Other examples of rapidly varied flow are flows through culverts and bridges and over weirs and spillways. Figure 3.22 illustrates the classification of flow according to changes in depth with space and time. Unsteady, uniform flow is, at best, rare (Chow 1959). It would require change in depth with time, but no change in depth with respect to distance at any instant in time.

<u>Steady Flow</u>. For steady flow analysis, the flow is known at all points along the channel. It should be noted that steady flow does not mean the velocity and acceleration are constant. Flow in an open channel bend or flow transitioning from a mild to a steep channel may be steady but the velocity and/or acceleration are not constant. The numerical solution to the energy or momentum equations is such that one boundary condition is known either downstream or upstream and the solution for the other boundary is computed through an iterative process (see Section 3.4.4). If the flow is subcritical, the computations are performed from downstream to upstream. For supercritical flow the direction of the solution is from upstream to downstream. The basic gradually varied flow equation in open channels can be written as:

$$\frac{dy}{dx} = \frac{S_0 - S_f}{1 + \frac{d(V^2/2g)}{dy}} \tag{3.66}$$

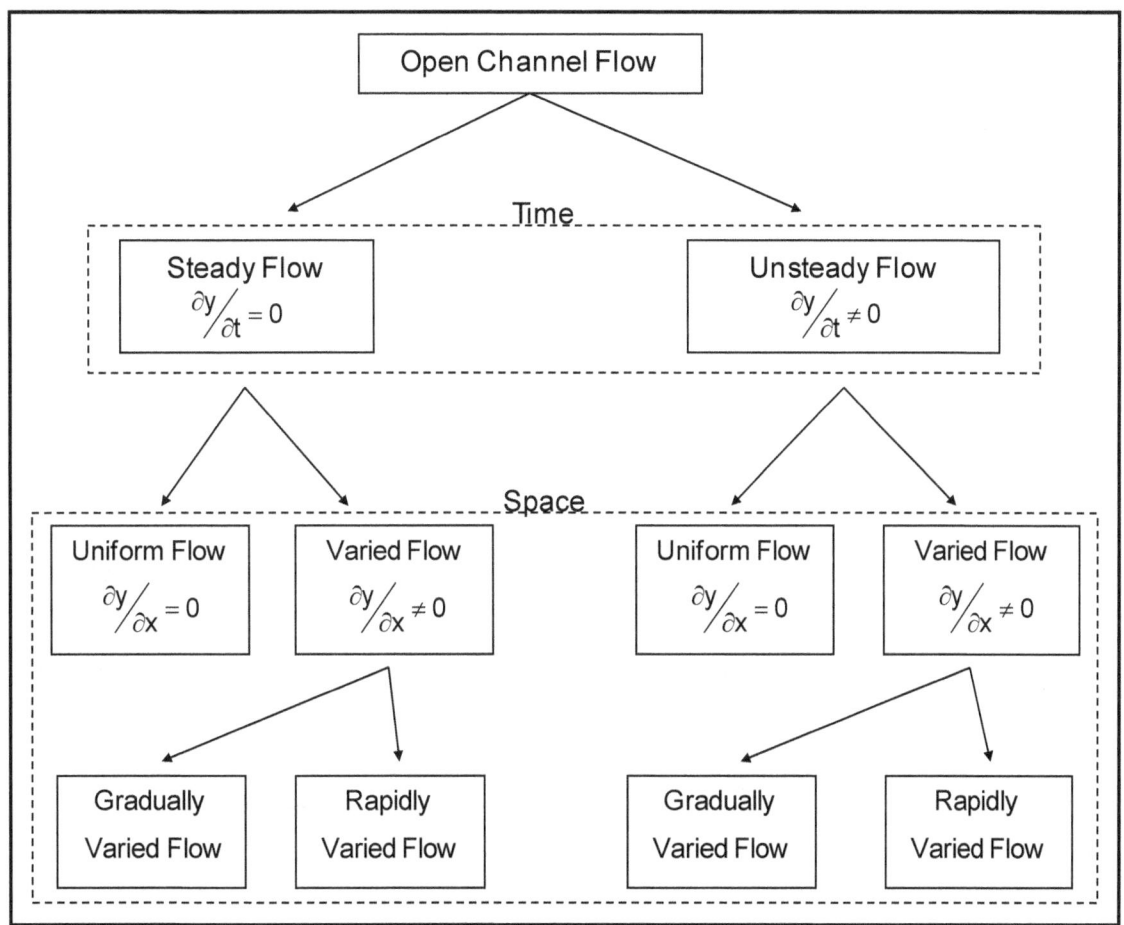

Figure 3.22. Flow classification according to change in depth with respect to space and time.

For a rectangular channel it can be shown that the change of the velocity head with respect to depth $\frac{d(V^2/2g)}{dy}$ is equal to $-F_R^2$ and the general differential equation for gradually varied flow can be written as:

$$\frac{dy}{dx} = \frac{S_o - S_f}{1 - F_R^2} \tag{3.67}$$

For a non-rectangular channel $F_R^2 = \frac{Q^2 T}{gA^3}$.

<u>Conditions When Steady Flow Modeling is Appropriate</u>. This consideration is of considerable practical interest, since unsteady flow is significantly more complex and requires more data and effort to analyze than steady flow. The flow is steady if the depth of flow at a particular point does not change or can be considered constant for the time interval under consideration. The flow is unsteady if the depth changes with time. Open channel flow in natural channels is almost always unsteady, although it is often analyzed in a quasi-steady state for channel design or floodplain mapping as well as the hydraulic design of bridges and culverts. In most open channel flow problems, practical flow conditions are typically studied under steady flow conditions.

Determining if a steady flow approximation is a reasonable assumption depends on the degree the flow variables change with time. During most rainfall-runoff events the water surface in the river will rise but the stage increases slowly with time and $\partial y/\partial t = 0$. Many natural open channel flow problems can be adequately analyzed with the steady-state approximation. On the other hand if a dam embankment were to fail, the resulting flood wave would create a relatively fast increase in water surface depth downstream with respect to time ($\partial y/\partial t \neq 0$) and the acceleration terms would need to be included in the computations for the water surface elevations downstream.

Under steady flow, the user inputs as boundary conditions a discharge upstream and a stage downstream. The numeric model calculates stages throughout the interior points, keeping the discharge constant. Under unsteady flow, the user inputs a discharge hydrograph at the upstream boundary and a discharge-stage rating at the downstream boundary. The model calculates discharges and stages throughout the interior points.

Unsteady Flow. In unsteady flow analysis, two governing equations must be explicitly solved because the discharge and the elevation of the water surface are both unknown. The two governing equations are the conservation of mass and the conservation of momentum. Unsteady flow is presented in detail in Chapter 7. In steady flow the conservation of mass can be written as $Q = AV$ where the discharge is constant and the only unknown is the water surface. In unsteady flow the discharge must be explicitly solved for flows and elevations. The continuity equation for unsteady flow is given by:

$$\frac{\partial Q}{\partial x} + \frac{\partial A}{\partial t} = q_l \tag{3.68}$$

Where q_l is the lateral inflow rate per unit length of channel. The momentum equation for unsteady flow can be written as:

$$\frac{1}{g}\frac{\partial V}{\partial t} + \frac{V}{g}\frac{\partial V}{\partial x} = S_f - S_o + \frac{\partial y}{\partial x} + q_l \tag{3.69}$$

Equations 3.68 and 3.69 are known as the shallow water equations developed for one-dimensional unsteady flow in open channels and are also called the Saint-Venant equations.

Storage Effects for Unsteady Flow. Steady flow computations ignore storage effects both within the channel and the overbank areas. In comparison, the unsteady flow equations account for both types of storage. The storage can significantly reduce peak flows giving a more realistic assessment of the surface flooding. If storage does occur in a floodplain and flow is treated as steady, then the storage can be accounted for by blocking that portion of channel by defining it to be ineffective.

3.4.2 Subcritical Versus Supercritical Flow

Various types of waves and surges may occur in open channels and cause a locally unsteady flow. The simplest is the small surface wave which progresses radially outward from a point as a rock would cause if thrown into a lake. The rate that this wave progresses outward is called its celerity. Subcritical flow is when the flow velocity is less than the celerity of a gravity wave, $c = \sqrt{gy}$, and supercritical flow is when the flow velocity is greater than the wave velocity.

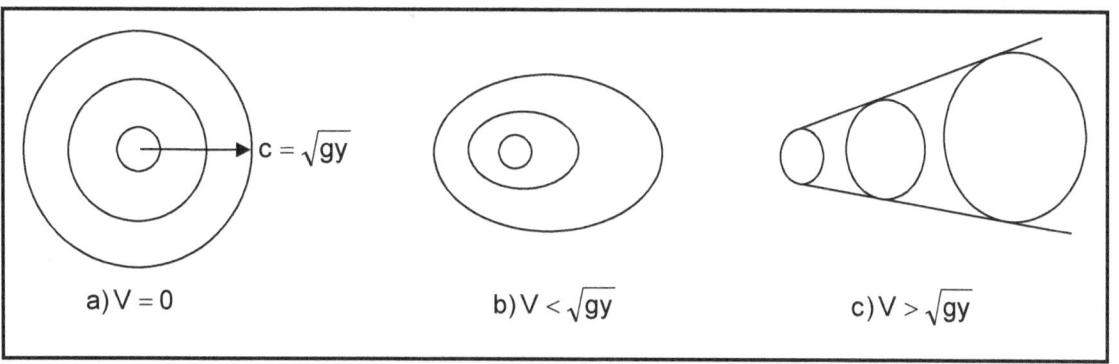

Figure 3.23. Propagation of a water wave in shallow water illustrating subcritical and supercritical flow.

Dropping a rock in a pond will cause a wave to propagate in all directions at the same velocity (Figure 3.23a). By dropping the rock in a stream and superimposing an average velocity V of the stream such that $V < \sqrt{gy}$ (i.e., $V/\sqrt{gy} < 1$), the wave will propagate upstream at a velocity of $\sqrt{gy} - V$ and downstream at $\sqrt{gy} + V$ (Figure 3.23b). This condition is defined as subcritical flow. If a higher velocity is imposed such that $V > \sqrt{gy}$ (i.e., $V/\sqrt{gy} > 1$), the wave will be washed downstream with no affect upstream (Figure 3.23c). This condition is defined as supercritical flow. The conclusion is that for subcritical flow any disturbance in the flow field will translate upstream (i.e., water surface computations must progress from downstream to upstream). On the other hand, for supercritical flow the computations must progress from upstream to downstream since any disturbance in the flow field will not translate upstream. In 1861 William Froude presented a paper where he defined the ratio of the characteristic velocity (average) V to a gravitational wave velocity $c = \sqrt{gy}$, which was later called the Froude Number.

$$F_R = \frac{V}{\sqrt{gy}} \qquad (3.70)$$

<u>Specific Energy</u>. Specific energy is defined as the energy per unit mass. Many practical problems of open channel flow are solved by application of the energy principle (Bernoulli's equation) using the channel bottom as the datum as shown in Figure 3.24.

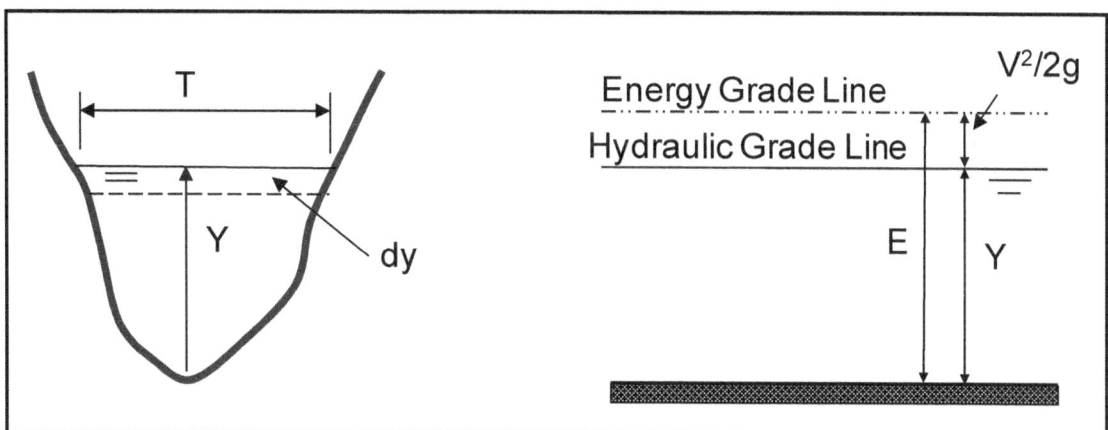

Figure 3.24. Specific energy description.

The concept of specific energy was first introduced by Bakhmeteff in 1912. The water surface is the same as the hydraulic grade line and the total energy head is represented by the energy grade line. Specific energy is a fundamental concept and widely utilized in solving problems of open channel flow. Flow represented in Figure 3.24 is essentially rectilinear (i.e., the flow has smooth parallel streamlines). The specific energy E is given by the total energy head consisting of the depth of flow and the velocity head and the total flow is given by the continuity equation Q = AV. Considering the special case where the head losses are negligible ($h_l = 0$), the Bernoulli equation can be applied between any two sections as:

$$\frac{V_2^2}{2g} + y_2 = \frac{V_1^2}{2g} + y_1 \tag{3.71}$$

The sum of the depth of flow and the velocity head is the specific energy and can be written as:

$$E = y + \frac{V^2}{2g} \tag{3.72}$$

Since the flow in channel cross section (Figure 3.24) is the product of the area and the average velocity, a change in the depth will result in a change in the Q unless the velocity changes inversely to keep the discharge the same. However, the redistribution of the velocity and the depth keeping the same discharge will result in different specific energies. Thus there are limitless combinations of velocity and depth that can have the same discharge. For very high depths the velocity would become very small and the specific energy would become equal to the depth. On the other hand if the velocity were to become very high then the depth would become very shallow and the specific energy would be approximately equal to the velocity head. These trends can be shown in the specific energy diagram shown in Figure 3.25.

It is obvious from the specific energy diagram that there are two depths for the same energy except at minimum energy where there is a unique depth. The two depths are referred to as alternate depths and the depth at minimum energy is typically called critical depth. Critical depth can be solved by taking the derivative of the energy with respect to depth at the minimum energy, which in this case will be equal to zero.

$$\frac{dE}{dy} = \frac{d}{dy}\left(y + \frac{V^2}{2g}\right) = \frac{d}{dy}(y) + \frac{d}{dy}\left(\frac{Q^2}{A^2 2g}\right) \tag{3.73}$$

$$\frac{dE}{dy} = 0 = 1 + \frac{Q^2}{2g}\frac{d}{dy}\left(\frac{1}{A^2}\right) \tag{3.74}$$

$$1 + \frac{Q^2}{2g}\frac{d}{dy}(A^{-2}) = 1 + \frac{Q^2}{2g}(-2)\frac{1}{A^3}\frac{dA}{dy} \tag{3.75}$$

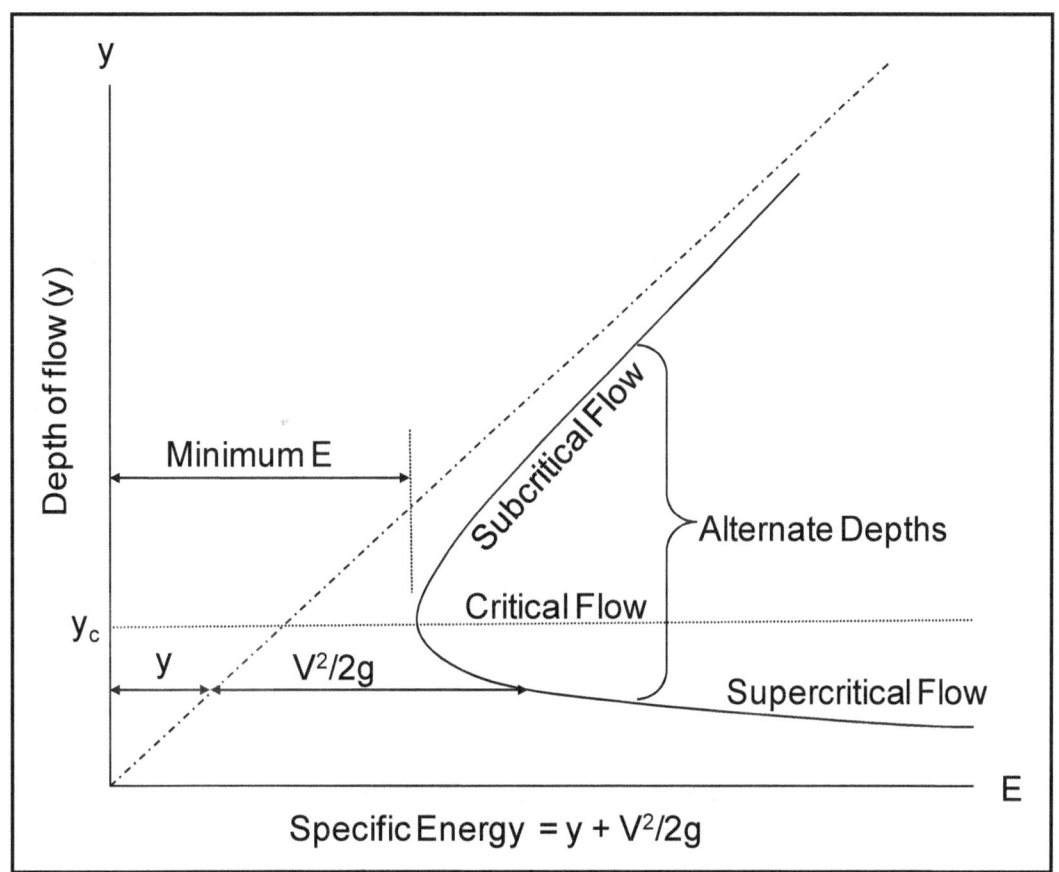

Figure 3.25. Specific energy diagram for a constant discharge.

Note from Figure 3.24 that dA = T dy where T is the topwidth. Substituting for dA/dy in Equation 3.75 gives us:

$$1 = \frac{Q^2 T}{gA^3} \tag{3.76}$$

Since both the area and the topwidth are functions of depth, Equation 3.76 defines a critical depth occurring at minimum energy. Substituting $V^2 = Q^2/A^2$ (i.e., continuity equation Q = AV), Equation 3.76 can be written in terms of the velocity and a hydraulic depth as:

$$\frac{V^2 T}{gA} = 1 = F_R^2 = \frac{V^2}{gD_M} \tag{3.77}$$

Where the hydraulic depth is defined as $D_M = \dfrac{\text{Area}}{\text{Topwidth}} = A/T$.

For a rectangular channel the hydraulic depth is equal to the depth of flow and by taking the square root the equation reduces to the Froude Number introduced at the beginning of this section (Equation 3.70).

$$F_R = \frac{V}{\sqrt{gy}}$$

<u>Dimensional Analysis and Similitude</u>. Many open channel flow problems can be solved only approximately by analytical or numerical methods. Physical model studies play an important role in verifying solutions or by providing results that can't be obtained through analytical solutions.

Similitude is a relationship between a full-scale flow and a flow involving smaller but geometrically similar boundaries. In 1861 Froude published a paper which dealt with identifying the most efficient hull shapes. He established a dimensionless number, later to be called the Froude Number that permitted small-scale tests to predict the behavior of full sized prototypes.

The three types of similitude involved in fluid mechanics are geometric, kinematic and dynamic similarity. Geometric similarity involves the length scales, x, y, and z. Ideally the ratios of the model geometry scales would all be the same. Kinematic similarity involves length and time ratios, which require that the streamline patterns be the same in the model as in the prototype. Dynamic similarity requires that the force ratios of model to prototype be the same from point to point.

$$\frac{F_{1M}}{F_{1P}} = \frac{F_{2M}}{F_{2P}} = \frac{F_{3M}}{F_{3P}} = \cdots = \frac{F_{nM}}{F_{nP}} \tag{3.78}$$

Typical forces that are encountered in open channel flow are inertia, gravity and viscous. If the model were to be in absolute similitude (holding to geometric, kinematic and dynamic similarity) then the model would be the same size as the prototype.

For model studies involving fluid motion, the inertial force has to be included for dynamic similarity. For open channel flow problems the other dominating force is that of gravity (i.e., viscous, surface tension and other forces are small and can be neglected). This requires that the dynamic force ratio of inertia to gravity for the model must equal that of the prototype.

$$\left(\frac{\text{Force}_{\text{Inertia}}}{\text{Force}_{\text{Gravity}}}\right)_{\text{Model}} = \left(\frac{\text{Force}_{\text{Inertia}}}{\text{Force}_{\text{Gravity}}}\right)_{\text{Prototype}} \tag{3.79}$$

In terms of open channel flow the inertial force can be represented as $F = ma = \rho l^3 \frac{V^2}{l} = \rho V^2 l^2$ and the gravitational force as $F = mg = \rho l^3 g$.

Therefore, the ratio of the inertia force to the gravity force for the model and prototype can be written as:

$$\left(\frac{\left(\frac{\rho l^3 V^2}{l}\right)}{\rho l^3 g}\right)_{\text{Model}} = \left(\frac{\left(\frac{\rho l^3 V^2}{l}\right)}{\rho l^3 g}\right)_{\text{Prototype}} \tag{3.80}$$

This reduces to:

$$\left(\frac{V^2}{gl}\right)_{Model} = \left(\frac{V^2}{gl}\right)_{Prototype} \tag{3.81}$$

The length scale can be represented by the depth of flow in many open channel flow situations. The square root of Equation 3.81 shows that the Froude Number of the model needs to be equal to the Froude Number of the prototype for open channel similitude. This is essentially what Froude found when he was performing his experiments in the 1800s.

$$\left(\frac{V}{\sqrt{gy}}\right)_M = \left(\frac{V}{\sqrt{gy}}\right)_P = (F_R)_M = (F_R)_P \tag{3.82}$$

3.4.3 Uniform, Gradually, and Rapidly Varied Flow

Uniform Flow. Uniform flow was discussed in Section 3.3.1 with the development of the Manning Equation. Uniform flow requires not only a longitudinally uniform cross section (prismatic channel), requiring the channel cross section, roughness, and slope to be constant throughout, but also that there be an equilibrium between the gravitational and frictional forces. Water flows down a channel by the force of gravity and is resisted by the boundary shear stress. The gravitational forces accelerate the flow while the frictional forces decelerate the flow. For a constant discharge there is only one velocity or depth that can satisfy these conditions. For most practical flow situations encountered in open channel hydraulics, cross sections, roughness, and/or slopes typically change from upstream to downstream and therefore, most flow situations are non-uniform (or varied) flow.

As illustrated in Figure 3.26, for the flow to be uniform the depth, water area, velocity, and the discharge at every section of the channel reach must be constant; and the energy slope, water surface slope, and channel bottom slope must be parallel (i.e., $S_f \equiv S_w \equiv S_0 \equiv S$). For practical purposes, the requirement of constant velocity may be interpreted as the requirement that the flow possess a constant mean velocity. Strictly speaking, however, this should mean that the flow possesses a constant velocity at every point on the channel section within the uniform-flow reach. In other words, the velocity distribution across the channel section is unaltered in the reach.

Uniform flow is considered to be steady state only, since unsteady uniform flow is practically nonexistent. In natural streams, a strict uniform-flow condition is unusual. Despite this, the uniform-flow condition is frequently assumed in the computation of flow in natural streams. The results obtained from this assumption are understood to be approximate and general, but they offer a relatively simple and satisfactory solution to many practical problems.

Hydrostatic Pressure. For a fluid at rest, the pressure at a point below the surface (assuming the surface is exposed to the atmosphere) is equal to the distance below the surface (depth of submergence) multiplied by the specific gravity of the fluid $P = \gamma y$. This is defined as the hydrostatic pressure for an incompressible flow having no shear stresses acting on the fluid since it is at rest. For uniform flow the pressure would also be hydrostatic since the streamlines in uniform flow are all parallel to one another, there are no vertical accelerations, and the velocity is everywhere the same (no shear stresses).

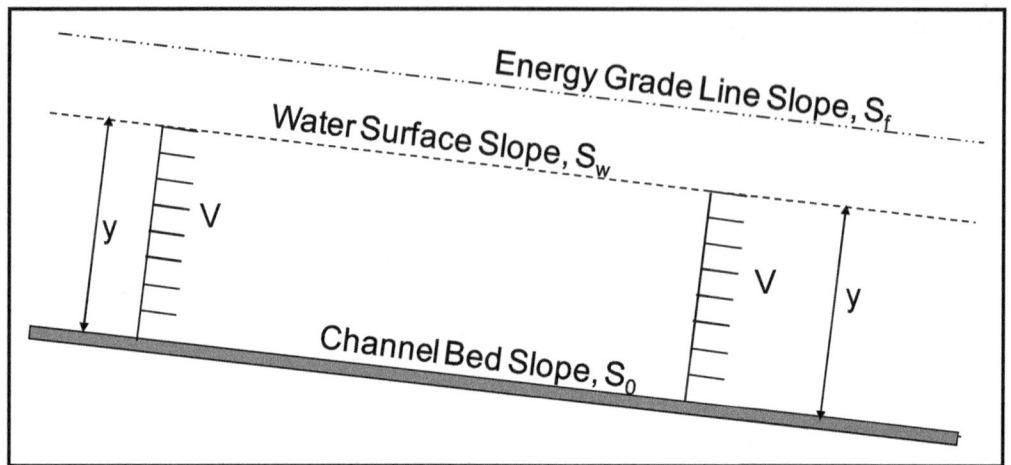

Figure 3.26. Definition sketch of uniform flow.

Gradually Varied Flow. Since most actual flows in open channels are non-uniform, the actual water surface elevations for any discharge need to be determined. As long as there is a gradual change in depth and velocity, vertical accelerations are small and the streamlines are essentially parallel. The conservation of energy principle is used to solve for the depth, which in turn allows calculation of the water surface elevation at any location along the stream. The hydrostatic distribution is still valid, which allows the assumption that $P/\gamma = y$. For rapidly varied flow, such as the hydraulic jump, the conservation of momentum equation is more effective. Rapidly varied flow will be discussed later in this section. Figure 3.27 shows the hydraulic conditions for gradually varied flow analysis using energy principles.

The value of the energy coefficient α is assumed to be equal to one for the following derivation, but must be included for better accuracy in practical application. The total energy of the flow at any location is shown in Figure 3.27 as:

$$H = \frac{V^2}{2g} + y + z \tag{3.83}$$

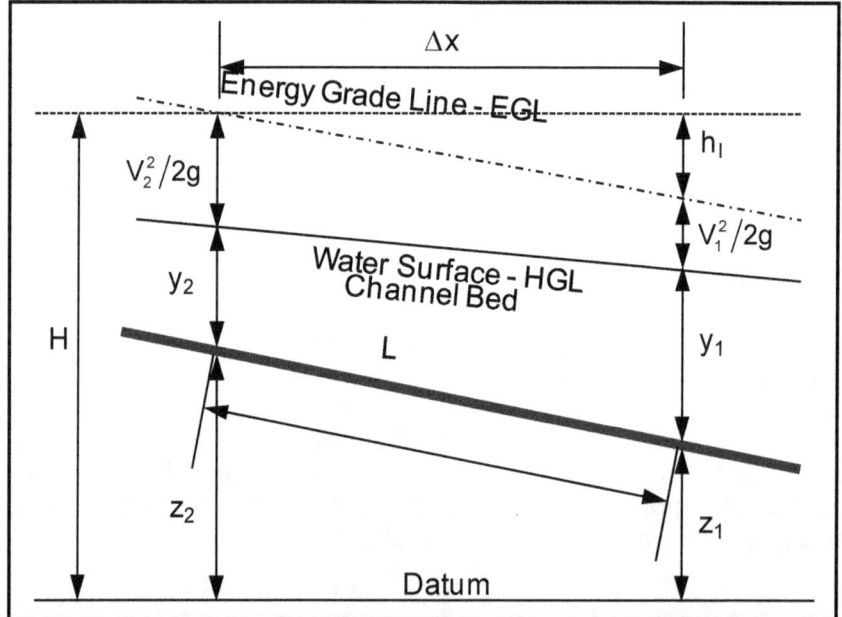

Figure 3.27. Definition sketch of a typical non-uniform water surface profile.

Differentiating Equation 3.83 with respect to x, the rate of energy change is:

$$\frac{dH}{dx} = \frac{d}{dx}\left(\frac{V^2}{2g}\right) + \frac{dy}{dx} + \frac{dz}{dx} \qquad (3.84)$$

The energy grade line $S_f = -dH/dL \cong -dH/dx$ for small slopes (i.e., less than 10%). The channel slope is given by $S_o = -dz/dx$ and the slope of the water surface (hydraulic grade line) is $S_w = -dH/dx - dy/dx$.

The energy equation can be written between sections 2 and 1 for steady flow as:

$$\frac{V_2^2}{2g} + y_2 + z_2 = \frac{V_1^2}{2g} + y_1 + z_1 + h_l \qquad (3.85)$$

Note that $S_o(\Delta x) = z_2 - z_1$ and $S_f(\Delta x) \cong h_l$ for small slopes. The energy equation can then be written as:

$$y_2 - y_1 + \frac{V_2^2}{2g} - \frac{V_1^2}{2g} = (S_f - S_o)(\Delta x) \qquad (3.86)$$

Dividing by Δx and noting in the limit $\Delta x = dx$ Equation 3.86 is written in differential form as:

$$\frac{dy}{dx} + \frac{d}{dx}\left(\frac{V^2}{2g}\right) = S_o - S_f \qquad (3.87)$$

This is similar to the equation developed in Section 3.4.1 where $\frac{d}{dx}\left(\frac{V^2}{2g}\right)$ was equal to $-F_R^2$.

Therefore, the gradually varied flow equation as presented in Section 3.4.1:

$$\frac{dy}{dx} = \frac{S_o - S_f}{1 - F_R^2} \qquad (3.88)$$

For gradually varied flow there is a gradual change in depth and velocity with distance and the conservation of energy with the losses being determined empirically using the Manning equation. Thus S_f is determined from the Manning equation.

$$S_f = \frac{Q^2 n^2}{(1.486)^2 A^2 R^{4/3}} \qquad (3.89)$$

Note that the coefficient 1.486 is 1.0 for SI units. For computations of the water surface elevations along the x-axis a finite difference scheme is used rather than solving the differential Equation 3.88. Equation 3.85 is written in terms of the water surface WSEL = y + z as:

$$WSEL_2 = WSEL_1 + \frac{V_1^2}{2g} - \frac{V_2^2}{2g} + h_l \qquad (3.90)$$

Equation 3.90 is solved iteratively and is presented in more detail in Section 5.2.

<u>Rapidly Varied Flow</u>. Rapidly varied flow has pronounced curvature of the streamlines; therefore the hydrostatic pressure distribution assumption is no longer valid. Often the flow curvature is so abrupt that the flow profile is broken and becomes discontinuous (such as a hydraulic jump). The energy equation is insufficient to solve for rapidly varied flow problems. Typically the momentum equation or hydraulic structure relationships are used. Sections 3.3.3 and 3.3.4 presented the weir, gates, and orifice equations that are used to solve for many rapidly varied flow situations.

<u>Hydraulic Jump</u>. The most important local rapidly varied flow problem is that of the hydraulic jump, which develops when supercritical flow transitions to subcritical. A hydraulic jump occurs when high velocity discharges flow into a zone of lower velocity either in natural rivers or over spillways or other hydraulic structures. Figure 3.28 is an illustration of the hydraulic jump phenomenon.

The equation relating the depths of flow up and downstream of the jump (y_2 and y_1) is developed by applying the conservation of linear momentum to the control volume assuming the channel bottom has a horizontal or small slope as shown in Figure 3.29. For channels with small slopes the gravity component (weight of water) in the downstream direction is relatively small and can be neglected. Also, since the length of the channel is short, the friction forces are small in comparison to the energy losses through the jump and can be neglected. Therefore the only significant forces are those caused by hydrostatic pressure (Figure 3.29).

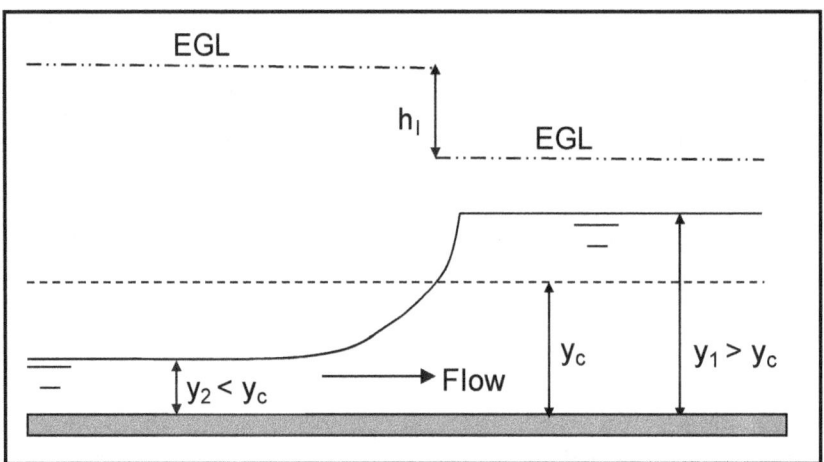

Figure 3.28. Hydraulic jump.

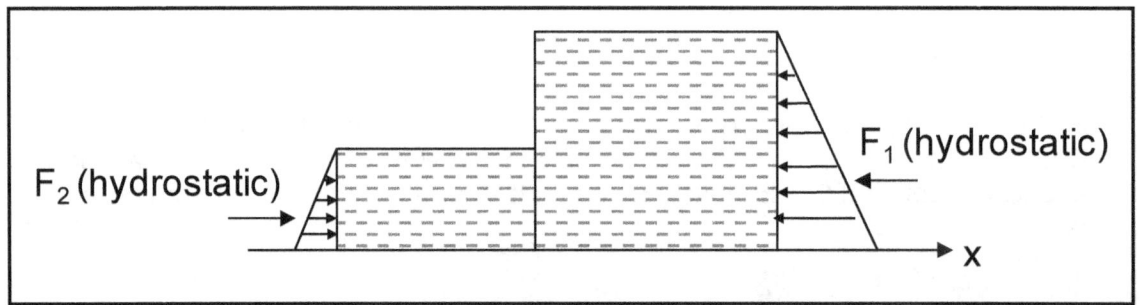

Figure 3.29. Control volume for the hydraulic jump.

By assuming hydrostatic pressure distribution at sections 2 and 1 where the flow streamlines are nearly parallel and summing the forces in the x-direction gives:

$$\sum F_x = F_{2H} - F_{1H} = m\bar{a} \tag{3.91}$$

The magnitude of the hydrostatic force is $F = \gamma \bar{y} A$, where $\bar{y}$ is the distance measured from the free surface to the center of gravity of the cross sectional area. Therefore Equation 3.91 can be written as:

$$\sum F_x = \gamma \frac{y_2}{2} A_2 - \gamma \frac{y_1}{2} A_1 = m\bar{a} = \rho Q (V_1 - V_2) \tag{3.92}$$

In the case of a rectangular channel $A = yW$ and $q = Q/W$, Equation 3.92 can be simplified to by dividing by γW:

$$\frac{y_2^2}{2} - \frac{y_1^2}{2} = \frac{q}{g}(V_1 - V_2) \tag{3.93}$$

Since $q = vy$, Equation 3.93 can be written as:

$$\frac{y_2^2}{2} - \frac{y_1^2}{2} = \frac{q}{g}\left(\frac{q}{y_1} - \frac{q}{y_2}\right) \tag{3.94}$$

The next series of equations are algebraic manipulations of the Equation 3.94:

$$(y_2 - y_1)(y_2 + y_1) = \frac{2q^2}{gy_1 y_2}(y_2 - y_1) \tag{3.95}$$

$$y_1 y_2^2 + y_1^2 y_2 = \frac{2q^2}{g} \tag{3.96}$$

Divide by y_1^3 and substituting yields:

$$\frac{y_2^2 y_1}{y_1^3} + \frac{y_1^2 y_2}{y_1^3} = \frac{2q^2}{gy_1^3} \tag{3.97}$$

$$\left(\frac{y_2}{y_1}\right)^2 + \left(\frac{y_2}{y_1}\right) = \frac{2q^2}{gy_1^3} \tag{3.98}$$

This is a quadratic equation for y_2/y_1 and knowing y_1, can be solved as:

$$\frac{y_2}{y_1} = \frac{1}{2}\left\{-1 \mp \sqrt{1 + \frac{8q^2}{gy_1^3}}\right\} \tag{3.99}$$

This can be reduced to the positive root as the negative root will give a negative depth.

$$y_2 = \frac{1}{2}y_1\{-1+\sqrt{1+8F_R^2}\} \tag{3.100}$$

3.4.4 Profiles for Gradually Varied Flow

To develop flow profiles the conservation of energy equation is used. The energy associated with section 1 and section 2 in Figure 3.27 can be stated as follows: the water surface elevation (potential energy) plus the velocity head (kinetic energy) at section 2 is equal to the water surface elevation and the velocity head at section 1 plus the energy losses through the reach. The total energy or head above an arbitrary datum was given in Equation 3.83.

Gradually varied flow is steady flow whose depth varies gradually along the length of the channel. This definition requires that two conditions be met: (1) the flow is steady (i.e., all of the variables remain constant for the time interval under consideration), and (2) the streamlines are approximately parallel (i.e., the pressure distribution over the cross section can be considered hydrostatic). The theory of gradually varied flow assumes that the head loss through the reach can be approximated using uniform flow equations (the Manning equation will be used to calculate the friction losses). Other assumptions for the development of the flow profiles will include the following:

1. The slope of the channel is small (i.e., the depth of flow is the same whether the vertical or normal direction is chosen). This assumes that the cosine of the slope angle is sufficiently small (i.e., $\cos\theta \cong 1$).
2. No air entrainment occurs.
3. The channel is prismatic (i.e., the channel has constant alignment and shape).
4. The velocity distribution in the channel is fixed (i.e., the velocity distribution coefficients are constant and for our case equal to one).
5. The roughness coefficient is independent of the depth of flow and constant throughout the channel reach under consideration.

Equation 3.101 is the gradually varied flow equation when the Manning equation is used to define energy loss. It is used to obtain the shapes of the water surface profiles. Recall that dy/dx represents the slope of the water surface with respect to the bottom of the channel. If the water surface converges to the bottom of the channel (i.e., the depth of flow decreases with distance downstream) the slope will be negative. Alternatively, if the water surface diverges from the bottom of the channel (i.e., the depth of flow increases with distance downstream) the slope is positive.

$$\frac{dy}{dx} = S_0 \frac{1-(y_n/y)^{10/3}}{1-(y_c/y)^3} \tag{3.101}$$

<u>Types of Channel Slopes</u>. Equation 3.101 is used to define the type of slopes relative to the x direction (direction of flow). Since hydrostatic pressure distribution has been assumed in the development of the equation, the application will be limited to flows with streamlines essentially straight and parallel, and of small slope. Also, the depth of flow is measured vertically from the channel bottom, the slope of the water surface dy/dx is relative to this channel bottom. Figure 3.30 shows water surface slopes resulting from the change in depth along the channel.

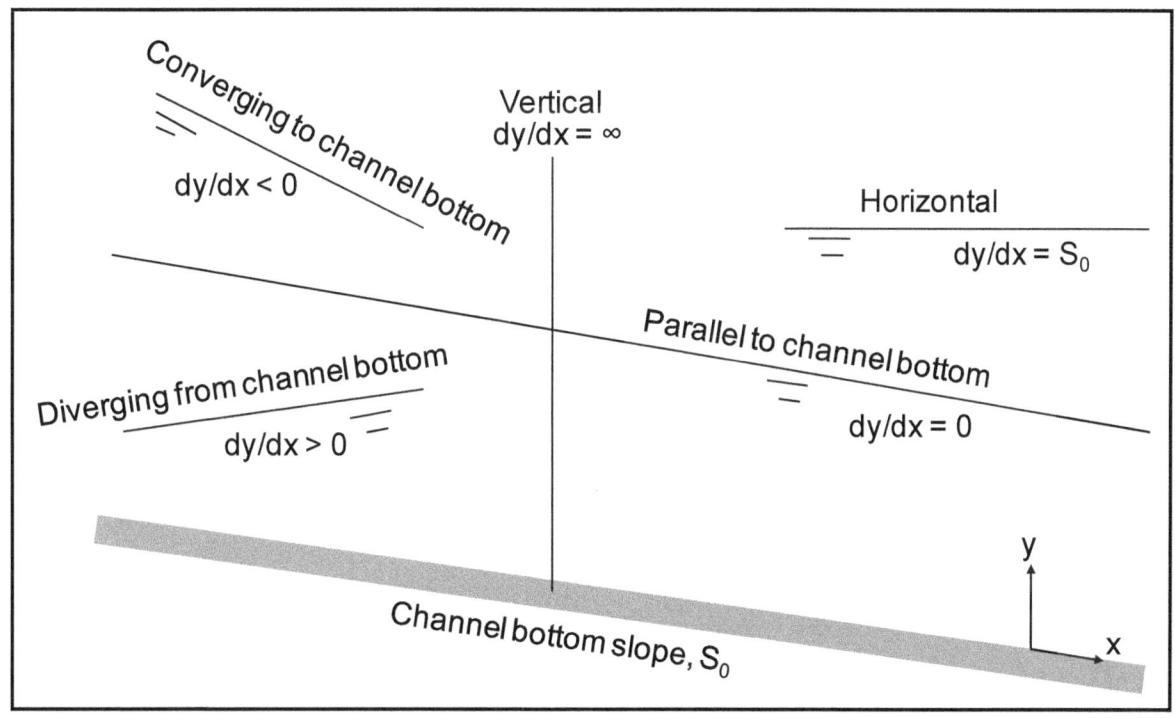

Figure 3.30. Water surface slopes versus channel bottom slope.

There are five types of bed slopes that are encountered: mild; critical; steep; horizontal; and adverse. All of the slopes are defined using the channel bottom slope S_0. To define mild, critical, and steep slopes the channel slope S_0 is compared to the slope if the channel were flowing at critical depth (i.e., solve for the critical slope S_c by substituting y_c into the Manning equation). Then we can define the slopes as follows:

Mild Slope	$S_0 < S_c$
Critical Slope	$S_0 = S_c$
Steep Slope	$S_0 > S_c$
Horizontal Slope	$S_0 = 0$
Adverse Slope	$S_0 < 0$

Figure 3.31 illustrates the 12 flow profile curves for the five slopes. A summary of the profiles is given in Table 3.4.

When there is a change in cross section geometry or channel slope, or there is an obstruction to the flow, the qualitative analysis of the flow profile depends on locating the control points, determining the type of curve upstream and downstream of the control points, and then sketching the backwater curves. When the flow is supercritical ($F_R > 1$) the control of the depth is upstream and the computations must proceed in the downstream direction. On the other hand, when the flow is subcritical ($F_R < 1$) the depth control is downstream and the computations must proceed upstream. Example flow profile curves that result from a change in slope are illustrated in Figure 3.32. For the M_1, M_2, S_1, C_1, H_2, and A_2 profiles the computations for water surface profiles start at the downstream control point and proceed upstream. For the M_3, S_2, S_3, C_3, H_3, and A_3 profiles the computations for water surface profiles start upstream at the control point and proceed downstream.

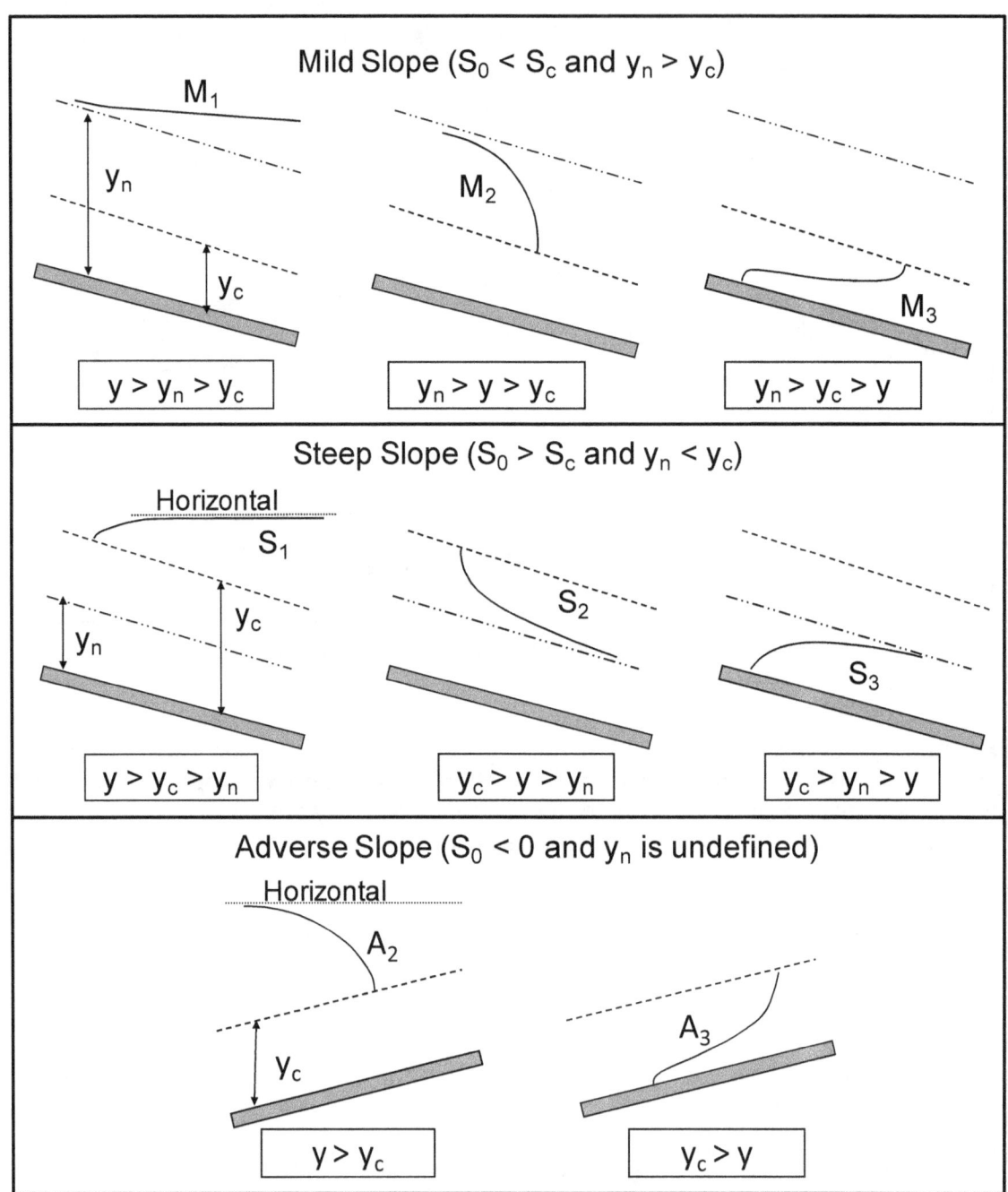

Figure 3.31. Flow profile curves.

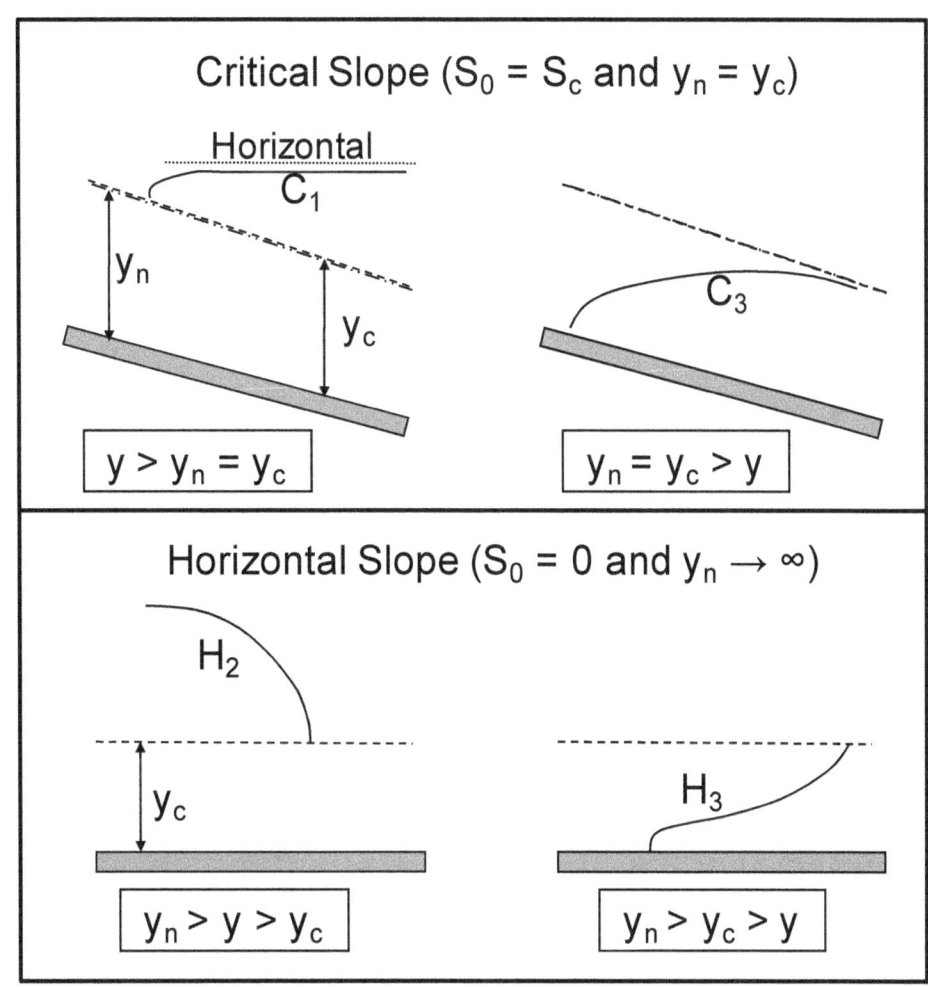

Figure 3.31. Flow profile curves (continued)

Table 3.4. Summary of the Flow Profiles				
Channel Slope	Designation	Relation of y to y_n and y_c	General Type of Curve	Type of Flow
Mild Slope	M_1	$y > y_n > y_c$	Backwater	Subcritical
	M_2	$y_n > y > y_c$	Drawdown	Subcritical
	M_3	$y_n > y_c > y$	Backwater	Supercritical
Critical Slope	C_1	$y > y_c = y_n$	Backwater	Subcritical
	None	$y = y_c = y_n$	None	None
	C_3	$y_c = y_n > y$	Backwater	Supercritical
Steep Slope	S_1	$y > y_c > y_n$	Backwater	Subcritical
	S_2	$y_c > y > y_n$	Drawdown	Supercritical
	S_3	$y_c > y_n > y$	Backwater	Supercritical
Horizontal Slope	None	$y > y_n > y_c$	None	None
	H_2	$y_n > y > y_c$	Drawdown	Subcritical
	H_3	$y_n > y_c > y$	Backwater	Supercritical
Adverse Slope	None	Not applicable	None	None
	A_2	$y > y_c$	Drawdown	Subcritical
	A_3	$y_c > y$	Backwater	Supercritical

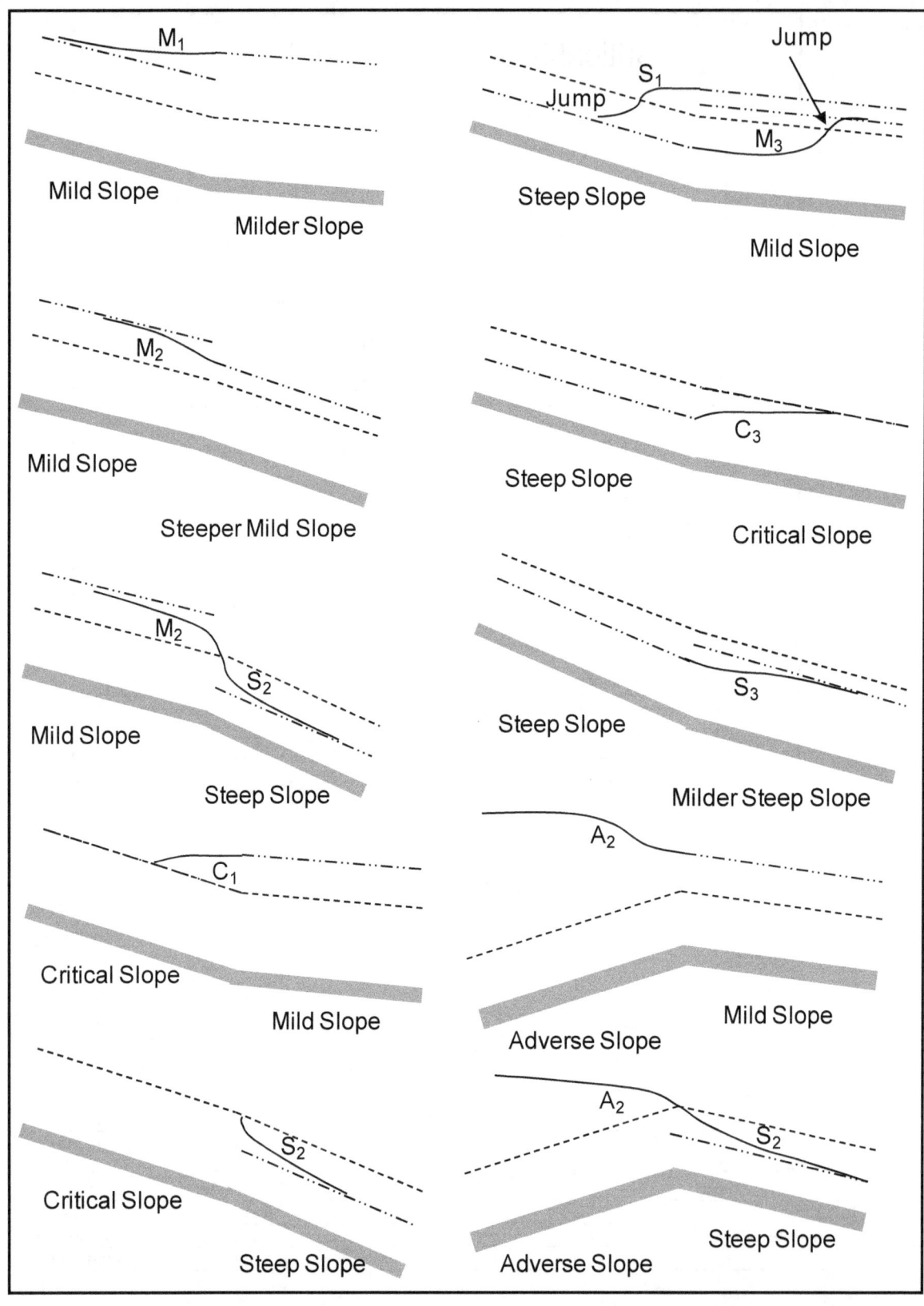

Figure 3.32. Example flow profiles for gradually varied flow with a change in slope.

CHAPTER 4

HYDRAULIC ANALYSIS CONSIDERATIONS

4.1 INTRODUCTION

Chapter 3 provides background on the fundamental open channel flow concepts that comprise the basis for the majority of the numerical hydraulic modeling and calculations encountered in open channel flow and bridge hydraulic analysis. The calculations are often complex and tedious, and many require iterative solution techniques due to interaction between variables. Therefore, computer programs have been the primary tool for hydraulic engineers ever since computers have become widely available. As computer technology has advanced, so has numerical hydraulic modeling. The primary analysis approach for bridge hydraulics is one-dimensional modeling, although two-dimensional modeling is becoming common and three-dimensional modeling is used to analyze complex flow fields. Chapters 5 and 6 provide information and guidance on the use of one- and two-dimensional numeric models for bridge hydraulic analysis. This chapter includes information on selecting the most appropriate approach whether it is one-, two-, or three-dimensional numerical modeling, steady or unsteady modeling, or physical hydraulic modeling. This chapter also provides background on developing input data and other considerations that are common to all bridge hydraulic problems regardless of the specific approach.

4.2 HYDRAULIC MODELING CRITERIA AND SELECTION

Any hydraulic model, whether it is numerical or physical, has assumptions and requirements. It is important for the hydraulic engineer to be aware of and understand the assumptions because they form the limitations of that approach. It is the goal of any hydraulic model study to accurately simulate the actual flow condition. Violating the assumptions and ignoring the limitations will result in a poor representation of the actual hydraulic condition. Treating the model as a black box will often produce inaccurate results. This is not acceptable given the cost of bridges and the potential consequences of failure. Therefore, the approach should be selected based primarily on its advantages and limitations, though also considering the importance of the structure, potential project impacts, cost, and schedule.

4.2.1 One-Dimensional Versus Two-Dimensional Modeling

One-dimensional modeling requires that variables (velocity, depth, etc.) change predominantly in one defined direction, x, along the channel. Because channels are rarely straight, the computational direction is along the channel centerline. Two-dimensional models compute the horizontal velocity components (V_x and V_y) or, alternatively, velocity vector magnitude and direction throughout the model domain. Therefore, two-dimensional models avoid many assumptions required by one-dimensional models, especially for the natural, compound channels (free-surface bridge flow channel with floodplains) that make up the vast majority of bridge crossings over water. Chapters 5 and 6 include detailed discussions of one- and two-dimensional model assumptions and limitations.

The advantages of two-dimensional modeling include a significant improvement in calculating hydraulic variables at bridges. Therefore FHWA has a strong preference for the use of two-dimensional models over one-dimensional models for complex waterway and/or complex bridge hydraulic analyses. One-dimensional models are best suited for in-channel flows and when floodplain flows are minor. They are also frequently applicable to small streams. For extreme flood conditions, one-dimensional models generally provide accurate results for narrow to moderate floodplain widths. They can also be used for wide floodplains when the degree of bridge constriction is small and the floodplain vegetation is not highly variable. In general, where lateral velocities are small one-dimensional models provide reasonable results. Avoiding significant lateral velocities is the reason why cross section placement and orientation are so important for one-dimensional modeling. Two-dimensional models generally provide more accurate representations of:

- Flow distribution
- Velocity distribution
- Water Surface Elevation
- Backwater
- Velocity magnitude
- Velocity direction
- Flow depth
- Shear stress

Although this list is general, these variables are essential information for new bridge design, evaluating existing bridges for scour potential, and countermeasure design. The Federal Emergency Management Agency (FEMA) also depends on numerical hydraulic models of extreme events to determine flood hazards. FEMA and NOAA (National Oceanic and Atmospheric Administration) commissioned the National Research Council (NRC 2009) to investigate the factors that affect flood map accuracy and identify ways of improving flood mapping. Among their findings, the NRC recommended greater use of two-dimensional models.

Two-dimensional models should be used when flow patterns are complex and one-dimensional model assumptions are significantly violated. If the hydraulic engineer has great difficulty in visualizing the flow patterns and setting up a one-dimensional model that realistically represents the flow field, then two-dimensional modeling should be used. One study that developed criteria for selecting one- versus two-dimensional models is "Criteria for Selecting Hydraulic Models" (NCHRP 2006). The recommendations from that study are summarized and expanded on below.

Multiple Openings. Multiple openings along an embankment are often used on rivers with wide floodplains. Rather than using a single bridge, additional floodplain bridges are included. Although one-dimensional models can be configured to analyze multiple openings, the judgment and assumptions that are made by the hydraulic engineer in combination with the assumptions and limitations of the software result in an extreme degree of uncertainty in the results. The proportion of flow going through a particular bridge and the corresponding flow depth and velocity are important for structure design and scour analysis. Because multiple opening bridges represent a large investment, two-dimensional analysis is always warranted.

Another type of multiple opening is multiple bridges in series. There are conditions when this bridge configuration should be analyzed using two-dimensional models. These include unmatched bridge openings or foundations that do not align. An upstream or downstream railroad or parallel road may significantly alter the flow conditions and warrant two-dimensional analysis.

Figure 4.1 shows two-dimensional model results (velocity magnitude) for the U.S. Route 1 crossing over the Pee Dee River in South Carolina. Flow is generally from top to bottom in this figure. This model illustrates several reasons for selecting two-dimensional modeling. The floodplain width ranges from 4,000 to 8,000 ft (1,200 to 2,400 m) and has highly variable land use and vegetation. The US 1 crossing includes a 2,000 ft (600 m) main channel bridge and two 500 ft (150 m) relief bridges. There is also a railroad crossing downstream. Although the railroad also has three bridge openings, they are shorter and not aligned with the US 1 bridges. The highest velocity, greater than 8 ft/s (2.4 m/s) occurs in the main channel. However, the center relief bridge has an average velocity of nearly 6 ft/s (1.8 m/s) and the eastern relief bridge has velocities of over 7 ft/s (2.1 m/s). The floodplain area under the main channel bridge, however, has velocities ranging from 1 to 3.5 ft/s (0.3 to 1.1 m/s). Therefore, overall conveyance would be improved and backwater would be reduced by shortening the main channel bridge and lengthening the relief bridges. If changing the bridge lengths would adversely impact the downstream railroad bridges, the two-dimensional model results would also quantify those impacts.

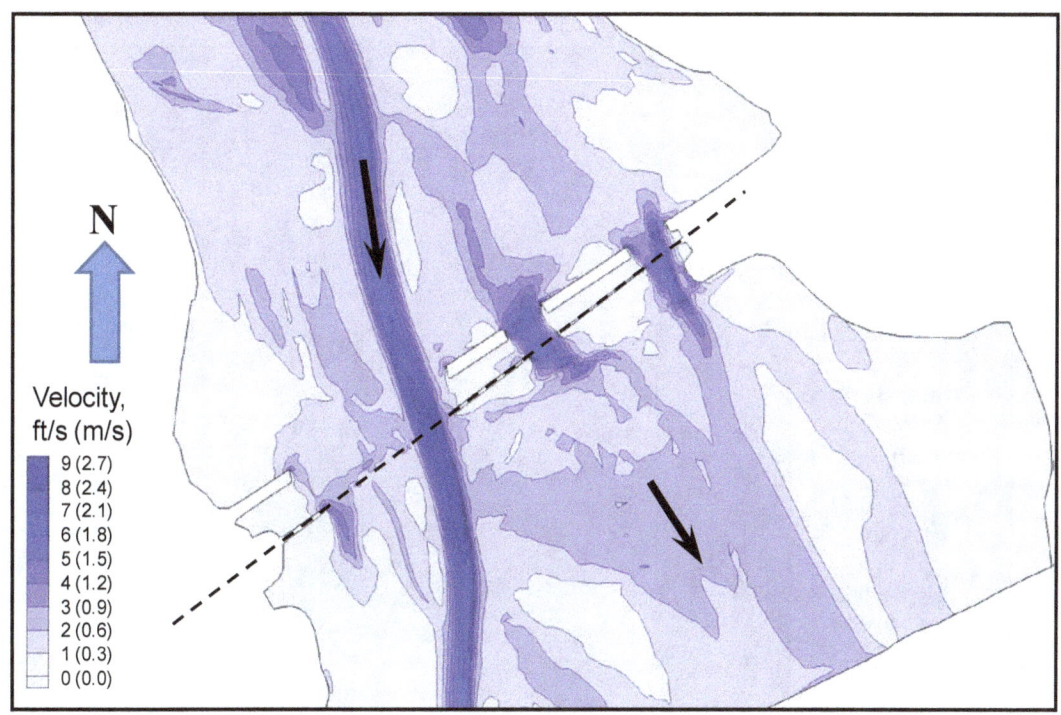

Figure 4.1. Two-dimensional model velocities, US 1 crossing Pee Dee River.

Wide Floodplains. Floodplains often include features that significantly impact flow conveyance and flow distribution. Historic channel alignments and changes in land use or vegetation affect floodplain flow distribution. In a one-dimensional model, two cross sections that are a short distance apart may have significantly different vegetation, such as wooded versus cleared, or may have significantly different topography due to land use activities. If the hydraulic engineer uses these cross sections exactly as they exist, the one-dimensional model will depict a sudden change in flow distribution that is not physically possible. To better depict the flow conditions, the hydraulic engineer would need to adjust the cross section locations or alter the Manning n values, although this is difficult to implement. The two-dimensional model avoids these difficulties because in the simulation all the flow is interconnected. Therefore, wide and complex floodplains benefit from two-dimensional analysis.

Skewed Roadway Alignment. Roadways should be aligned perpendicular to channel and floodplain flows. FHWA (1978) indicates that skewed crossings with angles of up to 20 degrees produced no objectionable flow patterns. The HEC-RAS Reference Manual (USACE 2010c) indicates that using the projected opening is adequate for skew angles of up to 30 degrees for small flow constrictions. Two-dimensional modeling is the recommended approach for higher skew angles or moderate amounts of skew combined with moderate to high flow contraction. Not only will the flow patterns and bridge conveyance be better defined, but potential problems with backwater will also be evident. Figure 4.2 shows a crossing with an approximate 25 degree skew to the floodplain with flow from top to bottom. This figure illustrates how floodplain impacts can vary greatly upstream of a skewed crossing. The colors represent the difference in water surface between natural (no bridge crossing) and existing conditions. The darkest color shows the greatest water surface increase and the opposite side of the embankment shows a decrease in water surface compared to natural conditions. The fact that this is also a multiple opening crossing also complicates the hydraulic conditions.

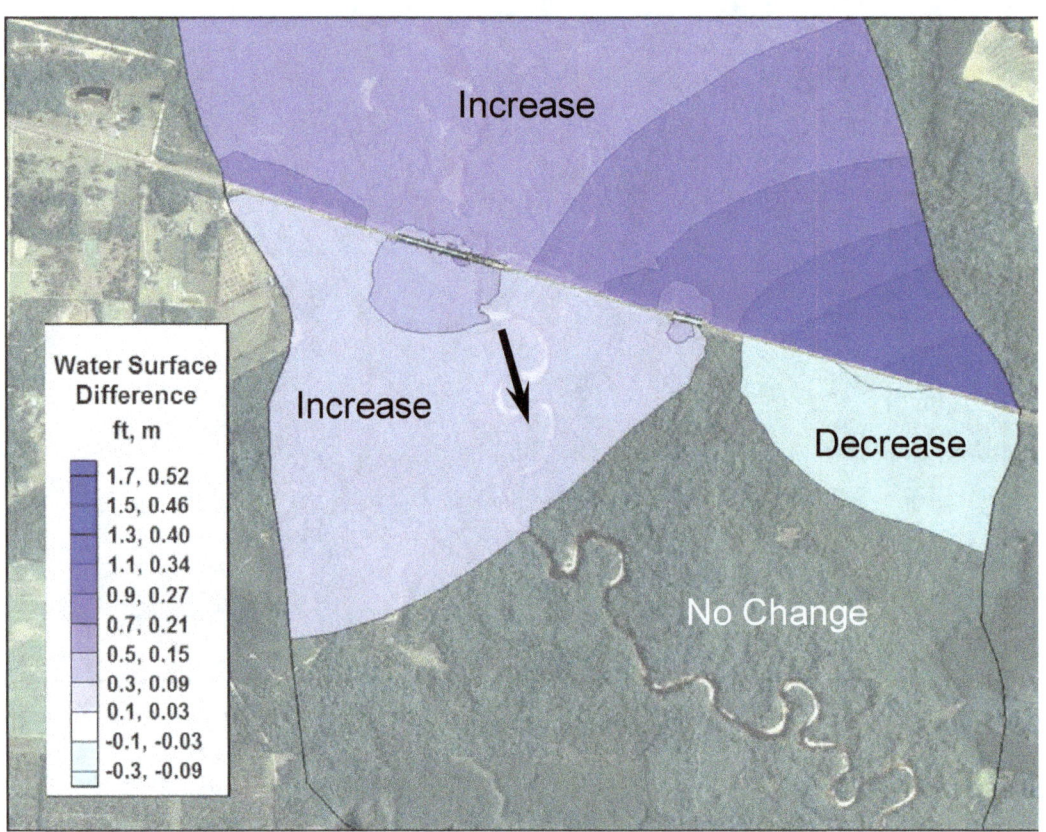

Figure 4.2. Backwater at a skewed crossing.

Road Overtopping. When computing road overtopping, the HEC-RAS model (USACE 2010c) uses the total energy grade line in the cross section upstream of the bridge as the head value in the weir equation. This assumption is reasonable for many conditions. Because standard use of ineffective flow areas can trigger full floodplain flow for any amount of overtopping, USACE (2010a) recommends comparing the road overtopping discharge to the floodplain flow and adjusting the Manning n to better maintain flow continuity. As illustrated in Figure 4.2, for roads crossing wide floodplains or skewed crossings, two-dimensional models offer a better approach. Road overtopping is still computed using the weir equation, but nodes on either side of the embankment are connected using a weir segment. The water surface and velocity at the two connected nodes are used to determine head and submergence. The head at the upstream node is used rather than the total energy grade line of the entire upstream cross section. Therefore, better estimates of the initiation of overtopping and overtopping discharges are achieved.

Upstream Controls. For sub-critical flow conditions, calculations progress from downstream to upstream. Locally, however, flow depth, velocity magnitude, and velocity direction can be controlled by upstream structures and obstructions. In one-dimensional modeling the usual, approximate approach is to incorporate ineffective flow areas to account for upstream obstructions. The overall flow area and conveyance are altered, but flow distribution is still based on the distribution of conveyance at the cross section. Therefore, upstream effects are not fully accounted for in one-dimensional models. Figure 4.3 shows velocity conditions at the I-35W crossing of the Mississippi River in Minnesota. This figure illustrates that two-dimensional models can be used to accurately determine whether an upstream condition impacts a downstream structure, even in sub-critical flow conditions. The I-35W Bridge is located downstream of St. Anthony Falls Lock and Dam, which concentrates the approach flow to the I-35W Bridge. During extreme events, the lock and dam could be operated with flow primarily through the three gates (as shown), or additional flow can be passed through the lock chambers. A range of upstream operating conditions was modeled for the I-35W new bridge design. For this situation flow is definitely not distributed in the downstream channel based on conveyance distribution. Another concern with this project was avoiding any adverse impact on the 10th Avenue Bridge immediately downstream. The 10th Avenue Bridge has a large pier in the center of the channel. The two-dimensional model was used to evaluate whether changes to the I-35W replacement bridge design would increase velocities approaching the 10th Avenue Bridge pier or change the flow angle of attack.

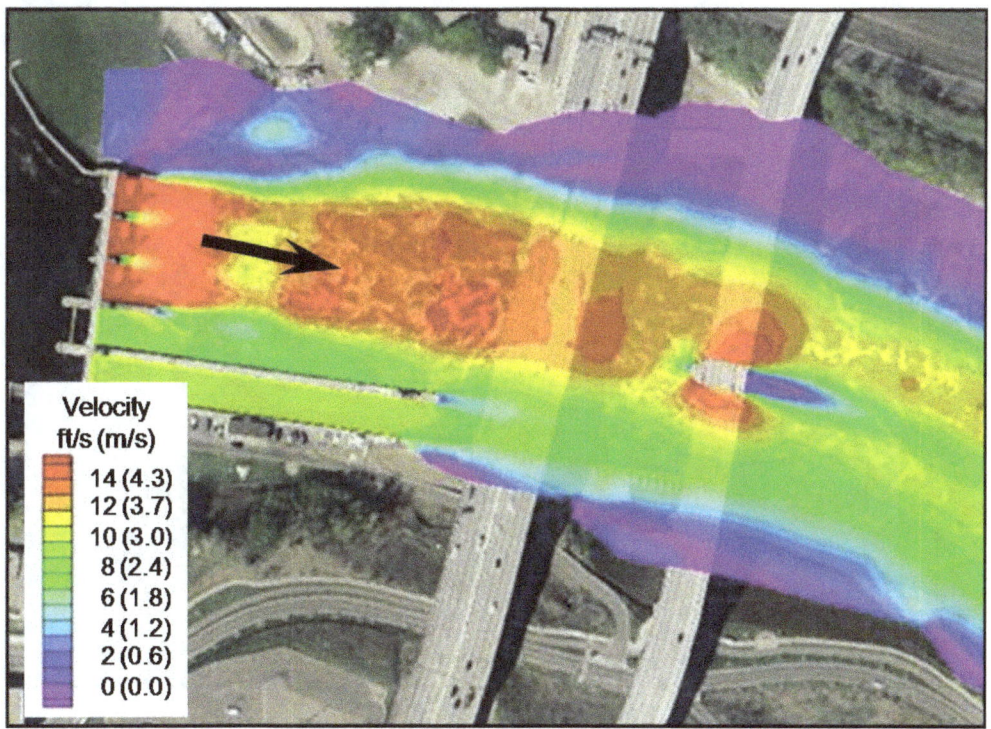

Figure 4.3. Two-dimensional model velocities, I-35W over Mississippi River.

<u>Bends, Confluences and Angle of Attack</u>. Highly sinuous rivers are, by definition, not one-dimensional, especially during floods when water in the floodplain flows more directly down valley and moves in and out of the channel. One-dimensional models must consider different channel and floodplain flow distances between cross sections and compute a discharge-weighted flow length. Two-dimensional models do not make any simplifying assumptions related to channel versus floodplain flow distance because the two-dimensional network directly incorporates flow paths. Flow conditions at confluences also vary depending on the proportion of flow in the main stem and tributary. With a one-dimensional model, determining the angle of attack for pier scour calculations is highly subjective in these situations and can be difficult for many other conditions. Two-dimensional models provide improved estimates of angle of attack because velocity direction is computed directly.

Multiple Channels. Anabranched and braided rivers have multiple channels and flow paths that complicate hydraulic calculations. Figure 4.4 shows an extreme example of multiple channels at Altamaha Sound in Georgia. The figure depicts channels in blue, flood-prone areas in green, and roadway alignments in red. The area is subject to riverine and tidal flooding. Not only are there nine crossings (five on I-95 and four on SR 17), but there are more than 20 individual channel segments, or reaches that would need to be included in a HEC-RAS split-flow model. The hydraulic engineer would also have to decide the amount of adjacent floodplain to assign to each channel segment and may well need to allow for lateral flow between floodplain segments. Two-dimensional models, while still a significant challenge, clearly have numerous advantages in this situation. Although many multiple channel situations are well simulated with the split-flow options in HEC-RAS, the effort in developing a two-dimensional model for these conditions may be less than an equivalent one-dimensional model.

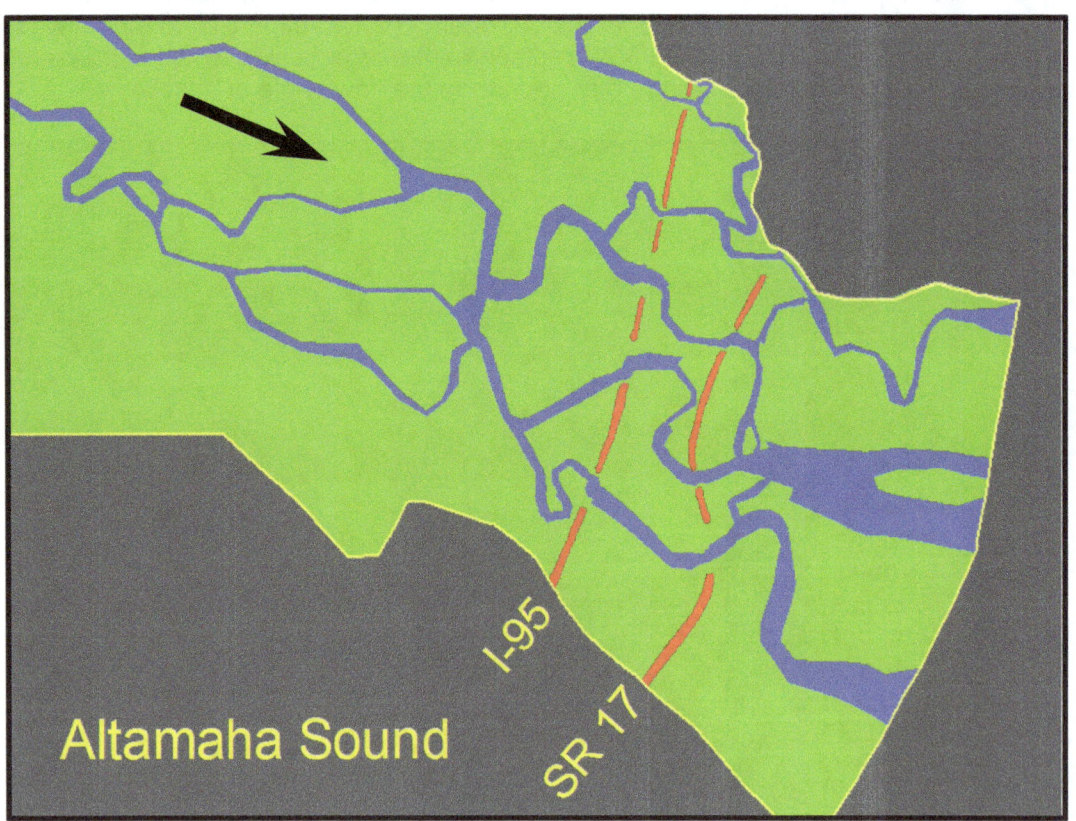

Figure 4.4. Channel network, Altamaha Sound, Georgia.

Tidal Conditions and Wind Simulation. Figure 4.4 is an example of the complex channel and hydraulic conditions that occur more frequently in tidal waterways than in upland rivers. Tidal waterways include inlets, estuaries, bays, and passages. Many bays and estuaries are crossed by causeways with multiple bridge openings and the potential for overtopping and wave attack. The HEC-25 manuals (FHWA 2004, 2008) include information and guidance on tidal and coastal conditions, including tides, storm surges, and wind, that impact transportation structures. Some coastal hydrodynamic conditions are dominated by wind-driven currents. Many two-dimensional models include wind stress acting on the water surface as a boundary condition. Therefore, two-dimensional models need to be used for many coastal bridge hydraulic analyses.

Flow Distribution at Bridges. The HEC-18 manual (FHWA 2012b) establishes scour evaluation procedures recommended by FHWA. Flow and velocity distributions are required within the bridge opening to calculate contraction, pier and abutment scour. One-dimensional models estimate flow and velocity distribution based on the incremental conveyance within a cross section (see Section 5.4). This assumption requires that each point in the cross section also have the same water surface elevation and energy slope. Figure 4.5 shows water surface and velocity vectors from a two-dimensional model. The model represents a relatively simple situation, but does not meet the one-dimensional criteria described above. In this figure, the thin lines indicate the channel banks and embankment. The water surface is relatively uniform along the upstream face of the bridge, varying by less than 0.3 ft (0.1 m), but the velocity vectors in the overbank areas in the bridge opening are not perpendicular to the bridge face. Although these are indicators that the flow is not one-dimensional, the most significant departure from one-dimensional assumptions is the velocity in the overbank areas under the bridge. A one-dimensional model would estimate much lower velocity in the overbanks based on conveyance and equal energy slope at the bridge cross section. The average energy slope in the overbank areas under the bridge is over five times the energy slope of the channel area, resulting in velocities more than twice what is computed from one-dimensional conveyance-weighted calculations.

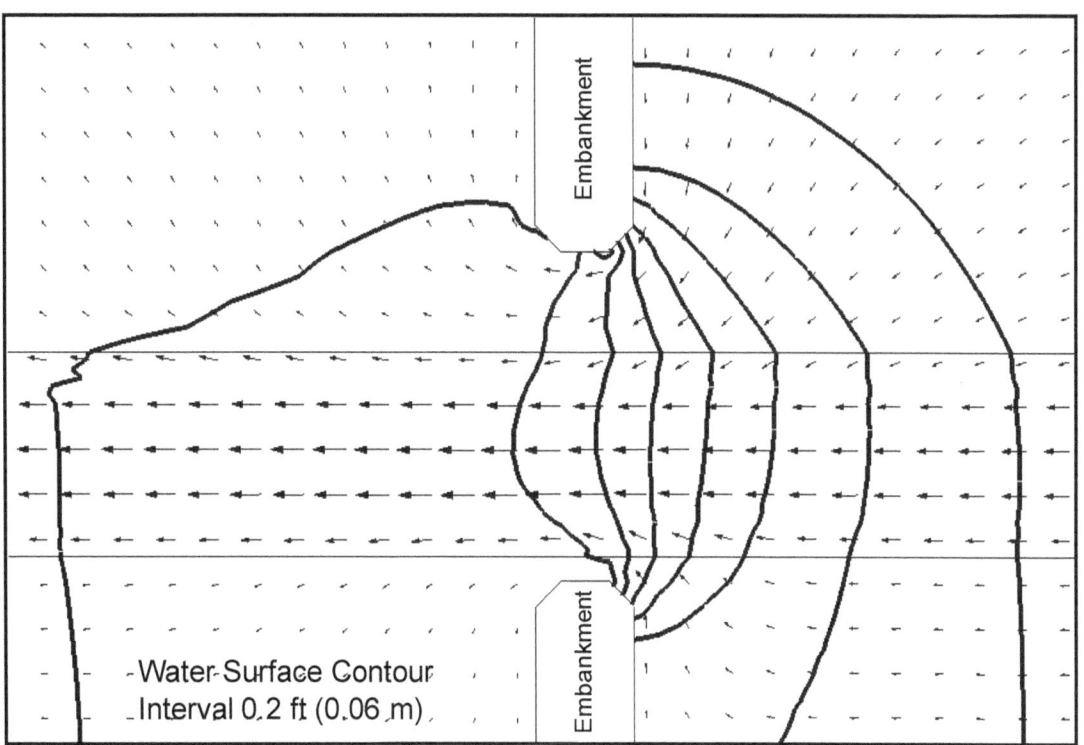

Figure 4.5. Two-dimensional flow within a bridge opening.

Countermeasure Design. The HEC-23 manual (FHWA 2009a) provides guidance on designing countermeasures for stream instability and scour. Many countermeasures, including spurs, guide banks, and transverse dikes, significantly alter flow paths and flow distributions. Two-dimensional models that are set up with a complete three-dimensional representation of the channel and countermeasure provide an accurate simulation of the flow field in the horizontal plane including locations of high velocity, flow separation and flow circulation. Three-dimensional models, CFD modeling, and physical hydraulic modeling may be required for analyzing extremely complex flow fields with large vertical velocity components that can occur at countermeasures.

Summary. No numerical model provides an exact representation of the complexities of an actual flow condition. This is especially true where roadways cross natural water courses with variable channel bathymetry, floodplain topography, land use, and vegetation. The assumptions that are required for one-dimensional models are often violated to a greater degree than is commonly thought, though in many cases experienced hydraulic engineers can compensate for some of the limitations of one-dimensional models. Because two-dimensional models avoid many assumptions required by one-dimensional models, they better represent the physics of the flow and provide more realistic hydraulic results, especially at highway encroachments. Therefore, two-dimensional models should be used for many bridge hydraulics and scour problems. Table 4.1 provides guidance on selecting one- versus two-dimensional modeling for bridge hydraulic and scour analyses. Two-dimensional models provide more accurate results for hydraulically complex conditions. Table 4.1 does not include three-dimensional numerical modeling, computational fluid dynamics (CFD), or physical hydraulic modeling because these methods are used primarily to simulate individual piers or other bridge elements and are rarely used to analyze the entire bridge reach.

Table 4.1. Bridge Hydraulic Modeling Selection.		
Bridge Hydraulic Condition	Hydraulic Analysis Method	
	One-Dimensional	Two-Dimensional
Small streams	I	●
In-channel flows	I	●
Narrow to moderate-width floodplains	I	●
Wide floodplains	●	I
Minor floodplain constriction	I	●
Highly variable floodplain roughness	●	I
Highly sinuous channels	●	I
Multiple embankment openings	●/O	I
Unmatched multiple openings in series	●/O	I
Low skew roadway alignment (<20°)	I	●
Moderately skewed roadway alignment (>20° and <30°)	●	I
Highly skewed roadway alignment (>30°)	O	I
Detailed analysis of bends, confluences and angle of attack	O	I
Multiple channels	●	I
Small tidal streams and rivers	I	●
Large tidal waterways and wind-influenced conditions	O	I
Detailed flow distribution at bridges	●	I
Significant roadway overtopping	●	I
Upstream controls	O	I
Countermeasure design	●	I

I   well suited or primary use
●   possible application or secondary use
O   unsuitable or rarely used
●/O possibly unsuitable depending on application

4.2.2 Steady Versus Unsteady Flow Modeling

The majority of bridge hydraulic analyses for upland rivers are performed with steady-state conditions where the peak flow conditions for various design events are used for hydraulic design and scour computations. Chapter 7 provides guidance on modeling unsteady flows. There are several conditions when unsteady modeling should be performed. These include nearly all tidal applications (FHWA 2004, 2008). An exception for tidal models is when a peak discharge and the corresponding water surface elevation can be established by other means. Unsteady flow analysis can also be beneficial when the available base modeling for floodplain regulation was developed as an unsteady model or when storage effects need to be evaluated.

There is often the perception that increasing the size of a bridge will increase downstream flooding by decreasing the amount of water stored upstream of the existing structure. This topic was investigated by McEnroe (2006), who concluded that few culverts and even fewer bridges are affected by structure-induced detention storage and, therefore, most do not increase downstream flooding when they are enlarged. Roads that overtop are unlikely to have increased downstream flow when structure sizes are increased. To fully assess the potential for increased downstream flooding, unsteady flow modeling is required. However, the model extent must be increased upstream to capture available storage and the downstream extent should be increased to account for dynamic routing effects. The downstream boundary condition must be applicable to the full range of flows (rating curve or normal depth).

Although the potential downstream impacts to bridge enlargement are generally minor or negligible, the benefits of bridge enlargement are often considerable. McEnroe (2006) indicates that benefits include reductions in backwater, flooding, overtopping, and scour. He also indicates that even if downstream flows increase, the flow hydrograph will more closely resemble the natural conditions that existed before the roadway was constructed.

4.2.3 Three-Dimensional Modeling and Computational Fluid Dynamics

Three dimensional modeling includes more variables in a given flow condition than one- and two dimensional modeling. While it requires more modeling effort and computational resources, three-dimensional Computational Fluid Dynamics (CFD) simulation often reveals more details of the flow at bridge elements. With the advancement of high performance computing facilities and computational algorithms, three dimensional CFD is becoming more feasible to use in hydraulic engineering. Computational facilities such as that at Transportation Research and Analysis Computing Center (TRACC) of Argonne National Laboratory house high-performance computational clusters with multi-core processing capability. In order to apply analytical formulation to complex stream conditions, three-dimensional modeling usually cuts the water body of interest into relatively small "cells," within which the flow condition is relatively simple so that the analytical formula is more readily applicable. Examples of numerical techniques for this purpose include Finite Difference Method (FDM), Finite Element Method (FEM), and Boundary Element Method (BEM). Depending upon the complexity and properties of the situation, general-purpose commercial software or custom-developed codes can be used. Most commercial software work with Computer-Aided Design (CAD) packages to streamline the modeling process of complex systems. Figure 4.6 shows the results from three-dimensional CFD modeling. The streamlines in Figure 4.6(a) illustrate the flow structure behind a rectangular pier with a 30 degree skew at the beginning of scour. Figure 4.6(b) shows the change in flow patterns at ultimate scour.

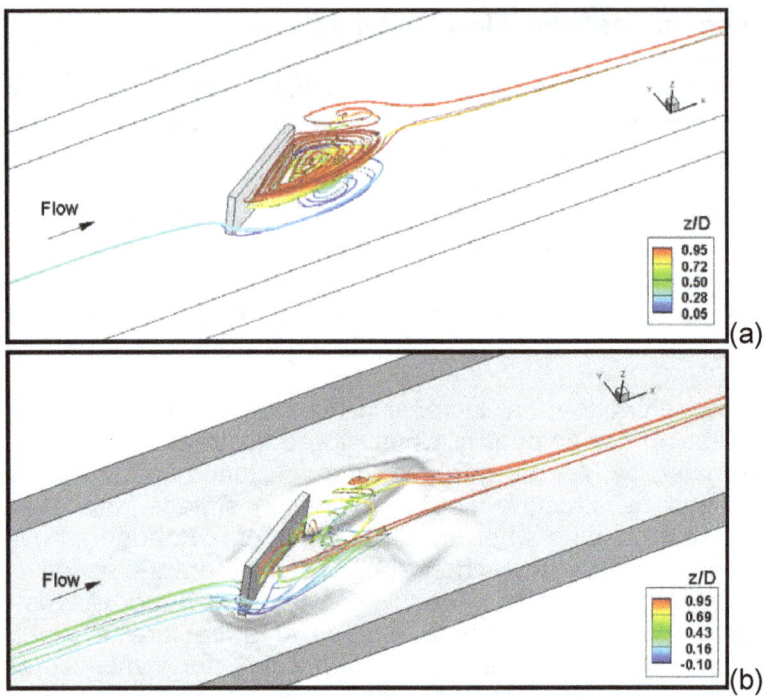

Figure 4.6. Three-dimensional CFD modeling (a) flow prior to scour (b) flow at ultimate scour.

Turbulence modeling is an important part of detailed three-dimensional CFD. It takes into account the fluctuation of velocity and energy transfer/dissipation in the simulation. The turbulence condition has a significant impact to bridge hydraulics and stream bed stability. Two important turbulent modeling approaches that are widely applied in bridge-related CFD simulations are large eddy simulation (LES) and Reynolds-averaged Navier-Stokes (RANS). The RANS method uses time-averaged equations of motion for fluid. Using Reynolds decomposition, the instantaneous velocity and pressure fields are decomposed into mean values and fluctuating components. An additional term compared to the original Navier-Stokes equations is a tensor quantity, known as the Reynolds stress tensor, in the resulting equations for the mean quantities. Reynolds stress tensor is modeled in terms of the mean flow quantities to provide closure of the governing equations.

Explicitly simulating eddies in all scales is extremely demanding on computer power and impractical. In LES, the turbulence over large scales is resolved by using filtered Navier-Stokes equations, which is a spatial averaging that eliminates the small scale turbulence. The small scale eddies are modeled based on the hypothesis that the smaller eddies are self-similar. LES allows the computation of instantaneous velocity distribution and hydrodynamic force, but requires a large amount of computational resources. Because of desire for information on temporal fluctuation in flow and because of continued advancements in computer power, the use of LES has increased rapidly in recent years. In bridge engineering, it is of great interest in scour development because the fluctuation of hydrodynamic force can significantly increase the erosion potential. In some past studies, it was found that the high fluctuation of bed shear may occur at a different location than high mean bed shear (see Figure 4.7).

LES can potentially provide additional temporal details to supplement RANS simulation and obtain more accurate dynamic measurements. There is not a one-size-fits-all optimal solution for turbulence modeling, so support from experiments is often needed. Once numerical modeling is calibrated by experiments, a large amount of additional conditions can be analyzed and expensive and time-consuming physical modeling can be greatly reduced.

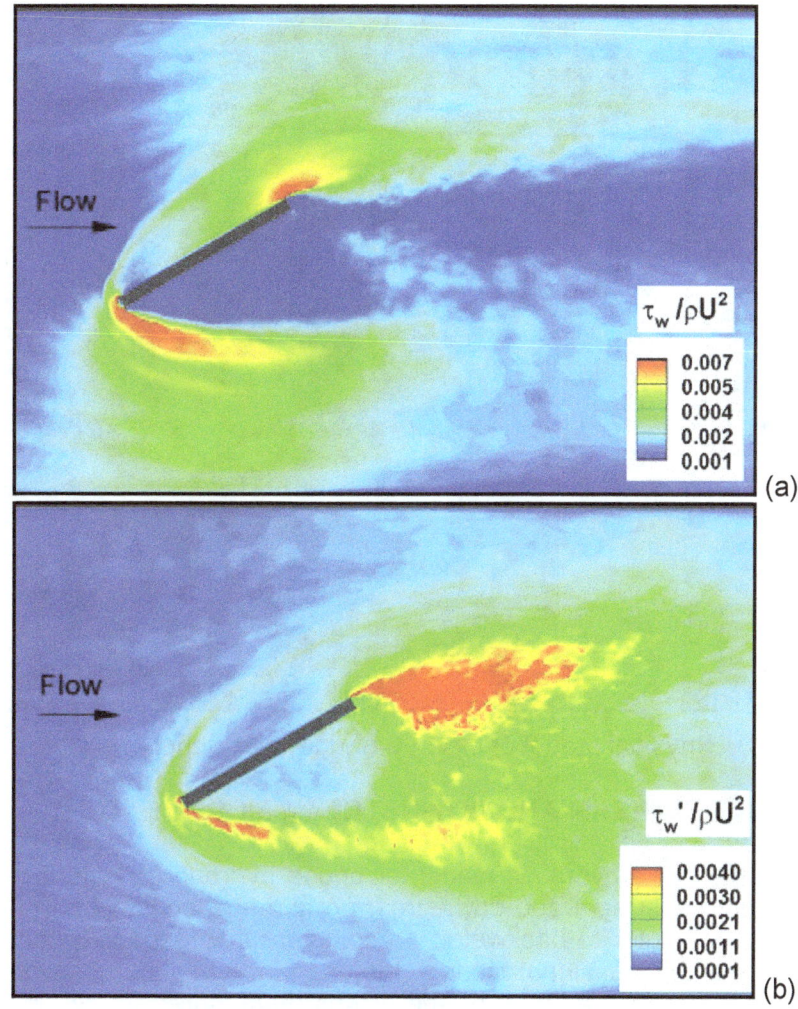

Figure 4.7. Bed shear from LES prior to scour (a) mean bed shear (b) fluctuation of bed shear.

4.2.4 Physical Modeling

Physical hydraulic modeling has and continues to be a valuable tool in fluid mechanics. Laboratory scale models provide direct experimental data for complex flow fields, flow-structure interaction, and erosion processes. Fluid mechanics textbooks (such as Munson et al. 2010) provide in-depth discussion of dimensional analysis and similitude requirements for laboratory scale models. Geometric similarity is the first requirement, although some conditions can be evaluated with distorted vertical and horizontal scales. For free-surface flow conditions, Froude number scaling (the ratio of inertial force to gravitational force) replicates the dominant hydraulic forces. When the Froude number is used for scaling, other force ratios, such as the Reynolds number, do not scale. Therefore, physical scale models are not a complete representation of the prototype conditions. Scale models range from three-dimensional fixed-bed models to fully three-dimensional moveable-bed models and moveable-bed models of individual piers or other structural elements to evaluate local scour (TAC 2004). For moveable-bed models, the sediment characteristics should also be scaled, though this is often difficult. Figure 4.8 shows a moveable-bed physical model of the I-90 crossing of Schoharie Creek in New York conducted at Colorado State University (CSU). The model was used to evaluate scour that caused the bridge to fail in 1987. ASCE (2008) provides a useful discussion of sediment transport scaling for physical models.

Figure 4.8. Physical model of the I-90 Bridge over Schoharie Creek, New York.

4.3 SELECTING UPSTREAM AND DOWNSTREAM MODEL EXTENT

The minimum extent of a hydraulic model for bridge hydraulics is the location where flow is fully expanded both upstream and downstream of the flow constriction. Flow constriction is often the major contributor to backwater, so complete flow expansion and contraction must be included. For one-dimensional models, the use of the minimum downstream extent does not detract from the results as long as the downstream water surface is known with a high degree of certainty. However, if the water surface is not know with confidence, then extending the model further downstream will decrease uncertainty at the structure. This is illustrated in Figure 4.9, which shows water surface profiles for a simple bridge model. The three profiles are all for the same discharge with the only difference being the downstream boundary condition. Each one of the profiles represents a valid solution to the equations of fluid motion. The downstream boundary is located far enough downstream so the profiles converge and the 4.0 feet (1.2 m) of initial difference is eliminated before reaching the bridge.

Thus an important principle of numerical modeling is that the farther downstream the model extends, the smaller the influence the boundary condition will have on the location of interest. The farther the boundary is from the bridge, the less uncertainty exists at the bridge because channel and floodplain geometry and roughness will dictate the results. This does not mean, however, that all uncertainty is removed. Inaccuracies or change in any of the input variables result in uncertainty in the results.

The minimum downstream extent for two-dimensional models is similar to one-dimensional models with flow fully expanded upstream and downstream. It is also desirable to select a location where flow is reasonably one-dimensional, especially at the downstream boundary. This is because the downstream boundary is usually specified as a constant water surface elevation along the boundary. One useful approach is to place the upstream and downstream boundaries at least one floodplain width up- and downstream of the crossing. As with one-dimensional models, the further the boundary is located away from the crossing the less influence the boundary condition will exert of the results.

When there are other structures or hydraulic controls either upstream or downstream that will influence or can be impacted by the project, then the modeling should be extended to include these structures. Figure 4.9 shows some backwater created by the crossing. Although the extent of the model probably captures the maximum water surface increase, extending the model upstream would be required to fully assess potential upstream impacts.

As indicated in Section 4.2.2, unsteady flow analysis also requires extending the model to account for storage-routing effects. Unsteady flow modeling of tidal waterways can require significant effort. Tidal models must extend far enough downstream to reach a well-defined tide or storm surge boundary condition and to account for storage and hydraulic controls between the downstream boundary and the structure. Tidal models must extend far enough upstream of the structure to account for storage because it is the storage that is the primary factor that determines tidal flow rates (FHWA 2004, 2008).

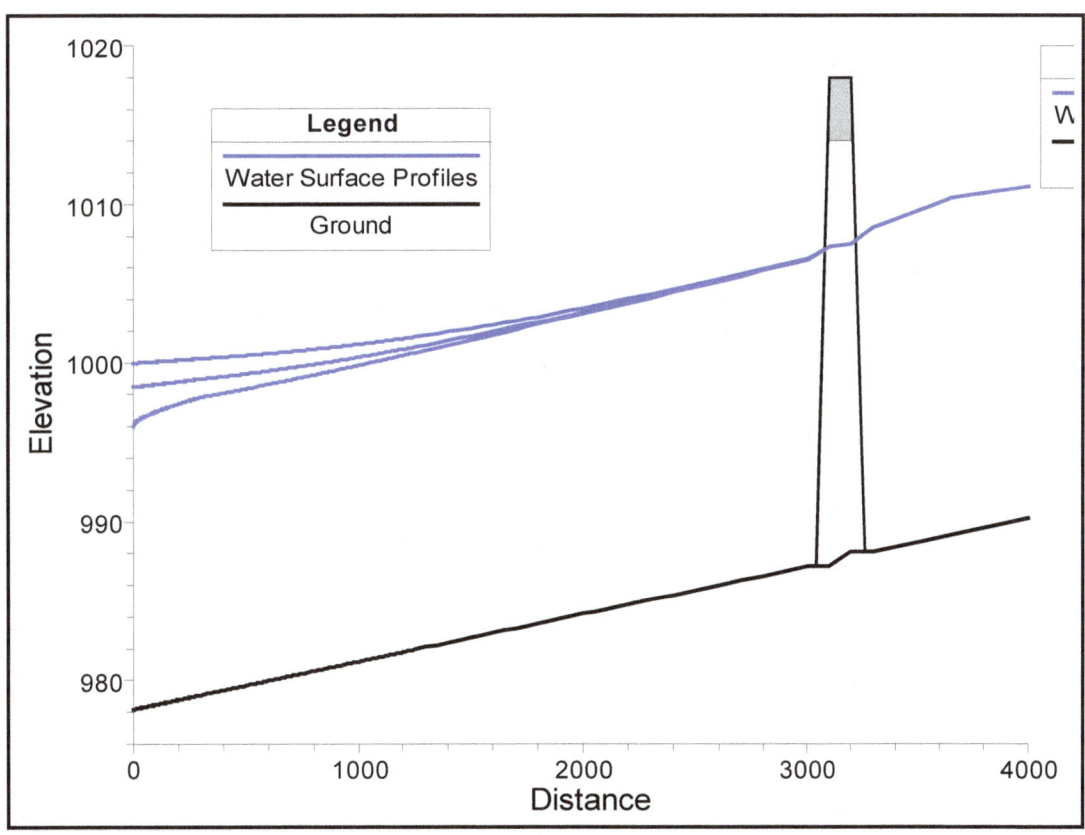

Figure 4.9. Flow profiles with downstream boundary uncertainty.

4.4 IDENTIFYING AND SELECTING MODEL BOUNDARY CONDITIONS

An important part of the hydraulic engineer's responsibility is to select representative boundary conditions for the hydraulic analysis. Peak discharge is one boundary condition that is commonly used for river projects and flood hydrographs are most frequently used for unsteady riverine modeling. For subcritical flow conditions, the downstream water surface must be specified or computed. For supercritical flow the upstream condition is specified and for mixed flow conditions the downstream and upstream condition is specified. The model extent (Section 4.3) and boundary condition should be selected based on identifiable hydraulic controls or on other reliable information. There are several types of hydraulic controls that can establish the boundary condition at a bridge. These include slope breaks where critical depth occurs (from flat to steep in the downstream direction), diversion dams, bridges, roads and other structures. The discussion below relates to a downstream boundary but also applies to upstream boundary conditions for supercritical or mixed flow models.

4.4.1 Water Surface

A known water surface is very commonly used in hydraulic modeling, where the hydraulic engineer specifies the elevation as the starting downstream condition. One common source for the known water surface is a Flood Insurance Study (FIS). The FIS profile may include many miles of river downstream of the bridge that is being analyzed. Starting the model relatively close to the bridge is more efficient and the water surface can be extracted from the appropriate location on the profile. Gage data or an observed high water mark can also be used to establish known water surface elevations as input boundary conditions.

4.4.2 Normal Depth and Energy Slope

Normal depth occurs when the bed profile, water surface, and energy grade line are all parallel, and the flow depth and velocity do not change along the channel flow path. This occurs relatively infrequently in natural rivers, though it can be a reasonable approximation for establishing boundary conditions in many situations. The use of the channel invert (thalweg) to compute bed slope should be avoided in natural channels because the channel bed elevation can vary widely over short distances. A better approximation is to use the floodplain slope as measured from a topographic map. The channel slope can, however, be much less than the valley slope for highly sinuous meandering channels. A conveyance weighted slope can be used, but this requires an assumed water surface to compute channel and floodplain conveyance.

When energy slope or normal depth is used, the model iteratively computes a water surface that produces the desired slope. Flow conditions are unlikely to actually satisfy normal depth criteria because of longitudinal topographic and roughness variations. The model downstream extent should be extended for this situation. The variability in channel and floodplain conditions is then incorporated into the model solution and uncertainty caused by the boundary condition is reduced.

There are situations when the FEMA Flood Insurance Study (FIS) only includes the 100-year flood profile but the bridge hydraulic study requires additional design flows. Similarly, the FEMA study may include several flood discharges but these do not match those desired for the hydraulic study. For this condition the energy slope can be computed from one discharge and applied to other discharges. This approach may also involve significant uncertainty and should be used in conjunction with extending the modeling downstream.

4.4.3 Rating Curve

A rating curve is a stage versus discharge table or curve relating stage and discharge. Gaging stations have published rating curves that are regularly checked and updated by the USGS. Gaging station data can also be used to establish rating curves. These data only apply to the specific gage location. Multiple profile data from FEMA studies can also be used to develop rating curves for a specific location and used as a model boundary condition. The same uncertainties can apply to the use of energy slope, so extending the model downstream is warranted.

4.4.4 Critical Depth

Critical depth is a relatively well defined boundary condition when a control structure produces a sudden drop in the channel. Critical depth in natural channels is unusual except in steep, bedrock or boulder-bed channels. In HEC-RAS (USACE 2010c) critical depth is defined as the minimum total energy. In a natural channel, total energy includes the energy correction coefficient, α, so roughness and flow distribution impact the determination of critical depth. Critical depth should be confirmed as reasonable before using it as a boundary condition in natural channels.

4.5 RIVERINE HYDROLOGIC INFORMATION

Peak discharge is the most commonly required input for river hydraulic models. For unsteady flow models the flow hydrograph is usually required. The HDS 2 manual (FHWA 2002) provides discussion and guidance on the analysis of aspects of the hydrologic cycle that are important to highway engineers. Of primary interest are statistical methods, regional regression equations and hydrologic modeling. Any one of these methods may provide the most reliable estimate of peak discharge based on basin conditions and availability of information.

When flood frequency information is available from a reliable and authoritative source then that information can be used, although it should be reviewed for reasonableness and applicability. Commonly used sources include FEMA Flood Insurance Studies, USACE, state agencies, and local agencies. One finding of the NRC (2009) study is that flood frequency analysis of gage records is the most reliable method for defining peak flood discharges. The reliability of gage analysis increases with the period of record. FEMA (2009) indicates that gaging station data are applicable to all studies if the record length is 10 years or longer. This does not necessarily mean that such a short period of record would provide the most reliable results, however. Where basin flow regulation is significant, hydrologic modeling is often required.

The sources of peak discharge in general order of preference are:

- Prior studies by authoritative sources
- Statistical frequency analysis of gage records at the site
- Transferring a gage analysis to a nearby, ungaged location
- Applicable regional regression equations
- Hydrologic models

The USGS StreamStats web-based application computes stream flow statistics for gaged and ungaged locations throughout the U.S. (USGS 2008). For states that do not have StreamStats fully implemented, USGS gaging station statistics are provided. States that have StreamStats implemented include gage analysis, gage transfer, and application of regional regression equations.

4.6 NUMERICAL MODEL EVALUATION

Numerical model verification, calibration and validation are all part of the evaluation process. Schwartz (2005) indicates that model verification involves testing to assure that the model solve the equations correctly. The verification process may include testing the model results against known analytical solutions to the same set of conditions. Although it can be assumed that widely used and accepted one- and two-dimensional models solve the appropriate equations correctly, errors in the programs do become evident from time to time. Therefore, it is the hydraulic engineer's responsibility to check results for reasonableness. Even though a program is correctly solving the equations, errors in data entry should also be checked.

Even though a model has been verified and all the input data are correct, it can produce erroneous results. This can occur in one-dimensional models if cross section spacing is too large and in two-dimensional models if the network is not sufficiently refined to solve the equations accurately. This type of error can be identified by reviewing model results. As discussed in Section 4.7, inaccurate or incorrect data of one type, particularly elevation, may require the use of unrealistic values of other parameters, such as roughness, to compensate.

A solution to a specific set of conditions also requires appropriate boundary conditions and, in the case of unsteady flow models, appropriate initial conditions. For a set of boundary and initial conditions, the model parameters (including roughness, turbulence, and other coefficients) must be adjusted to calibrate the model to match observed conditions within a desired degree of accuracy. If the calibrated parameters are not within the normal expected range, the model should be reviewed to determine if there are errors in the input data. In the case of hydraulic models, errors in geometry are often the source of unrealistic results or the need for unrealistic input parameters.

If possible, the calibration process should not use all available observed data. Part of the data, especially observations from another event, should be reserved for the validation process. The validation step tests the model and can improve the confidence in the model results, but may also identify deficiencies in the model.

Schwartz (2005) also includes sensitivity analysis and uncertainty analysis as part of numerical model evaluation. Sensitivity analysis, where each parameter is adjusted independently, is used to identify the parameters that have the greatest impact on the solution. Because numerical hydraulic models are used to simulate complex systems, any one parameter may dominate the solution. Uncertainty analysis is similar to sensitivity analysis, but is used to evaluate the overall uncertainty in the model results based on the uncertainties of the model input parameters. Monte-Carlo methods, which allow a set of input parameters to vary randomly based on expected probability characteristics, are very useful in determining modeling uncertainty.

4.7 DATA REQUIREMENTS AND SOURCES

There is a wide variety of information that is pertinent to bridge hydraulics and scour analyses. Table 4.2 provides a summary of the various types of information, their use, and sources. Although all of the data listed in Table 4.2 can be important in a bridge hydraulic study, geometric data are the greatest source of uncertainty and error. If the geometry is incorrect, then the flow, velocity, depth or water surface elevation must be incorrect. For example, if the floodplain elevation is several feet low and the modeled water surface is correct then the flow depth, and probably the velocity and floodplain discharge are incorrect. To obtain the correct velocity and discharge with an incorrect depth, then some other variable, probably flow resistance, must be adjusted incorrectly. That variable may well be outside the normal expected range. For these conditions, a model that has been calibrated for one flow is unlikely to produce accurate results for another event.

Table 4.2. Data Used in Bridge Hydraulic Studies.

Type of Information	Use	Sources
Floodplain topography	Hydraulic model geometry	Land survey, photogrammetry, LIDAR, USGS National Elevation Dataset (NED)
Channel geometry and bathymetry	Hydraulic model geometry	Land survey, hydrographic survey, LIDAR
Current/recent aerial photography	Land use and roughness, channel boundaries	Photogrammetry, web, city, county, and state agencies
Historic aerial photography	Land use and roughness change, channel migration	USGS, Farm Service Agency (FSA), web, city, county and state agencies
Existing structure information	Hydraulic model geometry	Bridge plans, as-built plans, roadway plans
FEMA Flood Insurance Studies and other flood hazard studies	Hydrology, flood history, channel and floodplain roughness information, flood profiles, coastal flooding range lines	Federal Emergency Management Agency (FEMA), U.S. Army Corps of Engineers (USACE), local floodplain administrator
Flood maps	Floodplain delineation, base flood elevations, floodway boundaries	FEMA and local floodplain administrator
Existing hydraulic models	Hydraulic modeling	FEMA, local floodplain administrator, USACE, other
Boring Logs	Sediment size, scour analysis, erodibility assessment	Geotechnical investigation
Core Samples	Sediment gradation, scour analysis, erodibility testing	Geotechnical investigation
Floodplain and channel roughness	Hydraulic model Manning n determination	Site visit
Bed and bank sediment surface and near-surface samples	Sediment gradation, scour analysis	Site visit
Coastal hydrographic survey maps and data and coastal DEMs	Tidal hydrodynamic model geometry	National Oceanic and Atmospheric Administration (NOAA) and National Ocean Service (NOS) Office of Coastal Survey
Existing bridge inspection reports	Channel stability assessment	Bridge owner
Gage data	Flood frequency analysis, historic flooding, hydraulic model calibration and validation	USGS, USACE, state water resources agencies
Tide gage data	Astronomic tide, water surface elevation frequency analysis	NOAA

The variable that has the greatest effect on accuracy is topographic data and the need for increased accuracy of elevation data increases for lower relief areas (NRC 2009). Geometric accuracy includes elevation, reach lengths, and bridge and roadway geometry. Improved elevation accuracy improves the results of all models. There is often the misconception that two-dimensional models require more accurate data and a larger domain. Two-dimensional models produce better results because they include more complete representations of the physical processes. If a topographic or vegetation feature needs to be incorporated in a two-dimensional model, it should also be incorporated in a one-dimensional model. Therefore, the complexity of the flow situation is the primary reason for selecting two-dimensional models, not data accuracy.

CHAPTER 5

ONE-DIMENSIONAL BRIDGE HYDRAULIC ANALYSIS

5.1 INTRODUCTION

The previous chapter describes many differences between one-dimensional and two- or three-dimensional hydraulic analysis. As stated, most bridge hydraulic studies employ one-dimensional analysis methods. This chapter provides information and guidance on the use of one-dimensional modeling techniques for bridge hydraulic analysis.

One-dimensional analysis encompasses a wide range of approaches from approximate methods requiring just a single waterway cross section to detailed water surface profile calculations involving many cross sections and multiple stream reaches. Approximate methods are frequently used for rapid assessment of flood inundation potential in support of FEMA floodplain mapping. They typically incorporate the assumption of uniform flow (see Chapter 3). If uniform flow is assumed, then the flow depth and corresponding water surface elevation at a particular cross section can be calculated using the Manning equation (Equation 3.46). The HDS 1 method described in the next section is an example of an approximate method. It includes an underlying assumption that flow conditions are essentially uniform downstream of the bridge, and it develops a backwater estimate using empirical equations based on energy loss principles.

The engineer must be cautious, however, in applying approximate methods to bridge hydraulics problems. Bridge-constricted floodplains and stream reaches usually exhibit significantly non-uniform flow characteristics. It is recommended, therefore, that engineers use methods employing water surface profile calculations for one-dimensional bridge hydraulic analysis.

5.2 HDS 1 METHOD

As explained in Chapter 1, the predecessor to this document is HDS 1 (FHWA 1978). HDS 1 presented a computational method of determining the backwater caused by a bridge crossing a floodplain. Chapter II of HDS 1 presented the basic expression for backwater as:

$$h_1^* = K^* \alpha_2 \frac{V_{n2}^2}{2g} + \alpha_1 \left[\left(\frac{A_{n2}}{A_4} \right) - \left(\frac{A_{n2}}{A_1} \right) \right] \frac{V_{n2}^2}{2g} \qquad (5.1)$$

where:

h_1^* = Total backwater, ft (m)
K^* = Total backwater coefficient
α_1, α_2 = Kinetic energy distribution coefficients at Cross Sections 1 and 2
A_{n2} = Gross water area in constricted bridge waterway measured below normal stage at Cross Section 2, ft² (m²)
V_{n2} = Average velocity in constriction (total discharge divided by A_{n2}), ft/s (m/s)
A_1 = Total flow area at Cross Section 1, including addition caused by backwater, ft² (m²)
A_4 = Total flow area at Cross Section 4, downstream of influence of bridge, ft² (m²)
g = Acceleration of gravity, ft/s² (m/s²)

The HDS 1 backwater expression (Equation 5.1) applies the energy equation between the location of maximum backwater upstream of the bridge and the point downstream of the bridge where flow is fully expanded, and including an empirical bridge loss coefficient. The expression is based on the assumption of steady, subcritical flow in the affected stream reach (classified in HDS 1 as Type I flow). Another significant assumption inherent in the expression is that the flow conditions are approximately uniform (a uniform water surface slope parallel to the stream bed slope) except for the backwater caused by the bridge. In the framework of HDS 1, Cross Section 1 is upstream of the bridge at the presumed point of maximum backwater, Cross Section 2 is at the upstream face of the bridge, and Cross Section 4 is downstream of the bridge at the presumed point of reestablishment of normal flow conditions (see Figure 5.1). Cross Section 3 is located at the toe of the downstream side slope of the road embankment, but it is not used in the calculation of backwater.

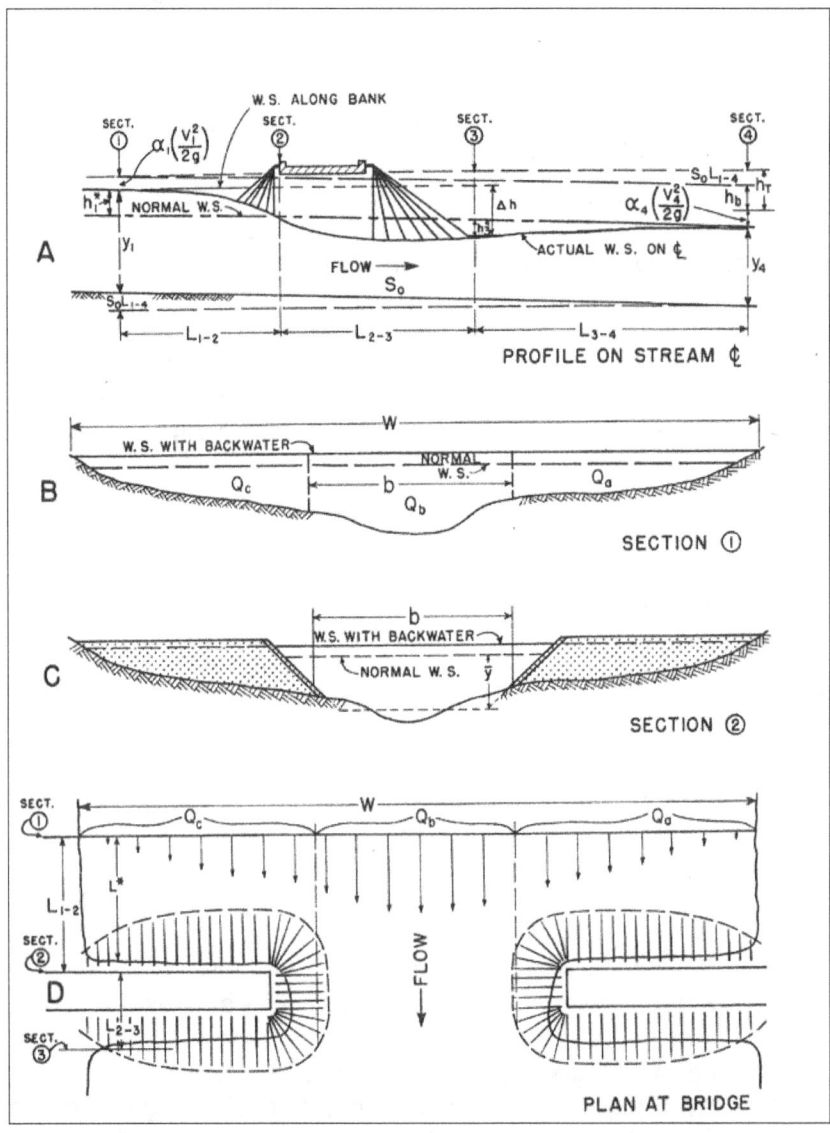

Figure 5.1. Sketch illustrating positions of Cross Sections 1 through 4 in HDS 1 backwater method (FHWA 1978).

The engineer applying the HDS 1 method would first compute the flow depth in a representative cross section under uniform flow conditions and without any constriction, for a given design discharge rate. Prior to incorporating the constriction caused by the bridge crossing, the representative cross-section properties would apply to each of the cross sections (1 through 4) because of the uniform flow assumption. Once the unconstricted flow depth is determined, the engineer computes A_4 and α_1. The values of A_{n2}, V_{n2} and α_2 are then computed based on superimposing the constriction caused by the road embankments and abutments onto the cross-sectional area and considering the area within the constriction and under the normal water surface (see Part C of Figure 5.1).

The engineer would then determine the bridge opening ratio (M), which represents the degree of constriction of the waterway. The value of M is computed by dividing the total cross section discharge (Q) into the discharge (Q_b) passing through the bridge without redirection by the encroaching road embankments or abutments.

$$M = \frac{Q_b}{Q} \tag{5.2}$$

where:

Q_b = Discharge that can pass through the bridge without redirection by the encroaching road embankments or abutments, cfs. Referring to Part D of Figure 5.1, Q_b is computed as the discharge contained in the portion of the representative unconstricted cross section that lies within the projected limits of the bridge opening, ft³/s (m³/s)

Q = Total cross section discharge, ft³/s (m³/s)

After computing the value of M, the engineer would develop the value of K*. The value of K* is found through a series of graphical charts that were derived empirically from physical modeling. K* is primarily a function of M, but is also affected incrementally by other factors including the skew angle (if any), the size and type of bridge piers, and the eccentricity of the bridge opening within the floodplain. Once the value of K* was obtained, the total backwater height h_1^* could be computed, which in turn would allow the upstream water surface elevation to be calculated.

Because the HDS 1 method incorporated the concept of uniform flow, the backwater could be estimated through relatively straightforward calculations, avoiding the complexity of the step backwater calculations associated with varied flow. The uniform-flow simplification, however, meant that the method would yield uncertain results when applied to situations involving highly varied flow conditions.

In addition to the basic backwater computation method, HDS 1 provided additional methods for bridges experiencing certain complex flow conditions, including:

- Flow passing through critical depth inside the constriction (Type II flow)
- Submerged-deck (pressure) flow conditions
- Flow overtopping the road
- Bridges with spur dikes (now called guide banks)

The methods presented in HDS 1 for submerged-deck flow and overtopping flow are still in common use at present, and are discussed in later sections of this chapter.

5.3 WATER SURFACE PROFILE COMPUTATIONS

The HDS 1 approach to backwater computation provided a useful analysis tool without overly cumbersome calculations. The assumption of approximately uniform flow, however, is an important limitation to the HDS 1 approach and other approximate methods. Natural streams and floodplains are typically characterized by variation in slope, cross section geometry and flow resistance arising from natural processes and from human activities and impacts. Methods that are based on the assumption of uniform or approximately uniform flow cannot accurately simulate the hydraulics of streams with significant variation.

A more accurate approach is required in order to ensure the protection of public safety and to meet the analysis demands of modern practice in bridge design. Thanks to the proliferation of powerful computers on the desks of engineers, more rigorous and accurate approaches are now not only feasible, but commonplace. This section describes significantly improved techniques that allow the analysis of natural streams without the assumption of uniform flow. These techniques are based on the concept of computing the water surface profile as a variable function along the length of the stream. All of the methods discussed in this section are governed by the continuity and energy equations discussed in Chapter 3.

5.3.1 Standard-Step Methods

Standard-step calculation methods allow the engineer to compute the water surface for a given flood scenario at cross section locations along the stream. Several widely used computer programs, including HEC-RAS, WSPRO, and HEC-2 use the standard-step approach to compute water surface profiles. The engineer chooses the cross section locations to capture transition points in the slope, width, geometry, flow rate and roughness in the stream channel and floodplain. All cross sections must be oriented perpendicular to the flow direction and have sufficient extent to contain the entire flow area. Unlike the direct-step method, described below, the standard-step methods do not require a prismatic waterway geometry or a pre-set distance between cross sections. Figure 5.2 shows the cross section layout for a floodplain model computed by the standard-step approach. Note the strategic locations of the cross sections to reflect the geometric changes.

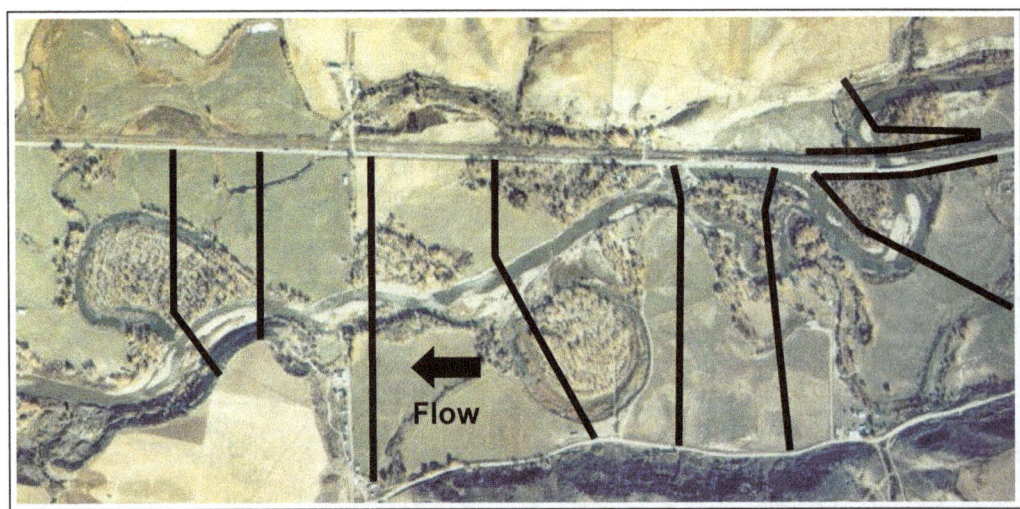

Figure 5.2. Cross section layout for a standard-step floodplain model.

Figure 5.3 graphically illustrates the computational framework for solving the energy equation using the standard-step approach. The calculations progress along the length of the stream segment, one cross section at a time. The hydraulic solution from the previously calculated cross section (Cross Section 1 in Figure 5.3) and the user-specified information about the cross section currently being calculated (Cross Section 2) are used to calculate the energy loss between the two cross sections. Once the energy loss has been determined, the energy grade line and water surface elevation at the current cross section can be computed. While the specific details of the implementation of the standard-step approach vary depending on the program used, the basic steps can be generally described as follows:

1. Set a trial water surface elevation for Cross Section 2. In HEC-RAS the initial trial water surface is determined by assuming the depth is the same as that calculated for Cross Section 1.

2. Use the trial water surface elevation to compute the conveyance, energy (friction) slope, kinetic energy distribution coefficient, and velocity head at Cross Section 2 (see Chapter 3).

3. Compute the average friction slope between Cross Sections 1 and 2, and multiply the average by the reach length between the two cross sections to estimate the friction loss.

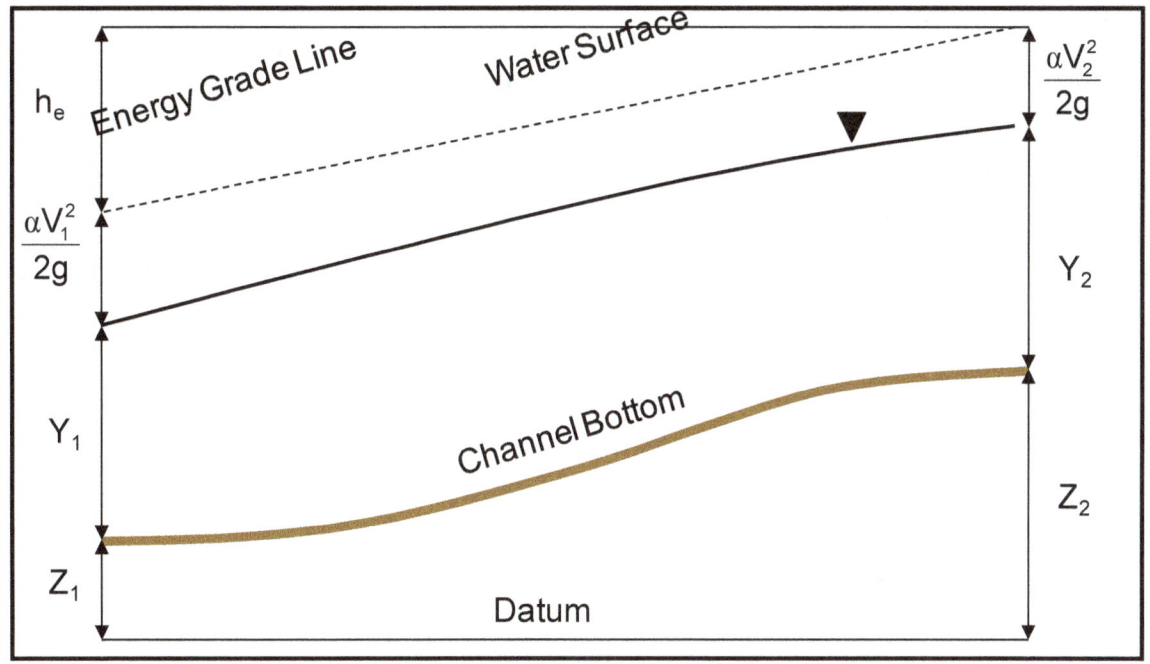

Figure 5.3. Illustration of water surface profile between two cross sections.

4. Multiply the absolute difference in the velocity head values between Cross Sections 1 and 2 by either a contraction loss coefficient or an expansion loss coefficient to estimate the transition loss.

5. Find the total energy loss (on a trial basis) from Cross Section 1 to Cross Section 2 by adding the friction loss and the transition loss.

6. Compute the energy grade line elevation (on a trial basis) at Cross Section 2 by adding the total energy loss to the energy grade line at Cross Section 1.

7. Subtract the velocity head from the energy grade line at Cross Section 2 to compute the resulting water surface elevation.

8. Compare the water surface elevation computed in step 7 to the trial water surface determined in step 1.

9. If the difference is within the specified tolerance, the calculation for Cross Section 2 is complete and the calculations progress to the next cross section. If the difference exceeds the tolerance, assign a new, adjusted trial water surface elevation and begin again at step 2. Iterate until the difference between the computed and trial water surface elevations is within the specified tolerance, then progress to the next cross section.

The procedure described above operates on two cross sections at a time. The calculation at any cross section (e.g., Cross Section 2) requires the information from the solution at the previous cross section (Cross Section 1). The analysis, therefore, must start with a known water surface elevation at the first cross section in the reach. For subcritical analysis, the downstream-most cross section is the first cross section. For supercritical analysis the calculation starts with the upstream-most cross section. This known water surface elevation at the first cross section is termed the "starting water surface" or the "boundary condition."

Most bridge-related studies involve subcritical flow and therefore require the engineer to specify a downstream boundary condition. This specified value is typically taken from a prior study (such as a FEMA Flood Insurance Study profile plot) or is calculated using the Manning equation, which requires an assumption of uniform flow conditions in the reach downstream of the analysis. If the downstream end of the model is located at a free overfall (such as a grade control structure) or a slope change from flat upstream to steep downstream, it may be appropriate to assign the water surface elevation at critical depth as the boundary condition.

Step 3 in the description above involves approximating the friction loss using the average friction slope. The friction slope at any cross section is:

$$S_f = \left(\frac{Q}{K}\right)^2 \tag{5.3}$$

where:

S_f = Friction slope in ft/ft (m/m)
K = Total cross section conveyance, a function of flow area, wetted perimeter and flow resistance, ft³/s (m³/s)

The friction loss between two cross sections is the integral of the friction loss function. An analytical solution would be highly complex. A simplified numerical integration is achieved by multiplying the average slope by the distance between the cross sections. The average friction slope between two cross sections is typically computed by one of four methods (USACE 2010c):

The Average Conveyance Equation

$$\overline{S}_f = \left(\frac{Q_1 + Q_2}{K_1 + K_2}\right)^2 \tag{5.4}$$

The Average Friction Slope Equation

$$\overline{S}_f = \left(\frac{S_{f1} + S_{f2}}{2}\right)$$ (5.5)

The Geometric Mean Friction Slope Equation

$$\overline{S}_f = \sqrt{S_{f1} \times S_{f2}}$$ (5.6)

The Harmonic Mean Friction Slope Equation

$$\overline{S}_f = \frac{2(S_{f1} \times S_{f2})}{S_{f1} + S_{f2}}$$ (5.7)

The friction slope is inversely related to the square of conveyance. In stream reaches with significant variation in cross section geometry, one can expect a corresponding variation in the conveyance values, which in turn could lead to large changes in the friction slope from one cross section to the next (see Figure 5.4). Since the standard step method uses the average friction slope between two cross sections to compute the friction loss, a large change in the friction slope can reduce the accuracy of the calculation.

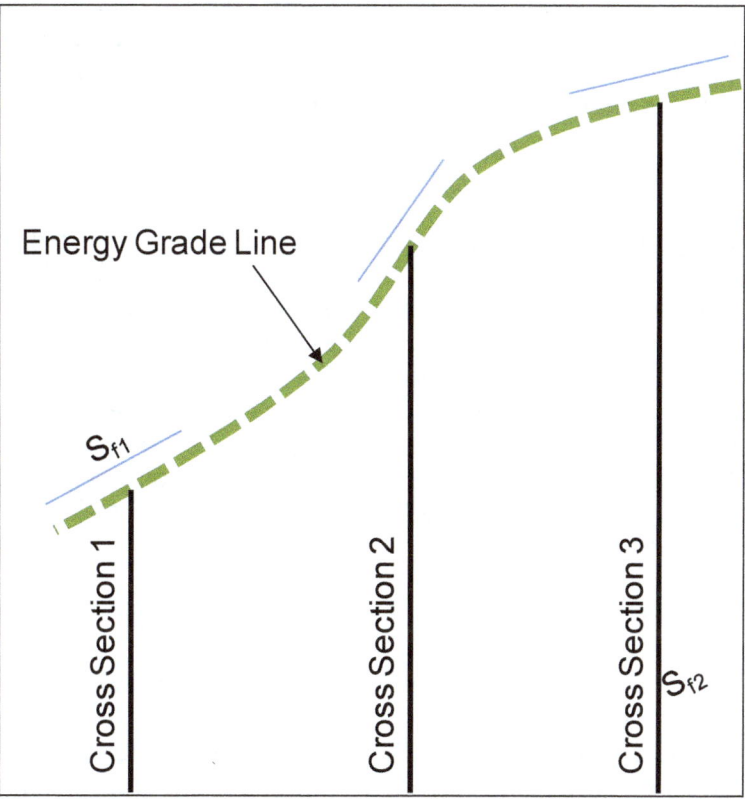

Figure 5.4. Illustration of the friction slope (the slope of the energy grade line) at each cross section in a stream segment.

Chapter 3 of this manual describes the various possible profile types on mild and steep slopes (M1, M2, S1, S2, etc). Each of the four friction slope averaging methods above is more suitable for some profile types than others. The HEC-RAS Hydraulic Reference Manual (USACE 2010c) quotes research by Reed and Wolfkill (1976) indicating that the average conveyance equation (Equation 5.4) gives the best overall results for a wide range of profile types. The average friction slope equation (Equation 5.5) is the most suitable method for M1 profiles. For M2 profiles the harmonic mean equation (Equation 5.7) was shown to be the most suitable. **Any one of the friction slope averaging methods will produce accurate results if the reach lengths are sufficiently short** (USACE 1986a).

As the reach length increases, however, so does the potential error in the computed average friction slope, regardless of the selected averaging method. Additionally, as the reach length increases, the error in the average friction slope is applied over a longer distance, thus having a greater effect on the total friction loss. The best remedy for the potential inaccuracy stemming from the variation of friction slope is to keep the reach lengths short from one cross section to the next.

As stated earlier, a water surface profile model using the standard step method should have cross sections located at all locations necessary to represent the major transitions in cross section geometry. Usually, however, additional cross sections are required to keep the reach lengths short enough to avoid significant error in the average friction slope and total friction loss calculations (USACE 1986a). The additional cross sections are often inserted using the interpolation function of the program being used. An advisable practice in developing a standard-step model is to shorten the reach lengths (e.g., add cross sections) in successive trials until the resulting water surface profile is insensitive to further shortening of the reach length. In other words, the number of cross sections is sufficient when inserting more cross sections does not significantly change the results.

5.3.2 Other Water Surface Profile Methods

Direct Step Method. The direct step method is similar to the Standard Step Method in that it uses average friction slope between two locations along the channel to compute the friction loss term in the energy equation. It is also similar in the fact that calculation of the water surface profile progresses from a known condition at one cross section (1) to another cross section (2). It is not a trial and error approach because the water surface at second cross section is determined in advance of the calculation, which is the flow depth at section 1 plus some increment in flow depth. The primary limitation of the direct step method is that the channel must be prismatic, meaning no change in the geometry, roughness, or discharge between the cross sections. Therefore, this approach is not applicable to natural channels. The steps in the Direct Step method are:

1. Calculate the specific energy ($E = y + \alpha V^2/2g$) and energy slope (Equation 5.3) for flow depth, y_1, at cross section 1
2. Based on a change in flow depth (Δy) and resulting y_2, calculate the specific energy and energy slope at cross section 2.
3. Calculate the average energy slope (S_f) between the two cross sections.
4. From the energy equation, the distance between the two cross sections is:

$$\Delta x = \frac{\Delta E}{S_0 - \overline{S}_f} \tag{5.8}$$

This method is limited to prismatic channels because the channel properties must apply to any location along the channel.

Integration Methods. The direct and exact integration of the energy equation for all types of channels and flow conditions is not possible, though many attempts have been made to solve the equation for specific cases or through the use of simplifying assumptions (Chow 1959, Henderson 1966, Chaudhry 2008 and others). The differential equation explicitly containing the independent variables (Chaudhry 2008) is:

$$\frac{dy}{dx} = \frac{S_o - S_f}{1 - \left(\frac{\alpha W Q^2}{g A^3}\right)} \tag{5.9}$$

Each of the variables on the right side of the equation can vary with distance along the channel. If one assumes that discharge and Manning n are constant along the channel reach, that conveyance is proportional to $y^{N/2}$, and that A^3/W is proportional to y^M, then Equation 5.9 can be written as:

$$\frac{dy}{dx} = S_o \frac{1 - \left(\frac{y_o}{y}\right)^N}{1 - \left(\frac{y_c}{y}\right)^M} \tag{5.10}$$

According to Henderson, if the channel is rectangular, very wide, and the Manning equation is used, then N = 3 and M = 10/3. The assumptions and difficulties in integration make this method limited to fairly short reach lengths.

Numerical Methods. The Standard and Direct Step Methods represent commonly used numerical integration of the energy equation. Other numerical approaches include single step methods, including Euler, Improved Euler, Modified Euler and Fourth-Order Runga-Kutta, and Predictor-Corrector methods (Chaudhry 2008). The single step methods represent successive improvements of computing average energy slope between cross sections. The Predictor-Corrector methods, including the Standard Step Method, involved iteration to arrive at a water surface computed to within and acceptable tolerance. Any of these methods is still limited by the fact that the mean (or integrated) energy slope must not be computed over too great of a distance. Therefore, the approach recommended in HEC-RAS (USACE 2010c) to limit cross section spacing is both necessary and robust. It should also be noted that in practical applications, discharge and Manning n will also vary longitudinally. Also the left floodplain, right floodplain, and channel distances between cross sections will not be equal. Therefore, many other assumptions are required for solution of water surface profiles using one-dimensional methods.

5.3.3 Mixed-Flow Regime

Natural streams and floodplains flow predominantly in the subcritical flow regime, and therefore most bridge hydraulic analyses are concerned exclusively with subcritical flow. Occasionally, however, supercritical flows are present within segments of the stream reach being analyzed. A water surface profile model that includes both subcritical and supercritical flow segments requires a mixed flow regime. The HEC-RAS Hydraulic Reference Manual

(USACE 2010c) describes the process used by the HEC-RAS program to analyze mixed-flow water surface profiles. The process is paraphrased here.

The program starts by computing a subcritical water surface profile, starting at the downstream most cross section and working in the upstream direction. After completing the subcritical profile, if the user has indicated that a mixed-flow profile is to be calculated, the program begins a supercritical profile calculation starting at the upstream end of the model. Working its way downstream, the program computes a supercritical profile wherever such a profile is possible. Some cross sections will be found to have valid solutions for both subcritical and supercritical flow. At such locations, the program determines which solution controls by computing the specific force for each solution. Whichever solution has the greater specific force is the controlling solution. The specific force is computed using the following expression:

$$SF = \frac{Q^2 \beta}{gA_m} + A_t \overline{Y} \tag{5.11}$$

where:

SF = Specific force, ft³ (m³)
β = Momentum distribution coefficient
A_m = Effective flow area in the cross section, ft² (m²)
A_t = Total inundated cross sectional area, including areas of ineffective flow, ft² (m²)
$\overline{Y}$ = Depth from water surface to the centroid of the total inundated area, ft (m)

The program completes the mixed-regime profile after identifying the controlling solution at each cross section. The upstream end of a hydraulic jump (abrupt transition from supercritical flow upstream to subcritical flow downstream) can be located through the mixed-regime profile calculations.

Figure 5.5 is an example of a mixed-flow water-surface profile computed by HEC-RAS. In this profile, a mild slope flows into a downstream steep slope, passing through critical depth at the slope break. Flow is subcritical at both boundaries, but there is an internal segment of supercritical flow (S_2 curve) upstream of the hydraulic jump up to the slope break. There is a subcritical profile (S_1 curve) at the downstream end of the steep slope controlled by a high water surface elevation at the downstream boundary. Had the downstream boundary been set as normal depth for the steep slope, the entire steep slope would have been an S_2 curve. There is an M_2 profile downstream of the bridge crossing and an M_1 profile upstream of the bridge crossing. Each of these mild-slope profiles converge on normal depth. The various types of flow profiles are described in Section 3.4.4.

5.4 CROSS-SECTION SUBDIVISION AND INEFFECTIVE FLOW

5.4.1 Cross Section Subdivision

Equation 5.3 shows that the friction slope is inversely related to the square of the conveyance, K. In a natural floodplain, the flow depth, roughness and velocity usually vary throughout the width of the cross section. Accurate conveyance calculations usually require that the cross section be subdivided into regions of similar flow properties as illustrated in Figure 5.6.

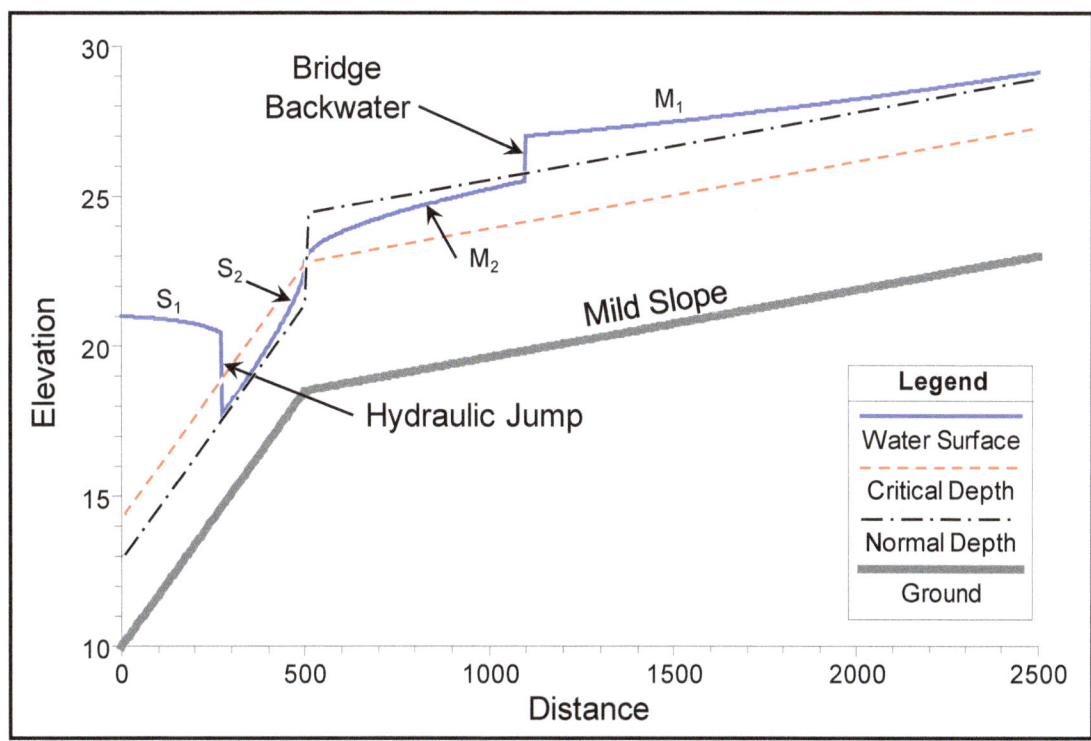

Figure 5.5 Example mixed-flow regime profile in HEC-RAS.

As an illustration of the inaccuracy that could result without cross section subdivision, consider the cross section in Figure 5.6 (but without the indicated subdivisions) for a condition in which the water surface is just below the top of the left bank of the main channel. The conveyance would reflect only the area and hydraulic radius of the main channel. Next consider a water surface elevation just above the top of the left bank. At this elevation the water surface would extend out onto the left overbank. If the cross section were not subdivided, the wetted perimeter might increase by hundreds of feet, but the area would increase very little because of the small depth of flow on the overbank. The much-increased wetted perimeter divided into the little-increased area would lead to a decrease in the hydraulic radius, which would lead to a decrease in the conveyance compared to the water surface just below the channel bank. This condition causes a discontinuity in the calculated conveyance as a function of water surface elevation (see Figure 5.7).

In reality, the small increase in water surface elevation would lead to a small increase in conveyance, but in a conveyance calculation without subdivision, the conveyance would appear to decrease in response to the increased water surface. This is an error that can be avoided through cross section subdivision.

The key concept to consider in subdividing a cross section is that the velocity inside any subdivision should be more or less uniform, even if the average velocities in two adjacent subdivisions are significantly different. The total conveyance of a cross section is the summation of the conveyance for each subdivision of the cross section.

$$K = \sum_{1}^{N} K_i \tag{5.12}$$

where:

K_i = Conveyance of a subdivision, ft³/s (m³/s)
N = Number of subdivisions in the cross section

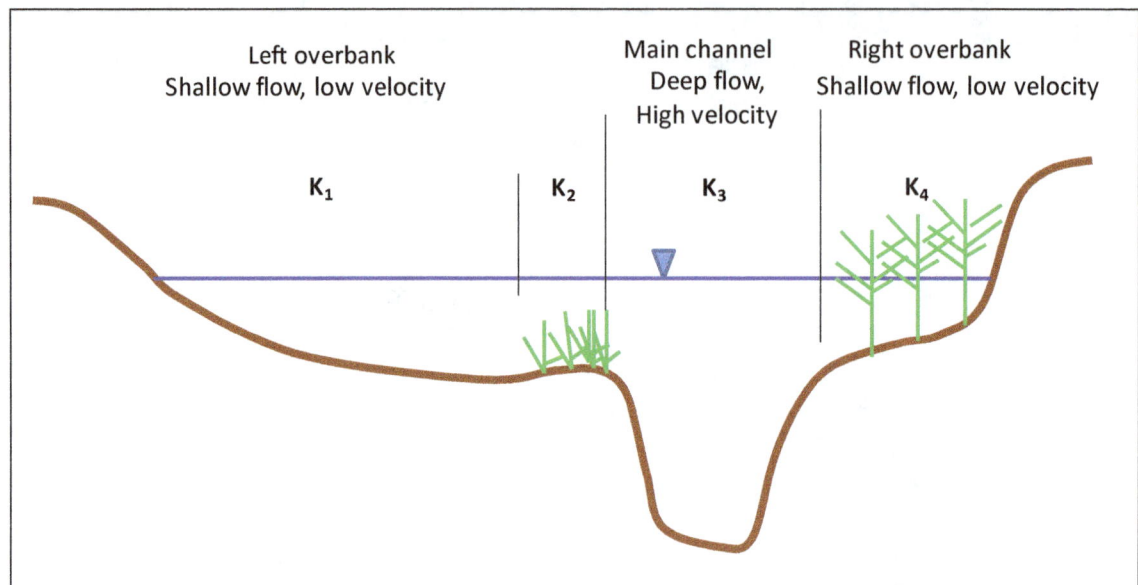

Figure 5.6. Example of a properly subdivided cross section.

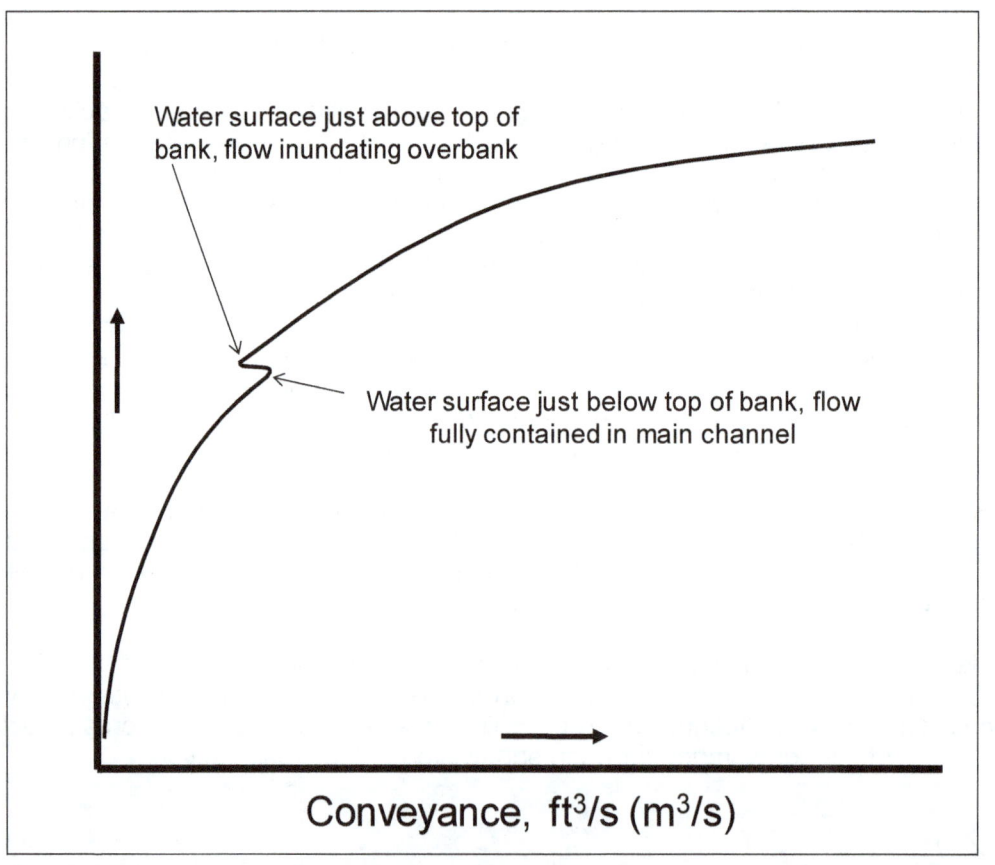

Figure 5.7. Discontinuity of computed conveyance for a non-subdivided cross section.

The conveyance of a subdivision is

$$K_i = \frac{1.49}{n_i} A_i R_i^{2/3} \qquad (5.13)$$

where:
- n_i = Manning roughness coefficient for the subdivision
- A_i = Effective flow area of the subdivision, ft² (m²)
- R_i = Hydraulic radius of the subdivision, ft (equal to A divided by the wetted perimeter), ft (m)

In the example shown in Figure 5.6, the cross section has four subdivisions, two in the left overbank and one each in the main channel and the right overbank. It is recommended to subdivide overbank areas at changes in roughness. In the main channel, however, it is more common to treat variable roughness (e.g., willows on the upper channel banks with an unvegetated channel bottom) by calculating a single composite Manning roughness value that applies to the entire channel width.

The subdivision of conveyance, in addition to refining the accuracy of the total cross section conveyance, also provides a rational means of distributing the total discharge within the cross section. Most one-dimensional analysis programs, including HEC-RAS, distribute discharge within a cross section such that it is proportional to the conveyance. If the left overbank has one sixth of the total conveyance, for instance, the program will assign it one sixth of the total discharge. Distributing the discharge is important in the calculation of the velocity distribution coefficient, the representative reach length between cross sections, and the average velocity in each subdivision. An approximate determination of the lateral velocity distribution is possible by further dividing the main channel and overbank regions into smaller subdivisions and distributing the discharge to each subdivision in proportion to conveyance.

5.4.2 Ineffective Flow

In a one-dimensional model, the program assumes that any area in a cross section below the water surface elevation is available for conveyance, and makes use of that conveyance unless the user specifies otherwise. In certain cases the engineer may want a portion of a cross section to be excluded from the conveyance, for one of several reasons including:

- That portion of the cross section is in the stagnant or eddying wake area downstream of a large obstruction, such as a building
- It is immediately upstream of an obstruction such that any water moving in the area is in a lateral direction rather than in the downstream direction
- It is an area where water can pond but cannot effectively convey flow from upstream to downstream, such as an area behind a levee that is connected to the flowing water downstream but not upstream
- It is outside of the effective contraction zone upstream of a constriction or the effective expansion zone downstream of a constriction, such as a road crossing (discussed in more detail in the next section)

Engineers have used various approaches to exclude portions of cross sections from the conveyance, including raising the ground elevation value or artificially increasing the roughness coefficient. In the HEC-RAS program, the user can specify areas within a cross section where the flow is ineffective (the Ineffective Flow setting) up to a user-designated water surface elevation. If the water surface in the cross section reaches the designated elevation, the ineffective flow specification is nullified. The use of ineffective flow specifications plays an important role in modeling bridge crossings with HEC-RAS, as discussed in later sections. Figure 5.8 is an example of the use of ineffective flow specifications at a bridge crossing.

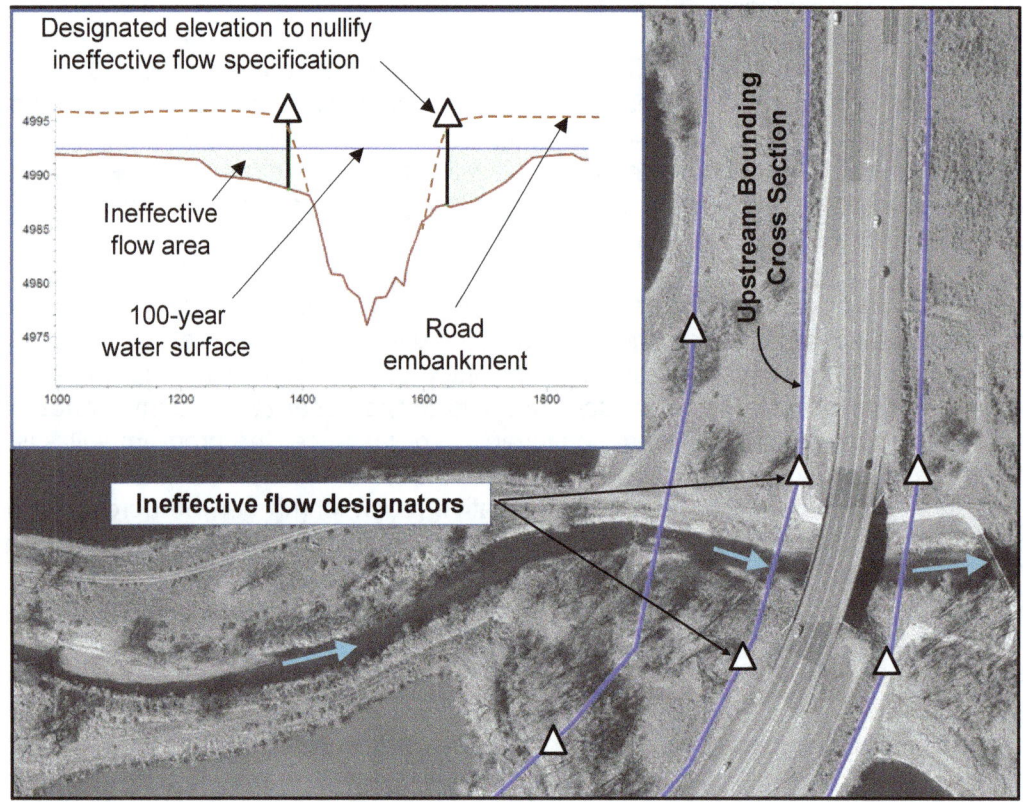

Figure 5.8. Example of the use of ineffective flow at a bridge.

5.5 FLOW CONSTRICTIONS AND CROSS-SECTION PLACEMENT

5.5.1 Effects of Highway Crossings

Both natural and manmade constrictions can affect streams and floodplains. Manmade constrictions are often the result of highway crossings. By economic necessity, a typical highway crossing consists of a cross-drainage structure (a bridge or culvert) together with earth-fill embankments encroaching into the floodplain from one or both sides. The encroachment forms a hydraulic constriction. Most bridge crossings are configured such that the abutments are at the channel banks or set back from the top of the banks, thereby avoiding significant constriction of the main channel flows. At such crossings the effects of the constriction are not appreciable until the flood discharges are high enough to exceed the main channel banks and inundate the overbank areas. Culvert crossings, by contrast, often involve encroachment by fill in the main channel as well as the overbanks.

When flow is constricted by a bridge or culvert crossing, the energy losses in the region upstream and downstream of the crossing are greater than they would be without the constriction. The flow upstream of the bridge is forced to contract from the full floodplain width to the structure opening width. On the downstream side the flow re-expands to occupy the full floodplain width. Both the contraction and expansion processes require some longitudinal distance from the crossing for fully established flow. This manual refers to the zones of establishment as the contraction and expansion reaches, or collectively as the transition reaches. Increased friction losses (due to decreased conveyance) and transition losses characterize both the contraction reach and the expansion reach.

5.5.2 Cross Section Placement

In one-dimensional hydraulic modeling, the engineer's placement of cross sections and the modification of the conveyance properties of certain cross sections drive the analysis of the transition reaches. Figure 5.9 illustrates the typical one-dimensional modeling framework for analyzing a bridge crossing. The key concept for modeling transitions is the narrowing of the effective flow width in the contraction reach and the widening of the effective flow width in the expansion reach. The outer streamlines in the transition reaches would naturally follow curvilinear flow paths. In one-dimensional modeling, however, the engineer typically simplifies the problem by assuming linear transition tapers as shown on Figure 5.9.

The hydraulic model's computation of the excess loss is directly related to the length of the contraction reach (the distance from the approach section to the upstream bounding section) and the length of the expansion reach (the distance from the downstream bounding section to the exit section). The longer the transition reach, the more excess loss is expected. The contraction and expansion reach lengths are directly determined by the locations of the approach and exit sections, which are assigned by the engineer. This situation illuminates one of the limitations of one-dimensional analysis in modeling constrictions. In two-dimensional modeling, the model's governing equations determine the length and configuration of the transition flow regions. In one-dimensional modeling, the engineer imposes the length and configuration of the transitions on the model through the placement of the cross sections and the modification of the conveyance properties.

The placement of the approach and exit sections depends on the engineer's assessment of the appropriate rates of taper for the contracting and expanding flow. The taper rates CR and ER on Figure 5.9 (contraction rate and expansion rate, respectively) vary depending on many site-specific factors. Chapter 5 of the HEC-RAS Hydraulic Reference Manual provides guidance on the assignment of CR and ER (USACE 2010c).

<u>Expansion Reach</u>. Table 5.1 below is taken from the Hydraulic Reference Manual. The table summarizes the results of research carried out by the USACE Hydrologic Engineering Center and documented in Research Document 42 (USACE 1995). It gives the ranges of expected values of ER for different combinations of degree of constriction, longitudinal slope, and ratio of roughness between the overbanks and main channel. A cell in the table is selected based on the degree of constriction, slope and roughness ratio that are closest to those of the site being analyzed. The selected cell gives the range of appropriate ER value.

The engineer decides on a value within the range in the cell and uses that value to estimate the length of the expansion reach. The expansion reach length is the distance required for the effective flow to expand to the edges of the floodplain at the ER taper rate. For example if an ER value of 2 is chosen, and the floodplain encroachment distance is 100 feet (30 m) on one side of the floodplain, the flow will take 200 feet (60 m) to fully expand on that side. For asymmetric encroachments, where the constriction is more pronounced on one side of the floodplain than the other, the expansion reach length can be based on the average encroachment distance.

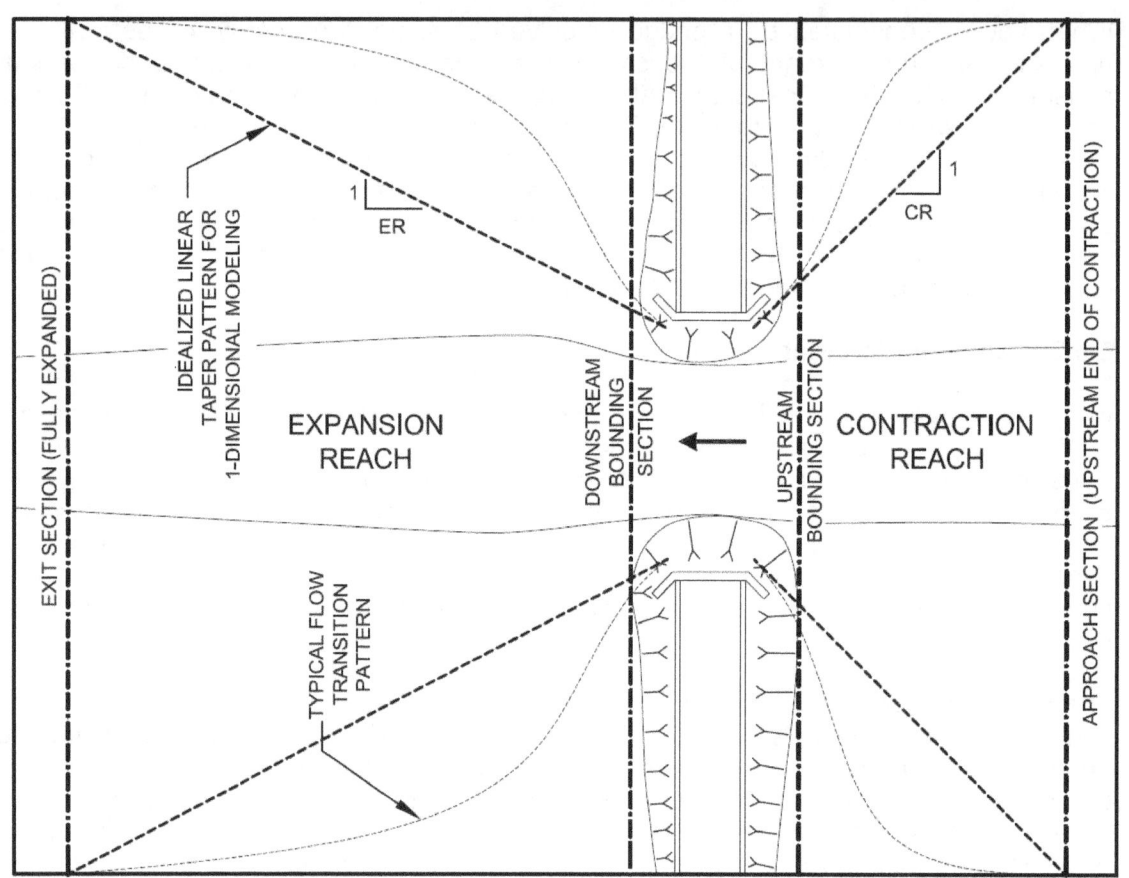

Figure 5.9. Illustration of flow transitions upstream and downstream of a bridge crossing.

Table 5.1. Ranges of Expansion Rates, ER (after USACE 2010c).				
b/B ratio	Slope	$n_{ob}/n_c = 1$	$n_{ob}/n_c = 2$	$n_{ob}/n_c = 4$
0.10	0.0002	1.4 – 3.6	1.3 – 3.0	1.2 – 2.1
	0.001	1.0 – 2.5	0.8 – 2.0	0.8 – 2.0
	0.002	1.0 – 2.2	0.8 – 2.0	0.8 – 2.0
0.25	0.0002	1.6 – 3.0	1.4 – 2.5	1.2 – 2.0
	0.001	1.5 – 2.5	1.3 – 2.0	1.3 – 2.0
	0.002	1.5 – 2.0	1.3 – 2.0	1.3 – 2.0
0.50	0.0002	1.4 – 2.6	1.3 – 1.9	1.2 – 1.4
	0.001	1.3 – 2.1	1.2 – 1.6	1.0 – 1.4
	0.002	1.3 – 2.0	1.2 – 1.5	1.0 – 1.4
Variables: b = bridge length, B = expanded flow width, S = slope, n_c = channel Manning n, n_{ob} = overbank Manning n.				

Once the expansion reach length has been estimated, the engineer locates the exit cross section and plots it on a topographic map or aerial photograph. Often the expansion reach is so long that the engineer will want to insert intermediate cross sections between the bridge and the exit section. Inserting intermediate cross sections is encouraged as long as ineffective flow specifications are used to represent the expansion taper, as discussed later in this section.

Contraction Reach. Table 5.2 below is taken from Appendix B of the Hydraulic Reference Manual. It summarizes the research published in Research Document 42 (USACE 1995) with regard to the contraction rate (CR). It is similar to Table 5.1 but involves fewer factors. A cell in the table is selected on the basis of longitudinal slope and roughness ratio. The values in the cell are the appropriate range of CR values. The engineer selects a value within the range and determines the length of the contraction reach. Every cell in the table includes a CR value of 1. For this reason, and because in the research study the overall data set had mean and median CR values both near 1, common practice is to routinely use a value of 1 for CR. This practice is usually acceptable, but for cases in which the bridge design is highly sensitive to the amount of backwater (e.g., to comply with FEMA floodway regulations) it may be advisable to use the table to select CR.

Table 5.2. Ranges of Contraction Rates, CR (after USACE 2010c).

Slope	$n_{ob}/n_c = 1$	$n_{ob}/n_c = 2$	$n_{ob}/n_c = 4$
0.0002	1.4 – 3.6	1.3 – 3.0	1.2 – 2.1
0.001	1.0 – 2.5	0.8 – 2.0	0.8 – 2.0
0.002	1.0 – 2.2	0.8 – 2.0	0.8 – 2.0

Variable: S = slope, n_c = channel Manning n, n_{ob} = overbank Manning n

Tidal Bridges. When bridges over tidal streams are analyzed with one-dimensional unsteady flow models, cross section locations must accommodate flow in both directions. The approach section during ebb tide conditions, in which flow is toward the ocean, will be the exit section during flood tide conditions, in which the flow is away from the ocean. In this case the CR and ER values should be the same, and should be a compromise between the ER and CR values that would have been selected if the bridge were not tidal. In many cases it is appropriate to use a value of 1.5 for both CR and ER.

5.5.3 Ineffective Flow Specifications at Bridges

Section 5.4.2 describes the Ineffective Flow feature in HEC-RAS. This feature is very useful in modeling bridge crossings. Referring to Figure 5.8, the upstream and downstream bounding sections are located beyond the side slopes of the embankment fills, and therefore the cross section geometry reflects the floodplain, not the roadway. Ineffective flow areas on the upstream and downstream bounding sections represent the presence of the highway embankments. Figure 5.8 is an example of an upstream bounding section at a bridge. The lateral position of the ineffective flow setting is set back from the abutment station by a distance equal to the CR or ER value multiplied by the distance of the cross section upstream or downstream from the bridge. The engineer assigns the elevation setting on the ineffective flow setting based on the water surface elevation at which a significant amount of discharge would flow over the top of the road embankment.

If intermediate cross sections are inserted in the transition reach between the bridge and the approach section or between the bridge and the exit section, then those sections must include ineffective flow specifications to reflect the taper of the contracting or expanding flow. It is strongly recommended that the engineer plot the cross section lines and the taper lines on a topographic map and/or aerial photograph in order to enter the ineffective flow specifications correctly.

5.6 BRIDGE HYDRAULIC CONDITIONS

This section provides a qualitative description of the various types of flow conditions that can exist at a bridge. Later sections explain the technical approaches to modeling the different conditions.

5.6.1 Free-Surface Bridge Flow

Free-surface bridge flow refers to the range of flow conditions at a specific bridge in which the bridge low chord is not submerged. The HEC-RAS Hydraulic Reference Manual classifies free surface bridge flow conditions as Class A, Class B or Class C, depending on flow regime in the stream reach being crossed and in the bridge waterway itself. Class A, B, and C flow can be considered roughly equivalent to Type I, II, and III flow, respectively, as described in HDS 1 (FHWA 1978).

Class A is the most commonly encountered free-surface bridge flow condition. In this class of flow the conditions are subcritical upstream of the bridge, downstream of the bridge, and throughout the bridge waterway. Class A flow generally satisfies the constraints of gradually varied flow throughout the reach of interest. HEC-RAS provides four available approaches to modeling Class A free-surface bridge flow at a bridge, described in detail in the next section.

In the Class B scenario, the flow passes through critical depth within the bridge waterway, which requires that supercritical flow exist at least for a short distance downstream of the critical depth control section. The potential for Class B flow to occur inside a bridge waterway stems from two causes. First, the elevation of critical depth is often higher in the constriction than upstream or downstream. Second, the water surface within the constriction is dropping rapidly. Most commonly the flow conditions upstream and downstream of a bridge in Class B flow are subcritical, and a hydraulic jump will often exist either within the bridge waterway or a short distance downstream of the bridge. Class B flow can sometimes occur in conjunction with a supercritical stream profile. In this case the bridge waterway is a control section with subcritical flow upstream and a hydraulic jump occurring some distance upstream of the bridge.

In Class C flow, the regime is supercritical upstream and downstream of the bridge and through the bridge waterway. Class C flow is an extremely rare condition because natural channels on steep grades, such as mountain streams, rarely support uninterrupted supercritical flow over long reaches (Jarrett 1984). Class C flow, therefore, would typically be expected only in engineered flood control channels on a steep slope. Figure 5.10 illustrates Class A, B, and C flow conditions.

5.6.2 Overtopping-Flow

Overtopping flow is the condition in which flow is crossing over the roadway approaches or the bridge deck itself. Overtopping flow conditions are appropriately represented by a broad-crested weir, since the road embankment is elevated above the floodplain grade, the dimension of the crest in the direction of flow (e.g. across the road) is broad, and the overtopping depth is comparatively shallow. Chapter 3 of this manual discusses broad-crested weirs. In a wide floodplain with a low road profile, the quantity of flow going over the road instead of through the bridge can be considerable. With just one foot (0.3 m) of overtopping depth, for instance, the weir flow could easily exceed 25 ft^3/s (0.7 m^3/s) for every 10 feet (3 m) of weir length.

Overtopping flow at bridge crossings is usually combined with either free-surface bridge flow or submerged-deck flow in the bridge waterway. When overtopping flow occurs, the engineer must determine how much flow is going through the bridge and how much over the bridge deck or roadway. This determination is accomplished by the principle that all flow paths from the upstream bounding section to the downstream bounding section should result in the same energy loss. Only one flow distribution between overtopping and bridge flow will result in equal energy loss.

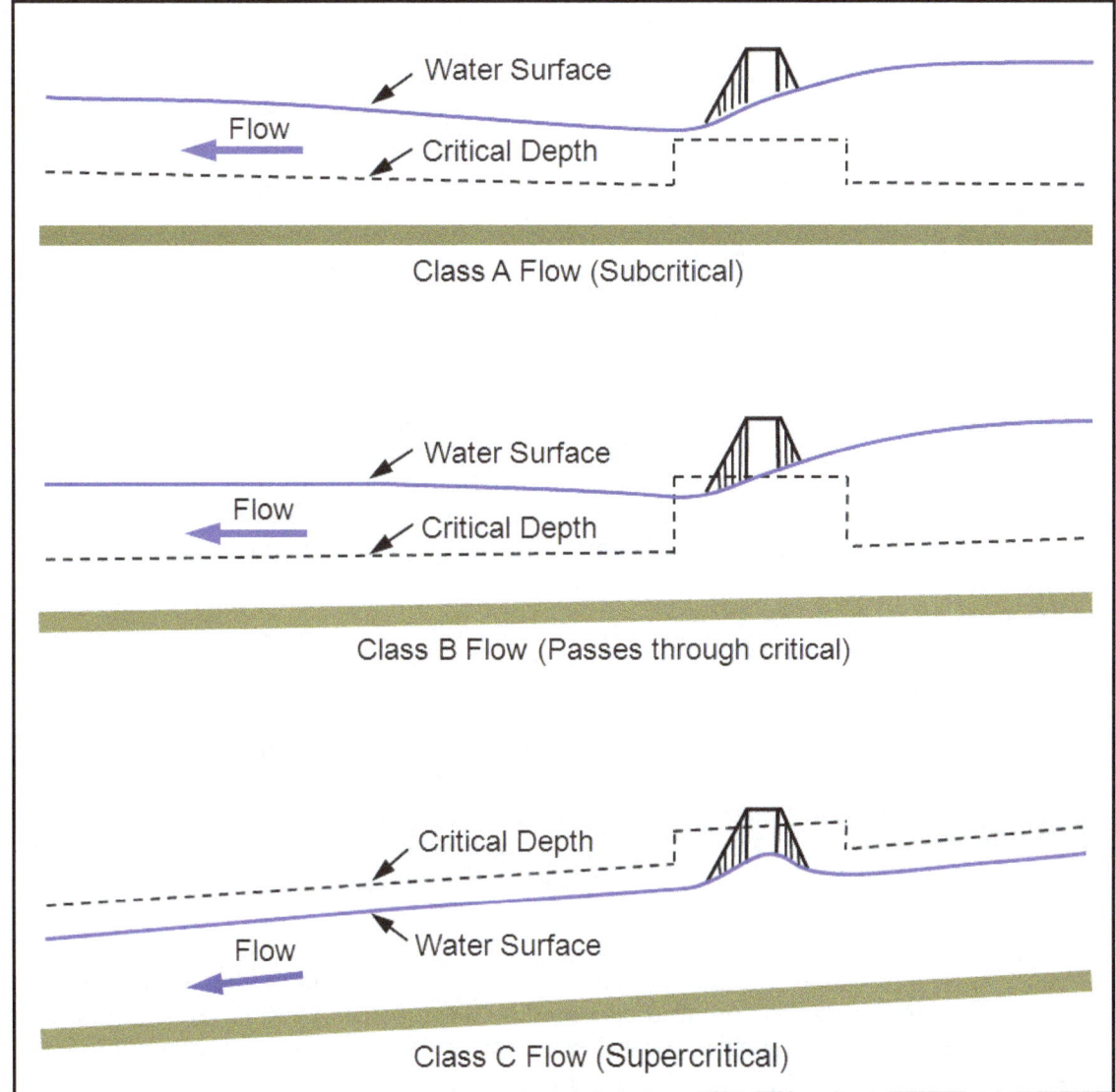

Figure 5.10. Illustration of free-surface bridge flow Classes A, B, and C (USACE 2010c).

In some cases a weir does not accurately represent the flow over the roadway approaches. This can occur either because the road is at or very near floodplain grade (in other words there is little or no embankment fill) or because the downstream water surface elevation is so high above the weir crest elevation that the weir control is drowned out.

5.6.3 Flow Submerging the Bridge Low Chord

A condition in which the water surface is above the highest point of the bridge low chord is usually representative of orifice flow. When the low chord is submerged only at the upstream edge of the superstructure, the orifice is considered free-flowing, and thus not affected by tailwater. This condition is analyzed using the same approach as for an orifice (FHWA 1978) and in this manual is referred to as "orifice bridge flow." Just as the headwater upstream of an inlet-control culvert is not affected by conditions downstream of the culvert entrance, the backwater upstream of a bridge operating under orifice bridge flow is not affected by conditions downstream of the upstream edge of the superstructure.

Another type of orifice flow exists when the highest point of the low chord is submerged at both the upstream and downstream edges of the superstructure. This type of flow is analyzed using a formulation for a tailwater-controlled orifice (FHWA 1978). Just as the headwater upstream of an outlet-control culvert is sensitive to conditions within and downstream of the culvert barrel, the backwater upstream of a bridge operating under full-flowing or tailwater submerged orifice conditions is affected by conditions within and downstream of the bridge waterway. For purposes of this manual, this condition is termed "submerged-orifice bridge flow."

5.7 BRIDGE MODELING APPROACHES

This section explains the approaches and equations that are used to analyze the various types of flow conditions that can exist at a bridge. The HEC-RAS Hydraulic Reference Manual explains the approaches described in this section in greater detail (USACE 2010c). The information presented in this section is predominantly taken from that source but with much of the detail omitted. Except in the case of the WSPRO method, the discussion below applies specifically to the region between the upstream bounding cross section and the downstream bounding cross section. Upstream and downstream of this region, the energy equation governs.

5.7.1 Modeling Approaches for Free-Surface Bridge Flow Conditions

The most commonly encountered free-surface bridge flow scenarios at bridges are Class A conditions. The HEC-RAS program makes four modeling approaches available to users for Class A flow: The Energy Method, the Momentum Balance Method, the Yarnell Equation and the WSPRO Method. The four modeling approaches are described below. Figure 5.11 shows the cross section identifiers for reference to the bridge flow equations.

<u>Energy Method</u>. Chapter 3 describes the energy equation in detail. Section 5.3 describes its general application to water surface profile calculations. When the energy equation is applied to bridge hydraulics, the area occupied by the road embankments, abutments, bridge deck and piers is subtracted from the effective flow area. The wetted perimeter is increased to account for the sides of piers (often a minor effect) and the low chord of the bridge if it is in contact with the flow. The low chord and pier sides can have a significant effect on the wetted perimeter. Since the area is decreased and the wetted perimeter is increased, the conveyance is often reduced significantly. The reduced conveyance, in turn, increases the friction slope which increases the friction loss.

<u>Momentum Balance Method</u>. As discussed in Chapter 3, a momentum-based formulation can be used to analyze open-channel hydraulics. The Momentum Balance Method is based on the principle that the sum of forces acting in a given direction on a control volume is equal to the mass of the water in the control volume multiplied by its acceleration. Hydraulics in the bridge waterway can be solved using this force-balance approach in three steps. The first step deals with the control volume between the downstream bounding section (designated with subscript 2) and the downstream face of the bridge opening (subscript BD):

$$A_{BD} \overline{Y}_{BD} + \frac{\beta_{BD} Q_{BD}^2}{gA_{BD}} = A_2 \overline{Y}_2 + \frac{\beta_2 Q_2^2}{gA_2} - A_{PBD} \overline{Y}_{PBD} + F_{f(2-BD)} - W_{x(2-BD)}$$

(5.14)

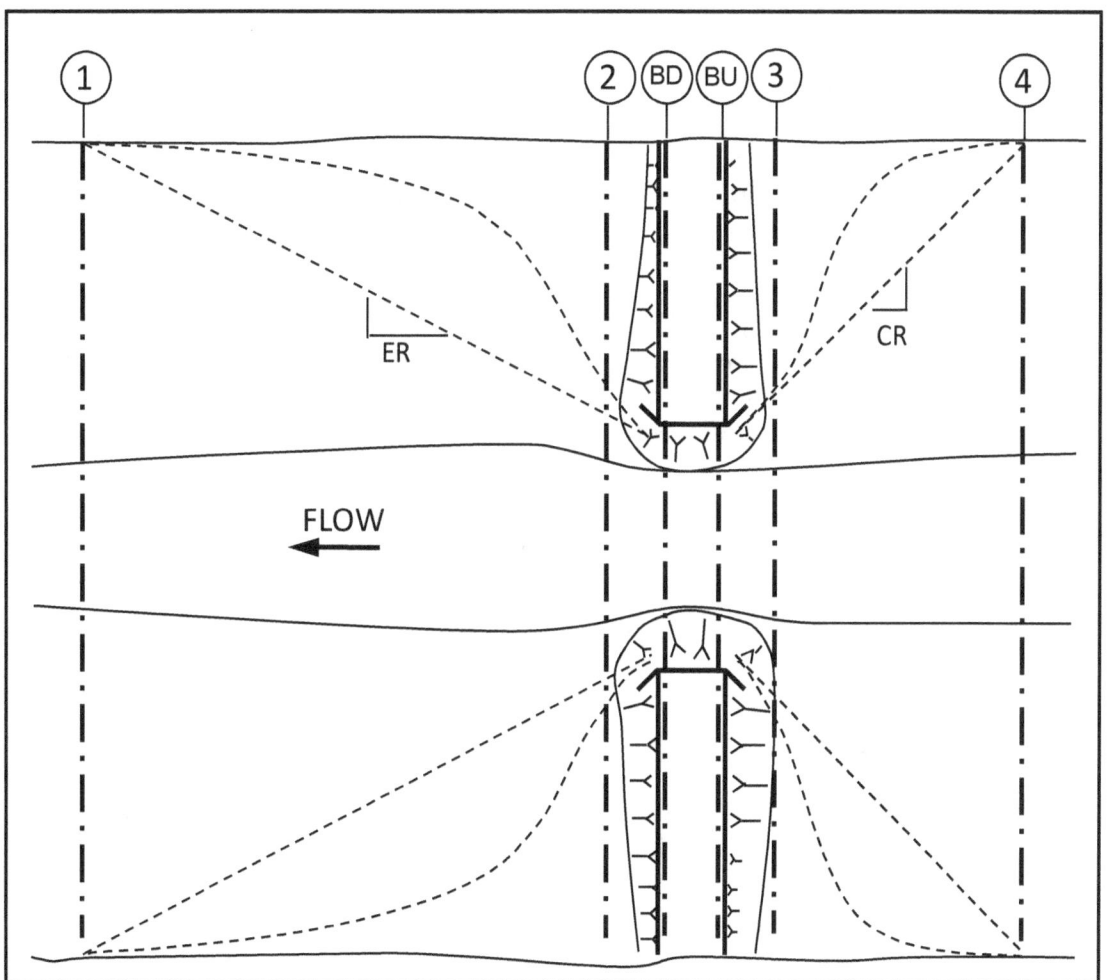

Figure 5.11. Plan view layout showing cross section identifiers referenced by the bridge hydraulics equations.

The second step operates on the control volume beneath the bridge superstructure:

$$A_{BU}\overline{Y}_{BU} + \frac{\beta_{BU}Q_{BU}^2}{gA_{BU}} = A_{BD}\overline{Y}_{BD} + \frac{\beta_{BD}Q_{BD}^2}{gA_{BD}} + F_{f(BD-BU)} - W_{x(BD-BU)} \qquad (5.15)$$

The third step analyzes the force balance on the control volume between the upstream face of the bridge opening (designated with subscript BU), and the upstream bounding section (designated with subscript 3):

$$A_3\overline{Y}_3 + \frac{\beta_3 Q_3^2}{gA_3} = A_{BU}\overline{Y}_{BU} + \frac{\beta_{BU}Q_{BU}^2}{gA_{BU}} = A_{PBU}\overline{Y}_{PBU} + \frac{1}{2}C_D\frac{A_{PBU}Q_3^2}{gA_3} + F_{f(BU-3)} - W_{x(BU-3)} \qquad (5.16)$$

In the equations above:

A_i = Active flow area at the cross section denoted by the subscript, ft² (m²)
A_{PBU}, A_{PBD} = Flow area obstructed by pier at the upstream and downstream faces of the bridge opening, ft² (m²) (see Figure 5.12)

$\overline{Y}_i$ = Vertical distance from the water surface to the centroid of the flow area at the cross section denoted by the subscript, ft (m)

$\overline{Y}_{PBU}, \overline{Y}_{PBD}$ = Vertical distance from water surface to the centroid of the pier area at the upstream and downstream faces of the bridge opening, ft (m) (see Figure 5.12)

Q_i = Discharge at the cross section denoted by the subscript, ft³/s (m³/s)

β_i = Velocity weighting coefficient for momentum at the cross section denoted by the subscript

F_f = External friction force acting on the control volume per unit weight of water, ft³ (m³)

W_x = Component of the weight of water acting in the direction of flow, per unit weight of water, ft³ (m³)

C_D = Drag coefficient for flow around the pier

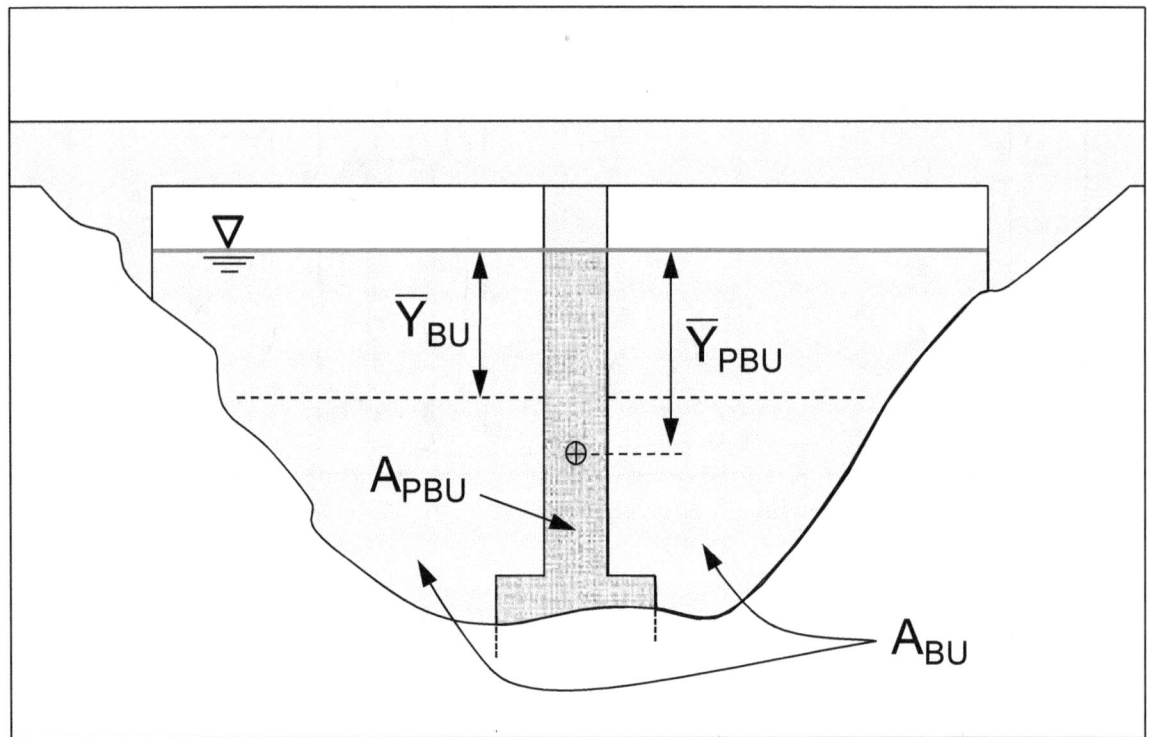

Figure 5.12. Cross section view illustrating the area and $\overline{Y}$ variables in momentum equation.

The user enters the drag coefficient, which is a function of the plan-view shape of the pier. Table 3.3 provides guidance on the values of the pier drag coefficient. Because of the pier drag coefficient, the Momentum Balance Method is sensitive to the hydraulic efficiency of the pier shape. This is an advantage over the Energy Method, which does not provide a way of accounting for streamlined pier shapes. The Momentum Balance Method is also the preferred approach to computing the bridge hydraulics in Class B flow, because it is not hindered by rapidly varied flow conditions.

Yarnell Equation. While the Energy Method and the Momentum Balance Method are theoretically derived, the Yarnell Equation is strictly empirical. It is based on the results of roughly 2600 flume experiments that were designed to test the relationship between the change in water surface elevation caused by a pier and the size, shape, and configuration of the pier in combination with varied flow rates. The resulting equation is:

$$H_{3-2} = 2K(K + 10\omega - 0.6)(\alpha + 15\alpha^4)\frac{V_2^2}{2g} \tag{5.17}$$

where:

H_{3-2} = Drop in water surface elevation from the upstream bounding section (Cross Section 3) to the downstream bounding section (Cross Section 2), ft (m) (see Figure 5.13)

K = Yarnell's pier shape coefficient (see below)

ω = Ratio of the velocity head to the depth at the downstream bounding section

V_2 = Velocity at the downstream bounding section, ft/s (m/s)

When using the Yarnell Equation, the engineer enters a pier shape coefficient. Table 5.3 provides appropriate values of the coefficient for various plan-view pier shapes. A disadvantage of the Yarnell Equation is that, because it is strictly empirical, its application should be limited to bridge sites that are similar in nature to the flume studies that were used in the development of the equation. In practical terms, this means that the equation is only appropriate for channels of generally regular cross section under approximately uniform flow conditions, where piers are the only significant source of losses.

WSPRO Method. Beginning in the 1980s the FHWA developed and supported a water-surface-profile computer program, called WSPRO, that became the standard bridge hydraulic analysis software for many state departments of transportation. The bridge hydraulics approach from that program is now included as an available method in HEC-RAS. The WSPRO Method is based on a standard-step solution of the energy equation, and is similar to the Energy Method in most respects. Unlike the other three free-surface bridge flow methods discussed here, the WSPRO Method works from the exit section to the approach section, and not just between the upstream and downstream bounding sections. In general:

$$WS_4 + \frac{\alpha_4 V_4^2}{2g} = WS_1 + \frac{\alpha_1 V_1^2}{2g} + h_L \tag{5.18}$$

where:

WS_1, WS_4 = Water surface elevation at the exit section (Cross Section 1) and at the approach section (Cross Section 4), ft (m) (see Figure 5.14)

V_1, V_4 = Velocity at the exit section and at the approach section, ft/s (m/s)

h_L = Sum of the energy losses between the exit section and the approach section, ft (m) (see Figure 5.14)

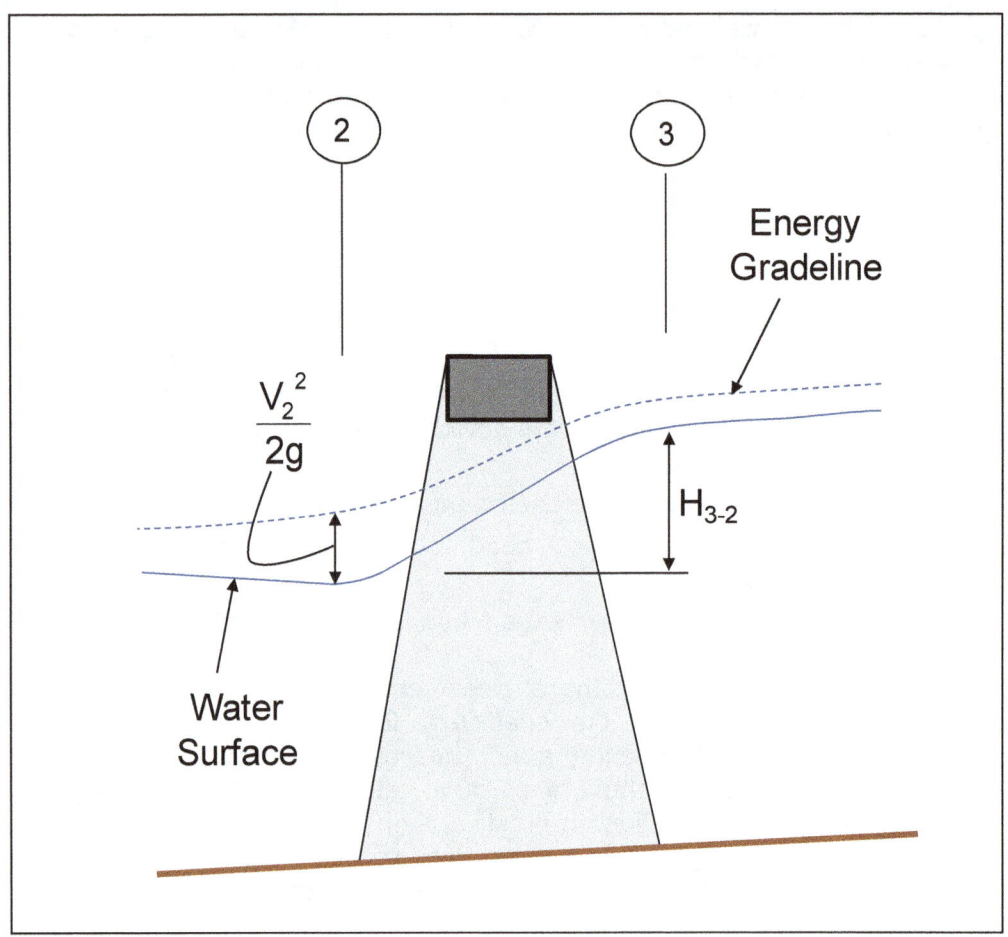

Figure 5.13. Profile view with definitions of variables in the Yarnell Equation.

| Table 5.3. Yarnell Pier Shape Coefficients (USACE 2010c). ||
Plan-View Shape of Pier	Drag Coefficient, C_D
Semi-circular nose and tail	0.90
Twin cylinders with a connecting diaphragm	0.95
Twin cylinders without diaphragm	1.05
Triangular nose and tail with 90-degree angle	1.05
Square nose and tail	1.25
Trestle bent with ten piles	2.50

The WSPRO Method computes the energy losses incrementally in six parts: five increments of friction loss and the expansion loss between the exit section and the downstream bounding cross section (Cross Sections 1 and 2).

The five increments of friction loss cover the segments between Cross Sections 1 and 2; between the downstream bounding section and the downstream face of the bridge opening (Cross Sections 2 and BD); the segment under the bridge deck (Cross Sections BD and BU); between the upstream face of the bridge opening and the upstream bounding section (Cross Sections BU and 3); and between the upstream bounding section and the approach section (Cross Sections 3 and 4). For each of the first four segments, the friction loss is calculated using the following general equation:

$$h_{f_{seg}} = \frac{L_{seg}Q^2}{K_d K_u} \tag{5.19}$$

where:

- $h_{f_{seg}}$ = Friction loss in the segment, ft (m)
- L_{seg} = Segment reach length computed as the discharge-weighted average reach length, ft (m)
- Q = Discharge, ft³/s (m³/s)
- K_d, K_u = Conveyance at the downstream and upstream cross sections, ft³/s (m³/s)

While not readily recognizable as such, the $Q^2/K_d K_u$ portion of Equation 5.19 is the geometric mean friction slope between the two cross sections (see Equation 5.6).

The friction loss for the upstream-most segment is slightly different:

$$h_{f3-4} = \frac{L_{av}Q^2}{K_3 K_4} \tag{5.20}$$

where:

- h_{f3-4} = Friction loss between Cross Sections 3 and 4, ft (m) (see Figure 5.14)
- K_3, K_4 = Conveyance at Cross Sections 3 and 4, ft³/s (m³/s)
- L_{av} = Effective average flow length in the approach reach, ft (m)

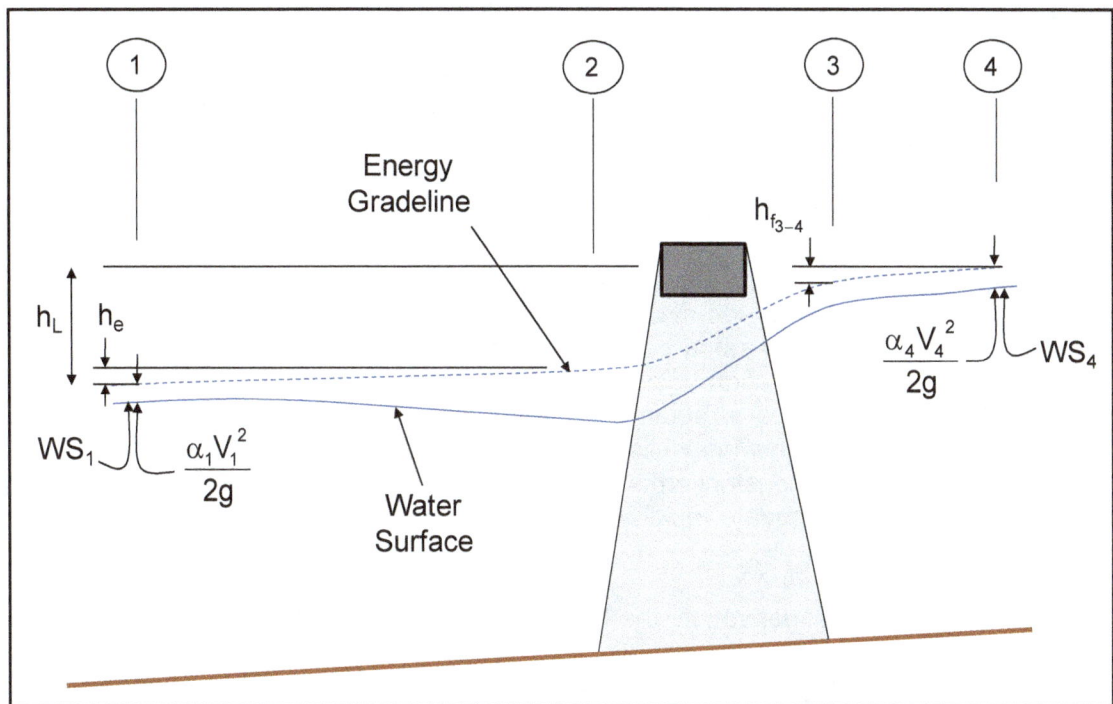

Figure 5.14. Profile view with definition of terms in the WSPRO Method.

HEC-RAS computes the effective average flow length in Equation 5.20 as the average length of 20 equal-conveyance stream tubes that flow from Cross Section 4 to Cross Section 3 on theoretical curvilinear paths. The details of the assumed stream tube flow paths are explained in Appendix D of the HEC-RAS Hydraulic Reference Manual and in the WSPRO User's Manual (FHWA 1986).

By default, the WSPRO Method does not include the standard contraction and expansion losses from the energy equation (e.g. the contraction or expansion coefficient multiplied by the absolute difference in velocity head between two cross sections) in the energy losses from the exit section to the approach section. Expansion losses between Cross Sections 1 and 2 are accounted for using the following equation:

$$h_e = \frac{Q^2}{2gA_1^2}\left[2\beta_1 - \alpha_1 - 2\beta_2\left(\frac{A_1}{A_2}\right) + \alpha_2\left(\frac{A_1}{A_2}\right)^2\right]$$

(5.21)

where:

h_e	=	Expansion loss in expansion reach, ft (m)
A_1, A_2	=	Flow areas at exit section and downstream bounding section, ft² (m²)
α_1	=	Kinetic energy distribution coefficient at exit section
β_1	=	Momentum distribution coefficient at the exit section
α_2, β_2	=	Factors related to discharge coefficient, C, which is a function of bridge geometry

Appendix D of the HEC-RAS Hydraulic Reference Manual (USACE 2010c) explains in detail the empirical discharge coefficient C mentioned in the definition of the variables α_2 and β_2.

5.7.2 Selection of Free-Surface Bridge Flow Modeling Approach

The Energy, Momentum Balance and WSPRO Methods are all suitable to a wide range of conditions. Among these three, the Momentum Balance Method is unique in accounting for the pier drag as a function of pier shape. Therefore the Momentum Balance Method is recommended in cases where piers are the predominant energy loss factor and especially when the pier geometry is somewhat streamlined.

The Energy and WSPRO Methods are both effective in conditions where friction loss and the effects of constriction are predominant. In most cases the results of the two methods, when applied correctly under the same conditions, are very similar in terms of the energy grade line and water surface elevation upstream of the bridge. Only the WSPRO Method, however, accounts for different types of abutments geometries (for instance spill-through abutments vs. vertical abutments with wing walls). The Momentum Method also typically performs well in situations where the constriction is the predominant loss factor, and has the advantage that it can better accommodate rapidly-varied flow, which is important in Class B and Class C free-surface bridge flow.

Because of its empirical derivation, the Yarnell Equation is suitable only for Class A cases in which the geometry of the waterway is generally uniform and regular, and without a great degree of constriction. It can be expected to perform well in analyzing bridges over man-made channels such as irrigation canals or engineered flood control channels.

Ideally engineers modeling Class B and Class C flow conditions should employ the Momentum Method for the reasons mentioned earlier. The Energy Method is also an acceptable approach, although less ideal.

5.7.3 Modeling Approaches for Overtopping and Orifice Bridge Flow

The HEC-RAS program makes two different approaches available to the user for modeling overtopping and orifice bridge flow conditions: Energy Method and Pressure and Weir Method. Note that in HEC-RAS terminology, orifice and submerged-orifice bridge flow conditions are termed "pressure flow."

Energy Method. Just as described above for free-surface bridge flow modeling, the Energy Method simply continues the standard-step solution of the energy equation through the bridge structure and vicinity. It accounts for the blockage caused by the road embankments, abutments, bridge deck and piers simply by reducing the conveyance. If the water surface is high enough to overtop the road, the program will treat the flow area above the road as conveyance area, but not as a weir. When the Energy Method is used, the quantity of overtopping flow will not be computed or reported. If the low chord is submerged, the added wetted perimeter will have a negative effect on conveyance, but the program will not attempt to compute orifice conditions.

Pressure and Weir Method. If the user has specified the Pressure and Weir Method, then the broad-crested weir equation is be used to compute the quantity of any overtopping flow. One of two orifice equations is used when orifice or submerged-orifice bridge flow is detected.

Overtopping (Weir) Flow. The technique for computing weir hydraulics in the case of flow over the road or bridge deck is very similar to the approach that was described and recommended in HDS 1 (FHWA 1978). Figure 5.15 depicts the condition of flow overtopping a roadway embankment. The broad-crested weir equation is:

$$Q_W = CLH^{3/2} \tag{5.22}$$

where:

Q_w = Discharge over the weir, ft^3/s (m^3/s)
C = Weir flow coefficient, see below
L = Length of weir overtopping, ft (m)
H = Head driving weir flow, which is the upstream energy grade line minus weir crest elevation, ft (m)

The discharge coefficient for a broad-crested weir generally ranges between 2.6 and 3.1. A bridge deck is not an ideal broad-crested weir and it is generally recommended that the lower value of 2.6 be used for the discharge coefficient where increased resistance to flow caused by obstructions such as bridge railings, curbs, and debris are present. HDS 1 provides a curve of the C value versus head on the roadway. That curve is reproduced here as Figure 5.16.

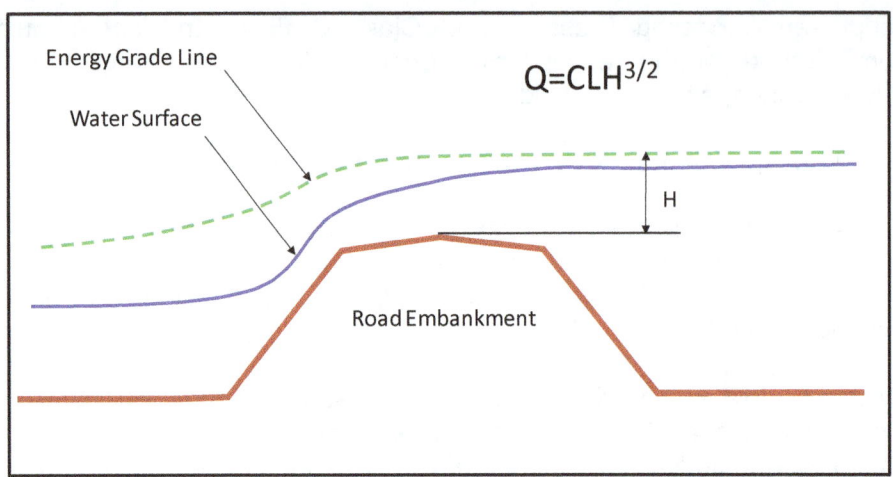

Figure 5.15. Illustration of flow overtopping a roadway embankment.

In the example depicted in Figure 5.15 the water surface downstream of the bridge is lower than the weir crest elevation (the crown of the road). In cases where the downstream water surface elevation is above the weir crest elevation, the weir flow is said to be submerged. The submergence doesn't start to affect the weir capacity until the degree of submergence reaches about 80 percent. The degree of submergence is defined as the tailwater depth divided by the head, and the tailwater depth is defined as the downstream water surface elevation minus the weir crest elevation (D in Figure 5.16). For submergence greater than about 80 percent, the discharge coefficient C should be reduced based on the relationship indicated on the graph along the right and top edges of Figure 5.16.

Overtopping flow at a bridge crossing is usually accompanied by free-surface bridge flow or orifice flow in the bridge. To compute an accurate water surface profile, it is necessary to determine how much flow goes through the bridge waterway and how much over the road or bridge deck. The HEC-RAS program uses an iterative approach to find the flow distribution. The solution approach is based on finding the amount of weir flow such that the head elevation driving the weir flow is the same as the energy grade line elevation upstream of the bridge resulting from the losses experienced by the non-overtopping flow passing through the bridge waterway.

<u>Orifice and Submerged-Orifice Bridge Flow</u>. Figure 5.17 is a sketch depicting the non-submerged orifice flow condition. The bridge opening acts as an orifice control section, with no influence from downstream conditions. The equation for orifice pressure flow is:

$$Q = C_D A_{BU} \sqrt{2g} \left[Y_3 - \frac{Z}{2} + \frac{\alpha_3 V_3^2}{2g} \right]^{1/2}$$

(5.23)

where:
- Q = Discharge under bridge deck, through the bridge waterway, ft³/s (m³/s)
- C_D = Orifice flow discharge coefficient
- A_{BU} = Net flow area at the upstream face of the bridge opening, under the low chord, ft² (m²)
- Z = Height of the bridge opening from the highest point on the upstream low chord to the mean riverbed elevation, ft (m)
- Y_3 = Hydraulic depth at the upstream bounding section (Cross Section 3), ft (m)
- V_3 = Velocity at the upstream bounding section, ft/s (m/s)
- α_3 = Kinetic energy distribution coefficient at the upstream bounding section

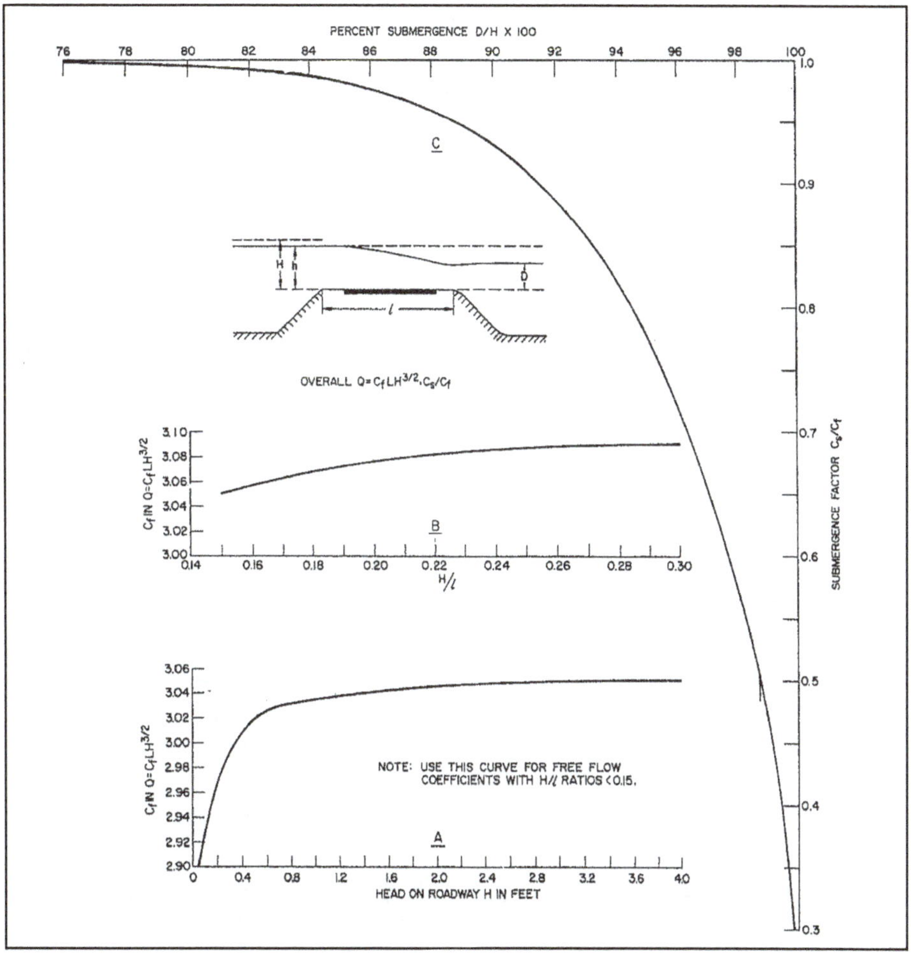

Figure 5.16. Guidance on discharge coefficients for flow over roadway embankments, from HDS 1 (FHWA 1978).

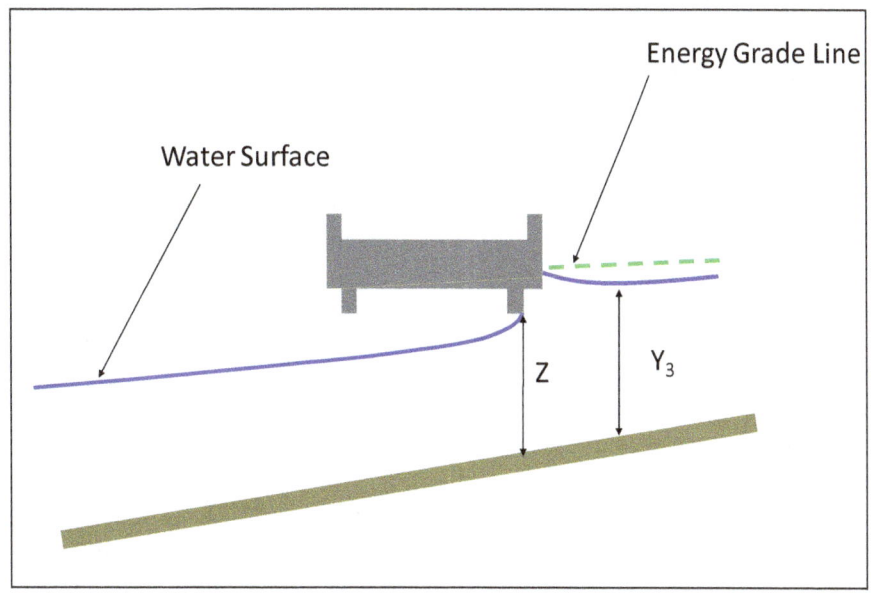

Figure 5.17. Sketch of orifice bridge flow.

The head forcing the flow through the orifice is defined as the vertical distance from the energy grade line upstream of the bridge to roughly the vertical center of the bridge opening height (the elevation halfway up the Z dimension). The value of the discharge coefficient C_D is related to the ratio of low chord submergence (Y_3/Z) as shown in Figure 5.18.

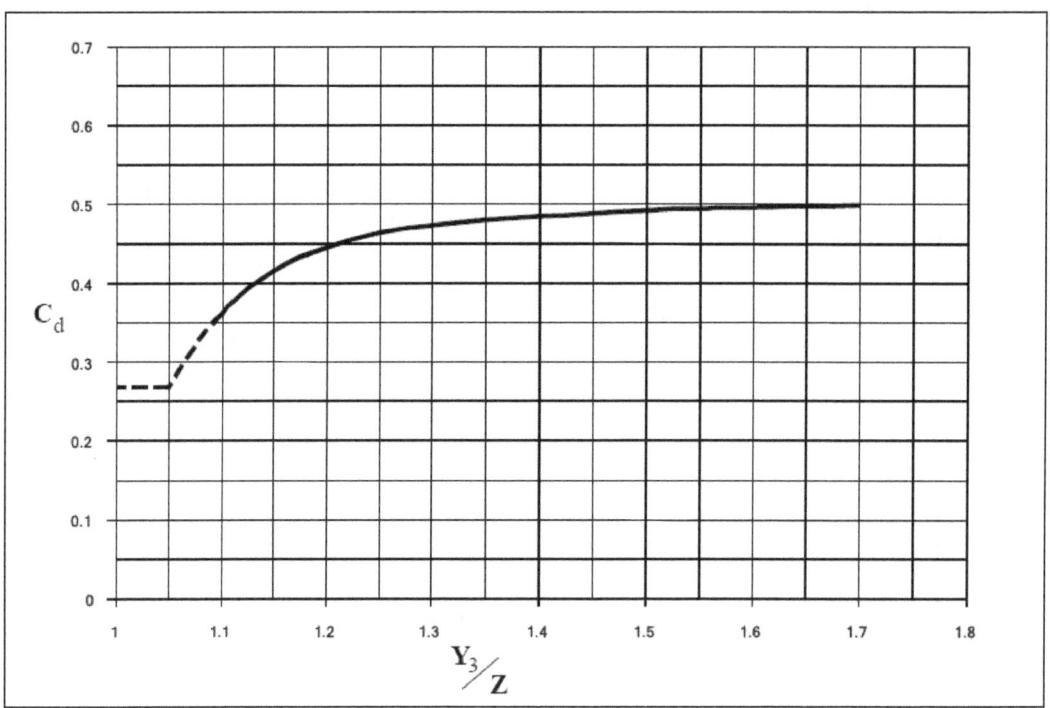

Figure 5.18. Relationship between orifice bridge flow discharge coefficient and submergence of the low chord, from HDS 1 (FHWA 1978).

Note that the curve is shown as a dashed line to the left of $Y_3/Z=1.1$. This region of the curve represents a transition zone in which orifice flow conditions have not been reliably established. At this submergence level, the flow could be expected to vary between open-channel and pressure flow conditions, and the orifice equation might not be reliable.

Figure 5.19 illustrates the case of submerged-orifice bridge flow. The equation for this case is:

$$Q = CA\sqrt{2gH} \qquad (5.24)$$

where:

- Q = Discharge under bridge deck, through the bridge waterway, ft³/s (m³/s)
- C = Discharge coefficient for submerged-orifice bridge flow (usually 0.8)
- A = Net flow area of the bridge opening, ft² (m²)
- H = Vertical distance between the upstream energy grade line and the downstream water surface, ft (m)

In the submerged-orifice case, the head is measured from the upstream energy grade line to the downstream water surface elevation, reflecting that the downstream conditions have a direct effect on the backwater. Field data reported in HDS 1 indicated that the values of C for submerged-orifice bridge flow range from 0.7 to 0.9. Common practice, encouraged by HDS 1, is to use a value of 0.8 for C.

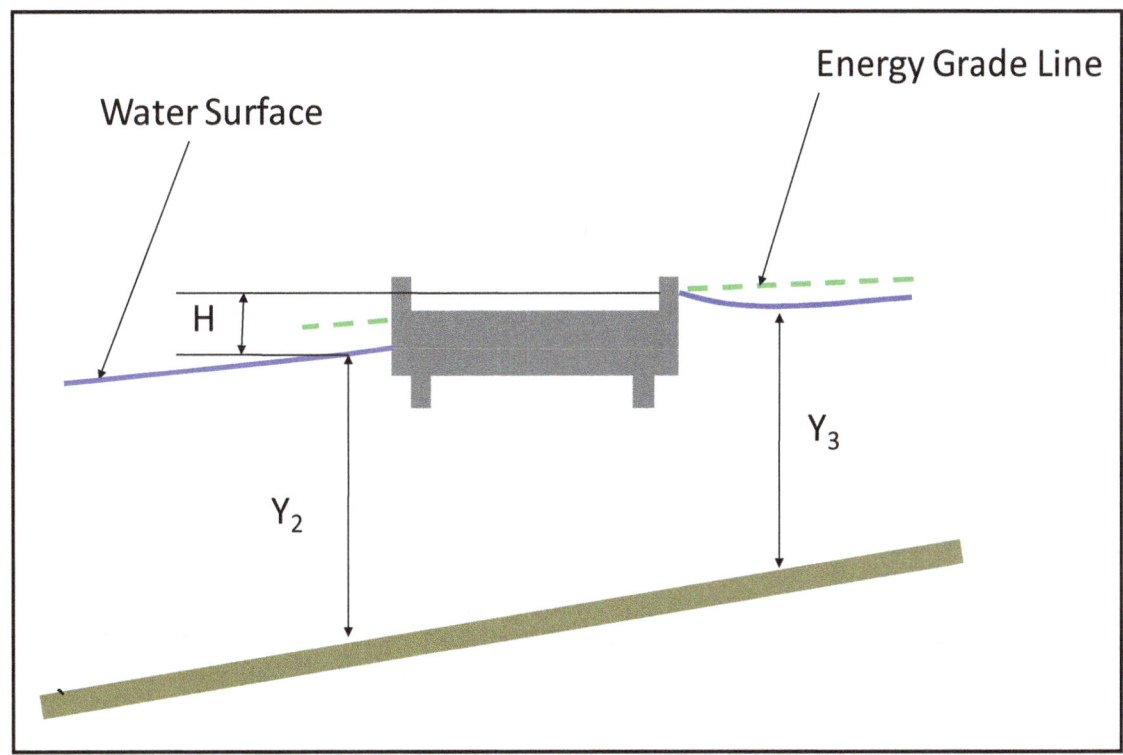

Figure 5.19. Sketch of submerged-orifice bridge flow.

5.7.4 Selection of the Overtopping and Submerged-Low-Chord Modeling Approach

The selection of the modeling approach for overtopping or submerged-low-chord conditions usually has much more significant consequences than the selection of the free-surface bridge flow modeling approach. The results of the Energy Method can be far different from those of the Pressure and Weir Method, which in turn can have major effects on the design of the bridge. This section provides guidance on identifying the better modeling approach for various situations.

The Pressure and Weir Method is the preferred approach for many scenarios, including the following:

- Overtopping with little or no tailwater submergence

    The embankment is truly functioning as a weir if there is flow over an elevated road embankment and there is no significant tailwater submergence. This is true even for a reasonably large depth of tailwater above the weir crest. Figure 5.16 shows that the ratio of tailwater depth to head must be greater than 0.80 to significantly reduce the weir capacity. In such cases the Energy Method typically overestimates the backwater caused by the crossing because it fails to acknowledge the high-efficiency flow conveyance provided by the weir. This effect is more pronounced with greater lengths of road overtopping. In addition to overestimating the backwater, the Energy Method also overestimates the amount of flow, and thus the velocity, inside the bridge waterway, which in turn could lead to overestimated scour depths.

- Overtopping with a significant change in water surface elevation

Similar to the previous case, if overtopping is occurring and the water surface difference is substantial across the road alignment, then true weir flow conditions are likely. Again the Energy Method overestimates the backwater caused by such cases.

- Orifice or Submerged-Orifice Bridge flow with $Y_3/Z \geq 1.1$

    The bridge waterway is truly functioning as an orifice when the above submergence ratio criterion is met, so the Pressure and Weir Method will yield appropriate results. In such cases the Energy Method usually tends to underestimate the head required to force the flow through the bridge waterway. Consequently the Energy Method underestimates the backwater caused by the crossing.

Scenarios that are better suited to the Energy Method include:

- Overtopping of a low or at-grade roadway

    The embankment will not act as a weir during overtopping flow if the road is at the floodplain grade or has a very low embankment. In such cases the Pressure and Weir Method could potentially underestimate the backwater by attributing too much capacity to the road overtopping segment. In addition to underestimating backwater, the Pressure and Weir Method could result in underestimating the flow through the bridge waterway, which would in turn lead to underestimated velocity and scour potential.

- Overtopping flow with very little water surface change

    A very small amount of drop in the water surface from one side of the road embankment to the other suggests that the weir crest is highly submerged such that the embankment is no longer functioning as a weir. In such cases the Energy Method is appropriate. Use of the weir equation would likely underestimate the backwater, similarly to the previous scenario. In HEC-RAS, the user can select the Pressure and Weir Method but specify that if the submergence ratio exceeds a threshold amount (often entered at 95%) then the program would revert to the Energy Method.

Engineers occasionally encounter situations that are borderline cases, where the decision between the two methods is not clear cut. One such example is a case in which the bridge low chord is submerged with a low degree of submergence ($Y_3/Z < 1.1$) and there is no overtopping flow. The flow conditions are not firmly in the realm of orifice flow at this degree of submergence and the Energy Method may be more appropriate. In this case it is recommended to conduct the analysis with both the Pressure and Weir and the Energy Methods. Use the more conservative of the two if the two results do not differ greatly. Use the Energy Method result if the two results differ by a significant amount.

The recommendations above call for making decisions based on the observed flow conditions. Usually the engineer will not be able to anticipate at the outset what types of flow conditions will be observed in the model results. The modeling process, therefore, requires some iteration by the engineer before arriving at a final analysis model. The recommended practice is to make an initial model run with the Pressure and Weir Method selected. If the model results show that overtopping is occurring, the engineer should identify whether the weir crest is submerged, and if so, by how much, and then decide whether the overtopping flow is truly functioning as weir flow. If so, then the Pressure and Weir Method is appropriate. If not, then the Energy Method should be used. If no overtopping is occurring but there is orifice or submerged-orifice bridge flow, then the engineer should decide between the Pressure and Weir and Energy Methods based on the degree of low chord submergence.

5.8 SPECIAL CASES IN ONE-DIMENSIONAL HYDRAULICS

The previous sections of this chapter have described the basic framework and technical approaches to one-dimensional hydraulic analysis of bridges. This section describes useful approaches to addressing special conditions that may exist at a bridge crossing site.

5.8.1 Skewed Crossing Alignment

Section 5.3 indicated that in one-dimensional modeling the cross sections should be oriented so that each part of the cross section is perpendicular to the anticipated direction of flow across that part. Often the highway crosses the floodplain and/or main channel on an alignment that is not perpendicular to flow. The crossing is said to be skewed to the flow direction. When a cross section has a skewed orientation, the cross sectional flow area is exaggerated, which leads to overestimation of the conveyance. The effects of skew are most important for flow conditions in which the overbanks are inundated and the road embankments are causing significant constriction. A skewed road crossing requires adjustment of the geometric input to the model in order to avoid overestimating the capacity of the crossing.

The recommended practice for skewed crossings with a skew angle less than or equal to 30° is to define the upstream and downstream bounding sections along the toes of the embankment side slopes, similar to a non-skewed crossing. The engineer initially enters the bounding cross section data, road embankment bridge deck, abutment and pier information without adjustment to account for the skew. Once the unadjusted input is entered, the engineer can use automated features in the program to make the required adjustment. The cross section points of the bounding sections require adjustment, along with the station locations of the bridge abutments and pier centerlines. The conveyance of the bounding sections and bridge opening must be calculated on the basis of the cross-sectional width and area that are projected to a line perpendicular to the flow (see Figure 5.20). The adjustment is accomplished by multiplying the cross-section station values by the cosine of the skew angle:

$$X_{Adjusted} = X_{Unadjusted} \times Cos(\theta) \tag{5.25}$$

where:

X = Station value of a cross section point, abutment, or pier centerline, ft (m)
θ = Magnitude of skew angle (the angle deviation from a line perpendicular to the flow direction)

If significant overtopping of the road embankments is anticipated at a skewed crossing, it may be advisable to leave the portions of the bounding cross sections adjacent to the embankments unadjusted, since weir flow tends to orient itself to be perpendicular to the weir crest.

Within the bridge opening of a skewed crossing, the bridge piers may be aligned with the flow or skewed. If the piers are skewed to the flow direction then the pier width entered into the model should be the width projected in the direction of the flow. HEC-RAS has a feature to calculate the projected width automatically given the actual pier width and length, and the pier skew angle as input. The projected pier width is calculated as follows:

$$a_{projected} = (L \times Sin\phi) + (a \times Cos\phi) \tag{5.26}$$

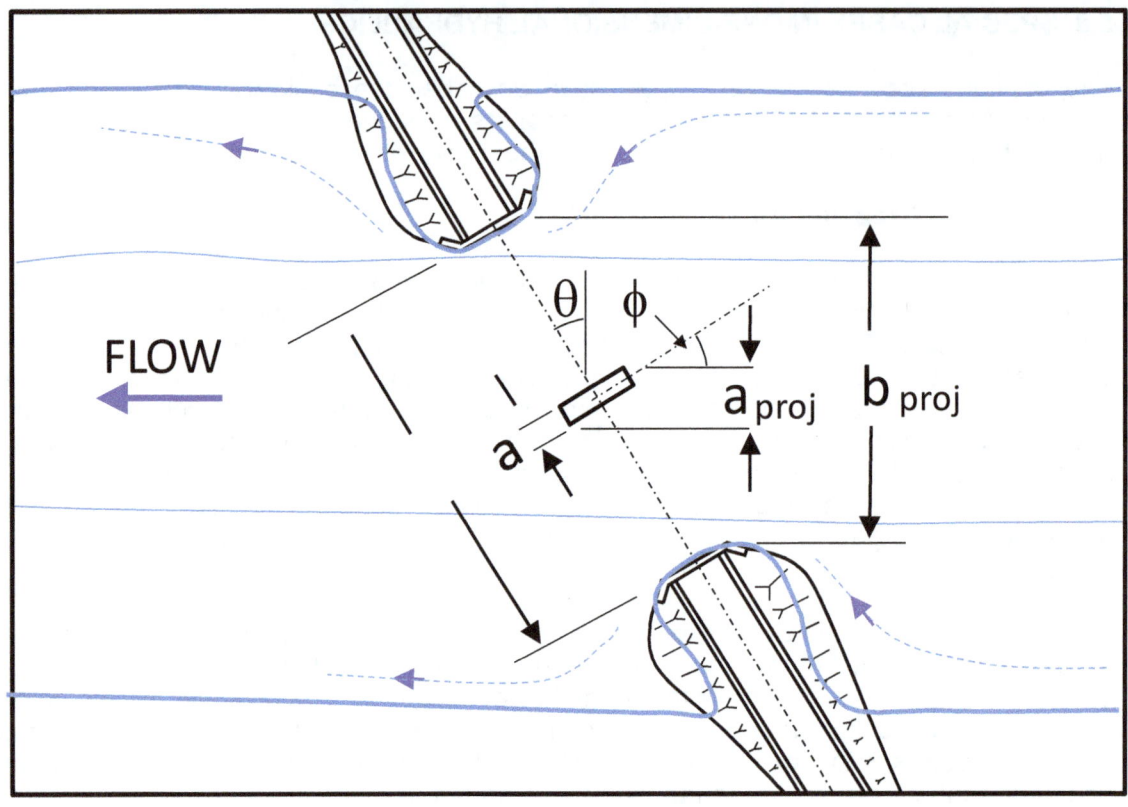

Figure 5.20. Illustration of a skewed bridge crossing.

where:

$a_{projected}$ = Pier width projected in the direction of the flow, ft (m)
L = Actual pier length, ft (m)
ϕ = Magnitude of pier axis skew angle (the angle from a line parallel to the flow direction)
a = Actual pier nose width, ft (m)

A skew angle of 30° is identified by HDS 1 as a practical maximum for analysis by the bridge opening adjustment concept described here. The engineer should consider a different modeling approach, such as two-dimensional analysis, when the skew angle exceeds 30°. Section 5.9 discusses some important limitations related to the one-dimensional treatment of skewed crossings.

5.8.2 Crossings with Parallel Bridges

Parallel bridges are a common occurrence when streams are crossed by divided highways. Figure 5.21 shows an example of a parallel bridge crossing. Hydraulically the two bridges are in series. Physical modeling results reported in HDS 1 show that two identical bridges in series and in close proximity to each other produce about 1.3 to 1.5 times the backwater caused by one bridge alone, depending upon the distance between the two bridges (see Figure 5.22). The maximum clear distance between the bridges in the study cited was 9 bridge deck widths (L_d/l equal to 11 in Figure 5.22). One likely reason for the total backwater being less than twice the single-bridge backwater is that the full contraction and re-expansion of the flow would have occurred only once (contracting upstream of the upstream bridge and re-expanding downstream of the downstream bridge).

Figure 5.21. Aerial image of Interstate 70 over the Colorado River.

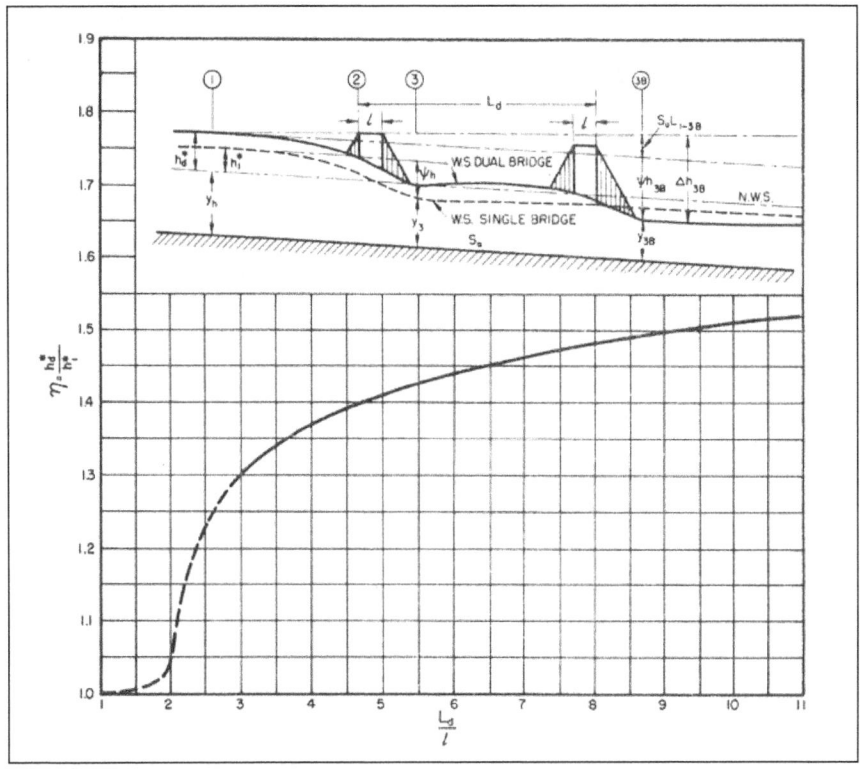

Figure 5.22. Backwater multiplication factor for parallel bridges (from FHWA 1978).

Except in very rare cases it is acceptable to model the two bridges as separate structures in series, but this approach does require additional effort by the engineer that may not be necessary. Depending upon the purpose of the analysis, it may be acceptable to model two parallel, identical bridges as a single bridge. If the two bridge decks are within just a short distance of each other, then treating them as a single bridge with the deck width equal to the sum of the two decks is appropriate. If the two bridges are farther apart, it might be more appropriate to enter the deck width as the total distance from the upstream edge of the upstream bridge to the downstream edge of the downstream bridge, so as not to completely neglect the losses that will be generated in the gap between the two structures.

Many scenarios make it advisable or necessary to model the two bridges as separate structures in series, each with its own bounding cross sections and bridge data specifications. Examples of such cases are:

- When the purpose of the model is to develop the hydraulic design of the two structures and the engineer needs refined hydraulic information to apply to each structure independently for freeboard determination, scour evaluation, etc.
- When the flow can re-expand and consequently re-contract between the two bridges, as could occur if there is substantial distance between the bridges and the ground between the divided roadways is low enough to allow it
- When submerged-orifice bridge flow is a possibility, since the backwater of the upstream bridge is very sensitive to conditions downstream in a submerged pressure flow condition
- When the two bridge structures are not identical in terms of span lengths, pier geometry, deck profile, etc.

5.8.3 Split Flow Conditions

Engineers performing one-dimensional hydraulic analysis occasionally find the need to model streams or floodplains that branch into two or more separate reaches. Often these split reaches remain separated and run generally parallel for some distance and eventually recombine downstream (see illustration in Figure 5.23).

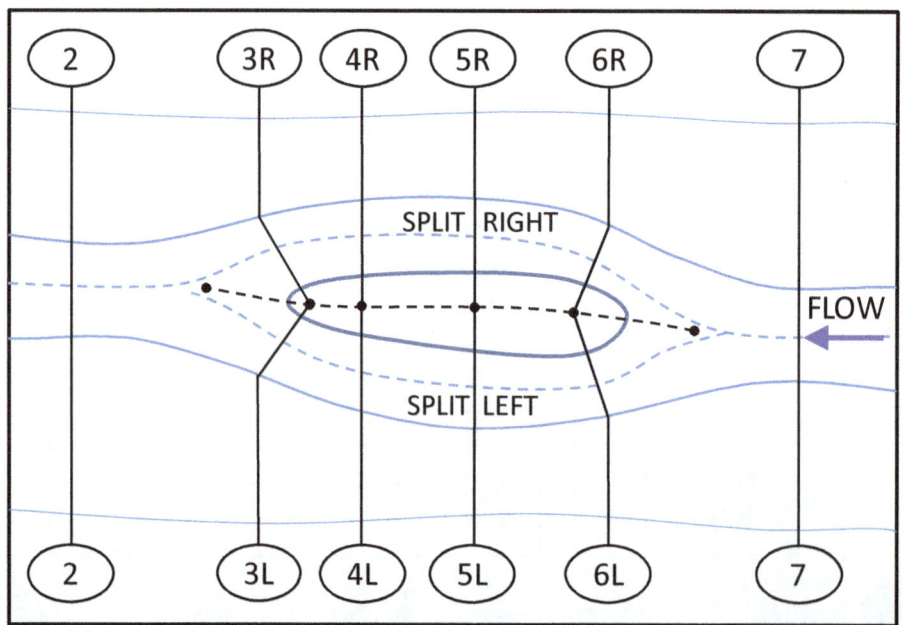

Figure 5.23. Illustration of a cross-section layout to model split flow conditions.

To model a split flow condition, the engineer defines a separate model reach for each flow path. Each reach has its own series of cross sections and its own flow rate. The flow rate for each reach is defined at the point where the flow paths diverge (usually defined by a branching junction in HEC-RAS). The correct apportionment of the total flow between the separate reaches is not known at the beginning of the analysis, but is determined through a trial-and-error process. The process of finding the correct flow apportionment is based on the principal that all separate reaches branching from a single point are expected to yield the same energy grade line elevation at the point of divergence. The analysis progresses as follows:

1. A trial flow rate is assigned to each reach, with the constraint that the sum of the individual reach flow rates must equal the total flow upstream of the point of divergence.

2. Using the assigned flow rates, a water surface profile is computed for each separate reach.

3. The resulting energy grade line elevations at the upstream ends of all of the separate reaches are compared.

4. If all of the resulting energy grade line elevations match each other within the desired tolerance, which is usually 0.05 feet (0.015 m) or less, then the correct flow apportionment has been found and the analysis proceeds beyond the split reaches.

5. If the disagreement between the resulting energy grade line elevations exceeds the desired tolerance, then the flow rates are reduced in the reaches that produced the highest energy grade lines and increased in those that yielded the lowest values, and the process begins again at step 2.

This process can be used whether or not the separate reaches recombine downstream. If they do recombine, then the total energy loss in all reaches must be the same from the branching junction to the confluence junction. If the reaches do not recombine, then the total energy loss in all reaches might not be the same, but in the final solution they must all produce the same energy grade line at the point of divergence.

The HEC-RAS program facilitates the modeling of split flow reaches. The Junction Optimization feature can be activated by the engineer at any branching flow junction. If the feature is activated the program automatically performs iterations and checks for convergence of the energy grade lines until the correct flow apportionment is found. The same principles and approaches used in solving the flow apportionment in split reaches are used in the analysis of multiple-opening crossings.

5.8.4 Crossings with Multiple Openings in the Embankment

Some crossings require relief bridge openings or culverts through the embankment in addition to the main bridge opening. Particularly wide floodplains and those with separate side channels are examples of sites that might require multiple openings. Figure 5.24 illustrates a multiple-opening crossing. Similar to split reaches, the multiple-opening scenario presents a special challenge in one-dimensional analysis. The analysis must correctly determine the apportionment of the flow to each opening in the embankment. The engineer is encouraged to consider the use of two-dimensional analysis for multiple-opening situations.

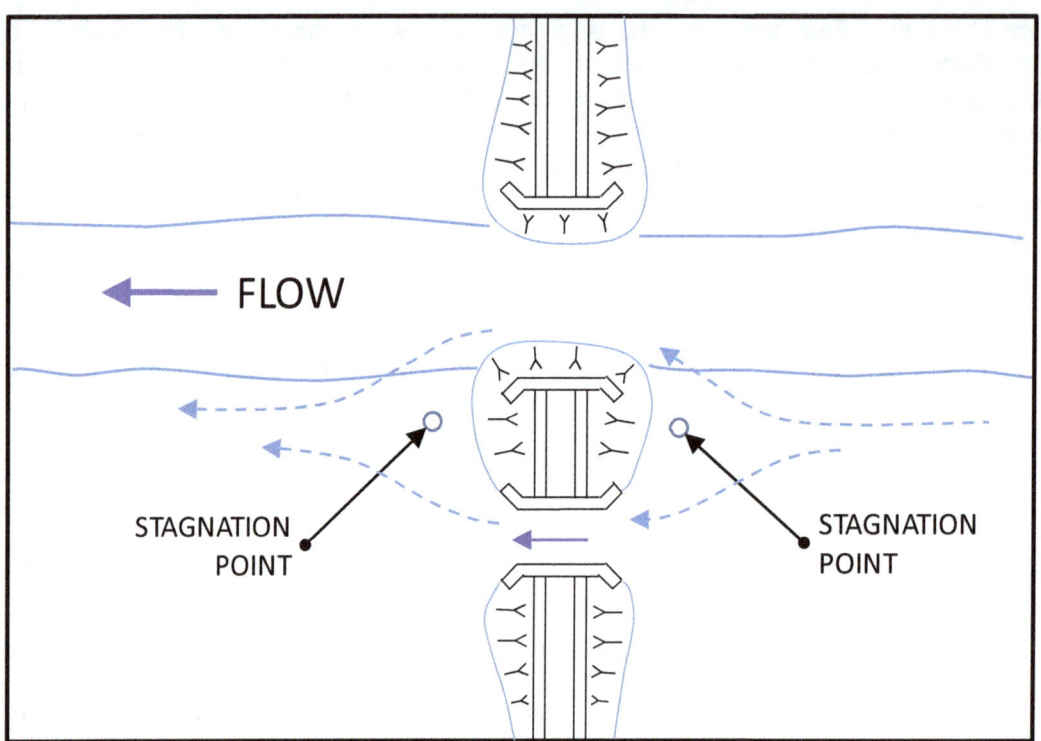

Figure 5.24. Plan view sketch of a multiple-opening bridge crossing.

In one-dimensional modeling, two types of approaches are available to engineers for analysis of multiple-opening crossings. One is the divided-flow approach, in which the engineer creates split reaches and attains correct flow apportionment through the method described above. The other is the automated multiple-opening approach, available in both HEC-RAS and WSPRO. In both programs, the multiple-opening feature automatically creates multiple reaches and iterates to find the correct apportionment of flow between the openings. Because the multiple-opening features of HEC-RAS and WSPRO differ, both are briefly explained below.

HEC-RAS Multiple Opening Approach. When HEC-RAS performs a multiple-opening simulation, it partitions the upstream and downstream bounding sections into as many partitions as there are openings. The engineer must specify stagnation limits at both bounding sections, which should fall into the ineffective regions of those sections, between the openings (see Figure 5.25 as an example).

The stagnation limits place constraints on the location of each partition boundary. Each opening in the embankment draws water only from its portion of the upstream bounding section. Once the partitions are established, HEC-RAS performs an automated split reach analysis, with the flow path for each bridge opening defined as a reach. The program iterates on varied flow apportionment between the openings until the resulting energy grade line elevations at the upstream bounding section are the same from all flow paths, within the specified tolerance.

WSPRO Multiple Opening Approach. A research report by the FHWA (1986) describes the approach to multiple-opening analysis in the WSPRO program. The WSPRO approach differs from HEC-RAS in some respects. Like HEC-RAS, the WSPRO program partitions the floodplain into "valley strips," one for each opening. Whereas the lengths of the divided partitions in HEC-RAS are from the upstream bounding section to the downstream bounding section, the WSPRO valley strips are longer, running from about one bridge opening width upstream to one bridge opening width downstream of the bridge.

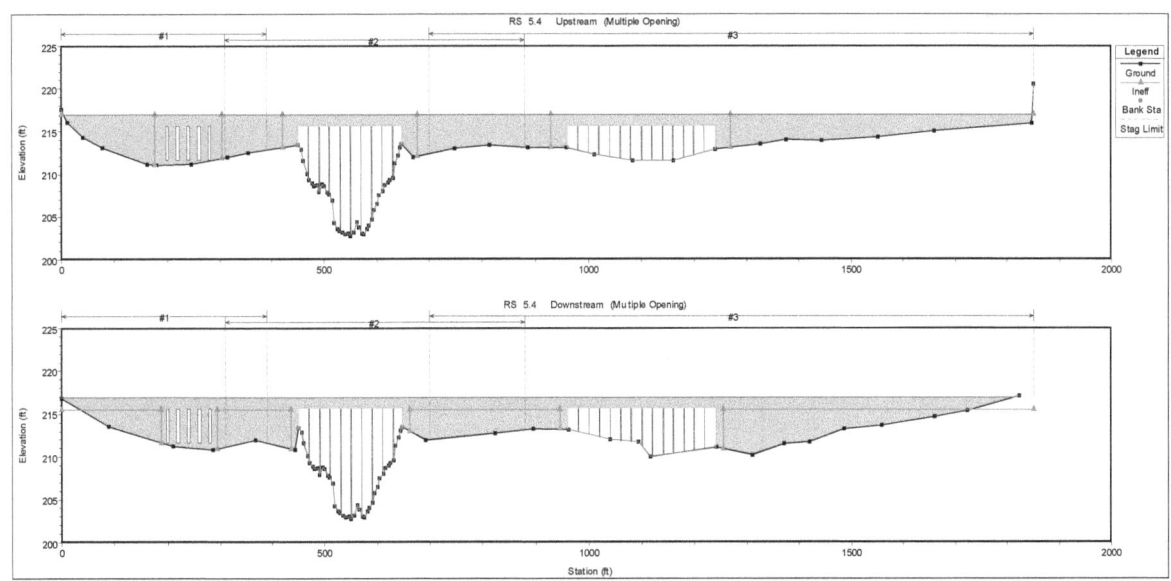

Figure 5.25. HEC-RAS multiple-opening example.

Like HEC-RAS, the WSPRO program performs iterations to determine the flow apportionment between openings. The iterations do not, however, attempt to converge on an equal upstream energy grade line. Instead, the program adjusts both flow apportionment and the stagnation limits at each iteration and computes the upstream water surface elevation for each valley strip. The program computes flow apportionment as a function of the relative opening size and the upstream conveyance in the valley strip. In each iteration, after the flow apportionment is calculated, the program computes the conveyance-weighted average upstream water surface of all of the strips. This value, along with the discharge apportionment, is compared to the results from the previous iteration. Adjustments and iterations continue until the results from two consecutive iterations match within a tolerance.

5.8.5 Lateral Weirs

Occasionally a raised embankment such as a levee, flood wall, railroad, or roadway defines an edge of a floodplain. If the flood profile is high enough, water will leave the floodplain by flowing over the embankment along the edge. The embankment in this case is functioning as a lateral weir. The flow in the floodplain decreases continuously in the downstream direction along the extent of the lateral overflow, as illustrated conceptually in Figure 5.26. In this example, flow is removed from the model in the region from cross section 7 to cross section 4 by flowing laterally over the road. The flow returns to the model at cross section 2.

The engineer should account for any significant loss of flow by lateral overtopping when performing one-dimensional analysis of a floodplain reach. HEC-RAS provides a lateral weir feature that automates the calculation of lateral overtopping flow. The engineer defines the lateral weir profile with respect to the floodplain model cross sections. The weir coefficient, C, must also be supplied as input. If the lateral weir feature is activated, HEC-RAS automatically computes the amount of lateral weir flow and removes the outgoing flow from the affected reaches as the water surface profile is being calculated. This is by necessity an iterative process, since the water surface profile that drives the lateral weir overflow is directly affected by the quantity of outflow.

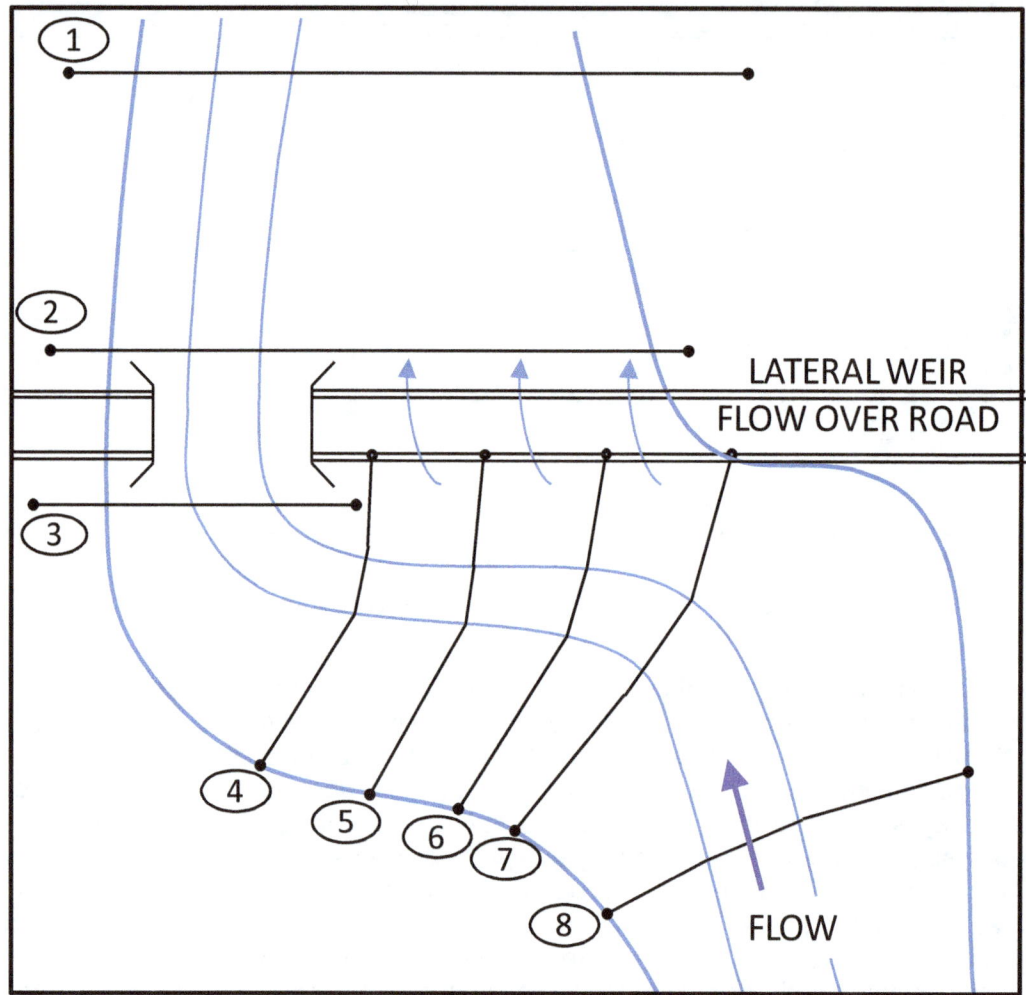

Figure 5.26. Illustration of a model incorporating lateral weir flow.

Lateral weir computations are more complex than those for a weir perpendicular to the floodplain flow. The head driving flow over the weir should be based on the water surface elevation rather than the energy grade line, because the flow's kinetic energy is generally parallel to the weir rather than over the weir crest. The variable water surface combined with the variable weir crest profile necessitated the derivation of a modified weir equation for HEC-RAS lateral weir computations (USACE 2010c):

$$Q_{x1-x2} = \frac{2C}{5a_1}[(a_1x_2 + C_1)^{5/2} - (a_1x_1 + C_1)^{5/2}] \qquad (5.27)$$

where:

Q_{x1-x2} = The weir overflow along the lateral weir segment, ft³/s (m³/s)

C = Weir discharge coefficient

a_1 = Slope of the water surface minus the slope of the weir crest along the direction of flow

x_1 = Location of the upstream end of the lateral weir segment, ft (m)

x_2 = Location of the downstream end of the lateral weir segment, ft (m)

C_1 = Height of the water surface above the weir crest at the upstream end of the lateral weir segment, ft (m)

The standard weir coefficients for a weir crest perpendicular to the floodplain flow are often inappropriate for lateral weir flow calculations because the momentum of the floodplain flow causes the overtopping flow to cross the weir crest at an oblique angle. HEC-RAS offers the option to use Hager's (1987) equation to compute the weir coefficient:

$$C = \frac{3}{5} C_0 \sqrt{g} \left[\frac{1-W}{3-2y-W} \right]^{0.5} \left\{ 1 - (\beta + S_0) \left[\frac{3(1-y)}{y-W} \right]^{0.5} \right\} \qquad (5.28)$$

where:

- C_0 = Base discharge coefficient, a function of the weir geometry
- W = The ratio of the height of the weir crest above the ground to the height of the energy grade line above the ground
- y = The ratio of the height of the water surface above the ground to the height of the energy grade line above the ground
- β = Main channel contraction angle in radians (zero if the weir is parallel to the main channel)
- S_0 = Average main channel bed slope

5.9 ONE-DIMENSIONAL MODELING ASSUMPTIONS AND LIMITATIONS

Flow in natural channels and floodplains is inherently three dimensional and unsteady. One-dimensional modeling requires simplifying assumptions to solve the equations of motion. It is the assumptions that establish the limitations of a particular numerical model. A model developer can derive better algorithms to solve the equations to improve the numerical model, but cannot fully overcome the limitations that are intrinsic to a specific approach.

The HEC-RAS Hydraulic Reference Manual (USACE 2010c) lists the steady state analysis program limitations as: (1) flow is steady, (2) flow is gradually varied, though alternative equations are used for some rapidly varied flow situations, (3) flow is one-dimensional, and (4) channel slopes are "small," generally less than 10 percent. Textbooks on open channel flow, (Chow 1959, Chaudhry 2008) expand on the assumptions related to one-dimensional gradually varied flow. The conditions of one dimensional, steady, gradually varied flow are: (1) that there is a single water surface at each cross section, (2) the flow is perpendicular to the cross section along its entire length, (3) the energy slope for the cross section applies to every point in the cross section, (4) hydrostatic pressure exists throughout the cross section, (5) channel slope is small, (6) energy slope is the same as for the corresponding normal depth, (7) the channel is prismatic with constant alignment and shape, and (8) roughness is constant through the reach. The first six conditions apply to individual cross sections and the last two apply to the reach. In practical applications these conditions will not exist, although the cross section conditions (1-6) are always applied within the computational framework of one-dimensional models.

Taking the textbook assumptions as absolute would preclude the use of one-dimensional models for many of the intended uses. For real-world application, the conditions can be interpreted such that sudden changes in alignment, shape or roughness are not fully simulated. Therefore, the hydraulic engineer should be aware of the limitations and use one-dimensional modeling when the accuracy of the results is not overly compromised. The following sections describe the assumptions and limitations of the one-dimensional modeling approach.

5.9.1 Water Surface, Velocity, and Cross Section Orientation

One-dimensional models use cross sections as the primary representation of the channel geometry. The other aspect of geometry is the distance between cross sections. A one-dimensional program treats each cross section as having a single water surface elevation and that flow is perpendicular to the cross section throughout its length. This is illustrated in Figure 5.27, which includes a plan view of a channel and floodplain, three cross section locations, and one cross section (A-A') showing ground, water surface, and energy elevation. The one-dimensional model computes a single water surface elevation that applies to the entire cross section and assumes all flow is perpendicular to the cross section.

The hydraulic engineer must select cross section locations and orientations that would have a consistent water surface elevation for the simulated flow condition. If the cross section cannot be reasonably oriented to achieve these conditions, then the numerical solution will differ from reality, potentially significantly. If the flow is uniform, these conditions are satisfied. The less uniform the flow, the more the model results will deviate from reality.

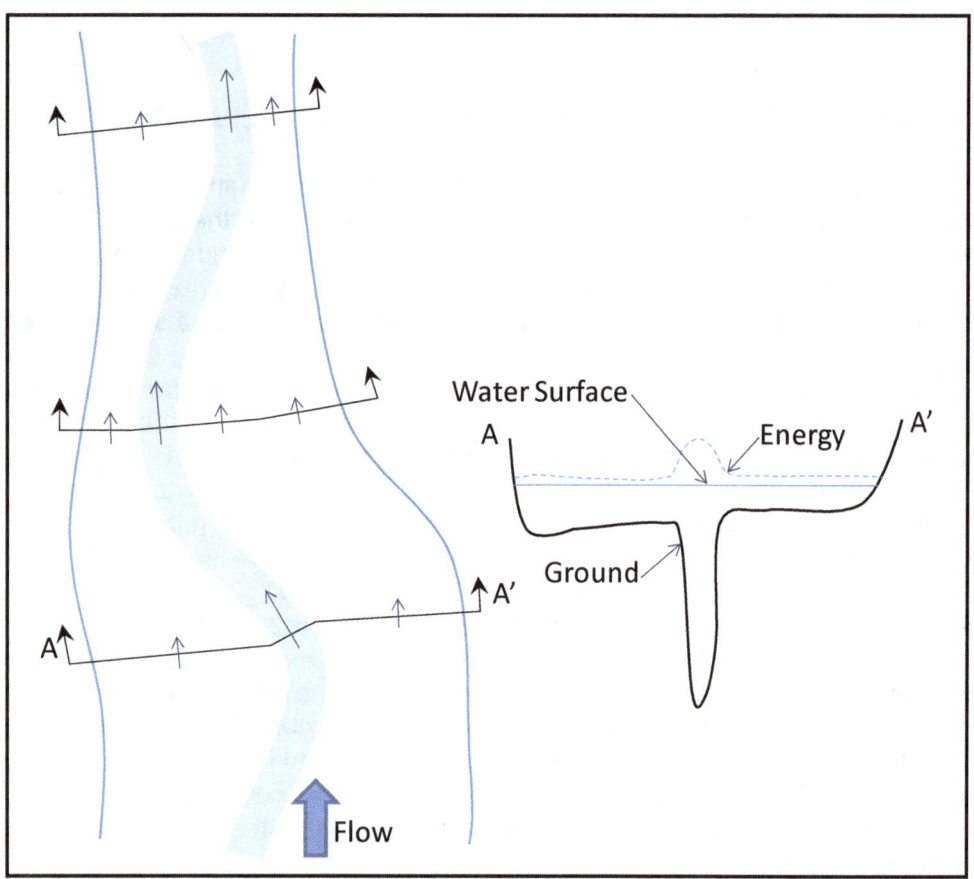

Figure 5.27. One-dimensional model cross sections.

5.9.2 Total Energy and Flow Distribution

Energy varies throughout the cross section because velocity is not constant throughout the cross section. Cross section A-A in Figure 5.27 shows local energy (WS + $V^2/2g$) computed from local water surface and velocity. Total energy is the local energy integrated for the entire cross section. Therefore, another assumption is that the total kinetic energy computed from $\frac{\alpha \bar{V}^2}{2g}$ is representative for the cross section. This implies that flow is distributed within the cross section proportional to conveyance. It also implies that every point within the cross section has the same energy slope. These are the assumptions that allow the program to estimate velocity and flow distribution between floodplain and channel, or between different locations within the floodplain or channel. The assumption is accurate for uniform flow and normal depth, but in locations of significant flow curvature the assumption breaks down.

5.9.3 Cross Section Spacing

One-dimensional numerical models assume that cross sections are spaced close enough that the numerical solution to the standard step equation is reasonable. They also assume that when flow distribution changes between cross section, that the flow can physically accomplish the redistribution within the allotted distance.

The standard step approach for computing water surface requires a calculation of head loss based on the friction slope between cross sections. The energy slopes at the two cross sections are used to compute the average friction slope. If the energy slopes are significantly different, the solution to the standard step method can be unreliable. Therefore, cross sections in one-dimensional models should not be spaced too far apart. Inserting additional surveyed cross sections or interpolating cross sections is recommended to avoid this problem.

Cross sections can also be too close together, resulting in physically impossible results. This is illustrated in Figure 5.28. Floodplains often contain abrupt changes in land use and hydraulic models must account for the variability of roughness to simulate flow conditions. Based on the assumptions described in Section 5.9.2, the one-dimensional solution would have significantly more flow in the right floodplain at cross section A-A' than in the right floodplain at B-B'. The opposite would be true in the left floodplain. Regardless of the distance between cross sections, the model would transfer the flow from one side to the other. In reality, the water would start shifting upstream of A-A' and complete the redistribution downstream of B-B'. Therefore the appropriate locations for these cross sections should be far enough up- and downstream of the vegetation break so that flow is reasonably redistributed. If a cross section is placed at the vegetation break, it would need to have an intermediate value of Manning n. However, because flow is actively redistributing at this location, the cross section would also need to be rotated to a better alignment with the flow direction and the realigned cross section would need to have a single water surface. Meeting all of these conditions is probably not possible.

The preceding discussion applies to any abrupt change in conveyance, not just from changing roughness. Any redistribution of flow caused by changing conveyance (area, depth, or roughness) should be reasonably possible within the distance between the cross sections. Otherwise, the redistribution is physically not possible. In summary, the cross section spacing should be small enough to avoid numerical inaccuracies and large enough to accurately represent flow redistribution.

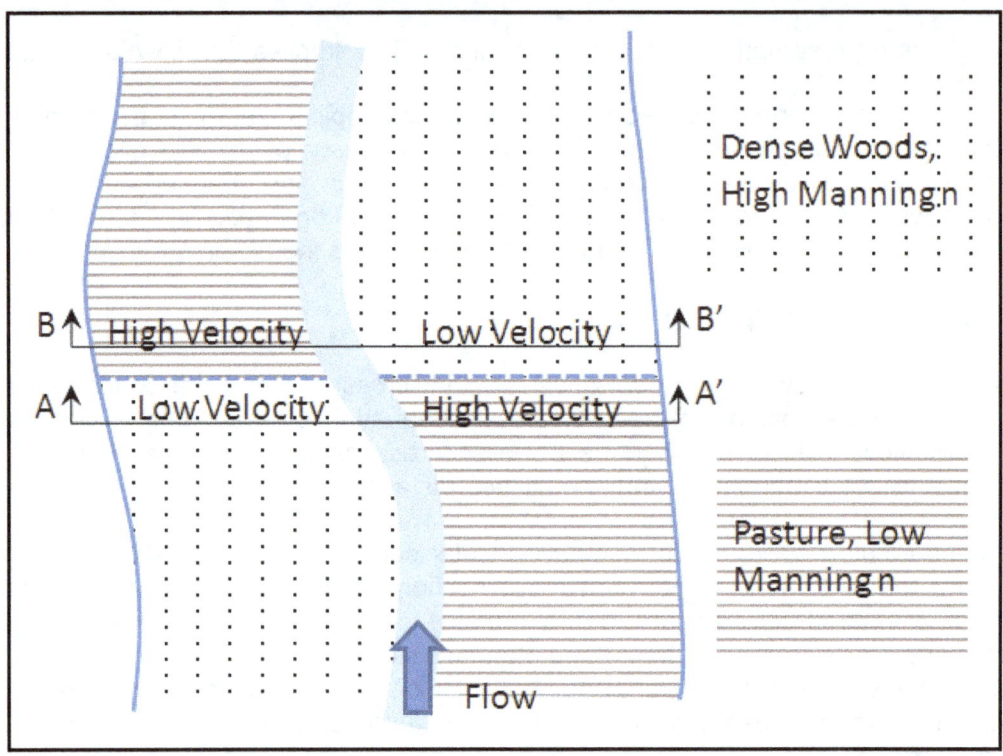

Figure 5.28. Cross section location at abrupt vegetation transitions.

5.9.4 Assumptions for Multiple Openings

The HEC-RAS Hydraulic Reference Manual provides detailed descriptions of the approaches used for calculating flow at multiple openings in an embankment. The two approaches are the multiple opening approach and the divided flow approach. The multiple opening approach apportions flow between the openings, which can include bridges, culverts, and conveyance paths. Apportioning the flow targets an equal energy at the upstream cross section, which is a condition that may well not exist. The HEC-RAS manual indicates that this method should not be used when water surface or energy vary significantly between the openings. Identifying the conditions when those variables would vary significantly is often very difficult.

The divided flow approach requires the hydraulic engineer to establish a separate reach for each opening. This requires that the flow paths must be readily apparent. The model apportions flow to the different reaches to establish an equal energy at the upstream junction. The divided-flow approach does allow for different water surfaces and energies immediately upstream of the bridge, but the results should be checked closely.

5.9.5 Cross-Section, Bridge, and Pier Skew

Cross sections, bridges and piers can be skewed to the flow within HEC-RAS. For bridges and cross sections, the cosine of the angle is used to adjust the distances along the cross section to compute flow area projected into the direction of the flow. This calculation is useful when the cross section is located along a skewed road embankment. However, the water surface is assumed to be level along the cross section, so large amounts of skew and skewed crossings at wide floodplains are unlikely to meet this requirement. Pier skew is used to calculate the projected blockage of the pier in the direction of flow. Flow does realign for long piers, so judgment should be used to avoid blocking too much of the bridge opening with skewed piers.

CHAPTER 6

TWO-DIMENSIONAL BRIDGE HYDRAULIC ANALYSIS

6.1 INTRODUCTION

Chapter 4 describes the differences between one-dimensional and two-dimensional hydraulic analyses. Most bridge hydraulic studies have used one-dimensional analysis methods that are described in detail in Chapter 5, though two-dimensional models are being used frequently, especially for complex situations. As the use of two-dimensional models becomes more commonplace, they will, inevitably, be used for all but the most straightforward bridge hydraulic conditions. This chapter provides information and guidance on the use of two-dimensional models for bridge hydraulic analysis.

In one-dimensional modeling, the standard step solution of the energy equation is most frequently used for hydraulic analysis. For two-dimensional modeling, the momentum equation (Newton's second law of motion, F = ma) is applied to a control volume in conjunction with the continuity equation. Figure 6.1 illustrates a control volume of flowing water in three-dimensional space and includes the primary forces acting on the control volume in two-dimensions. The calculated variables of velocity and depth are also shown. In two-dimensional models, vertical velocity components are considered as negligible and hydrostatic pressure is assumed. Velocity is a vector quantity that can be expressed as a magnitude and direction or as the x and y velocity components U (x direction) and V (y direction). The elevation of the bed (Z_b) and water depth (H) vary over the area. The force variables shown in Figure 6.1 are pressure (P) at the control volume horizontal surfaces, water weight (W), bed shear stress components (τ_b), and water surface shear stress (τ_s). For a set of unbalanced forces, the mass associated with the control volume will accelerate. As will be discussed in the next section, these variables are the primary set of forces acting on the control volume, though others are included depending on the problem.

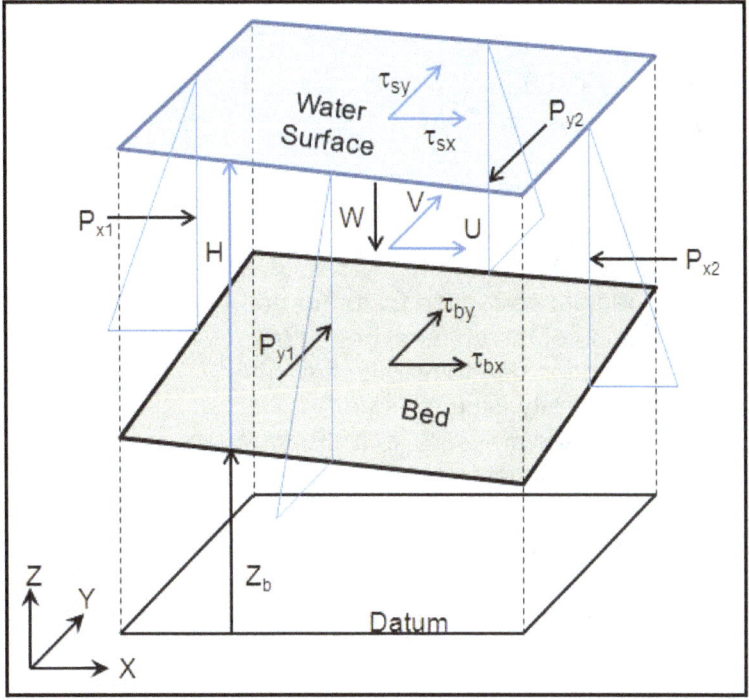

Figure 6.1. Three-dimensional coordinate system and two-dimensional hydraulic variables.

6.2 GOVERNING EQUATIONS

Although the representations of the variables may differ between the various two-dimensional modeling programs, and some variables may not be included in all the programs, the conservation of mass and momentum are used as the basis for hydraulic calculation in two-dimensional models. The following equations are presented for the FST2DH model (FHWA 2003). The conservation of mass (continuity equation) in two dimensions is:

$$\frac{\partial Z_w}{\partial t} + \frac{\partial q_x}{\partial x} + \frac{\partial q_y}{\partial y} = q_m \tag{6.1}$$

where:

Z_w = Elevation of the water surface, ft (m)
q_x = Unit discharge in the x direction (UH), ft²/s (m²/s)
q_y = Unit discharge in the y direction (VH), ft²/s (m²/s)
q_m = Inflow per unit area, ft/s (m/s)

The conservation of momentum equations in the x and y directions are:

$$\frac{\partial q_x}{\partial t} + \frac{\partial}{\partial x}\left(\beta \frac{q_x^2}{H} + \frac{g}{2}H^2\right) + \frac{\partial}{\partial y}\left(\beta \frac{q_x q_y}{H}\right) + gH\frac{\partial Z_b}{\partial x} + \frac{H}{\rho}\frac{\partial P_a}{\partial x} - \Omega q_y$$
$$+ \frac{1}{\rho}\left[\tau_{bx} - \tau_{sx} - \frac{\partial(H\tau_{xx})}{\partial x} - \frac{\partial(H\tau_{xy})}{\partial y}\right] = 0 \tag{6.2}$$

$$\frac{\partial q_y}{\partial t} + \frac{\partial}{\partial y}\left(\beta \frac{q_y^2}{H} + \frac{g}{2}H^2\right) + \frac{\partial}{\partial x}\left(\beta \frac{q_y q_x}{H}\right) + gH\frac{\partial Z_b}{\partial y} + \frac{H}{\rho}\frac{\partial P_a}{\partial y} + \Omega q_x$$
$$+ \frac{1}{\rho}\left[\tau_{by} - \tau_{sy} - \frac{\partial(H\tau_{yy})}{\partial y} - \frac{\partial(H\tau_{yx})}{\partial x}\right] = 0 \tag{6.3}$$

where:

H = Water depth, ft (m)
β = Momentum correction factor for non-uniform vertical velocity profile
Z_b = Elevation of the channel bed, ft (m)
g = Acceleration due to gravity, ft/s² (m/s²)
ρ = Water density, slug/ft³ (kg/m³)
P_a = Atmospheric pressure, lb/ft² (Pa)
Ω = Coriolis parameter, (1/s)
τ_{bx}, τ_{by} = Bed shear stress in the x and y directions, lb/ft² (Pa)
τ_{sx}, τ_{sy} = Water surface shear stress in the x and y directions, lb/ft² (Pa)
τ_{xx}, τ_{xy} = x-direction shear stresses due turbulence, lb/ft² (Pa)
τ_{yy}, τ_{yx} = y-direction shear stresses due turbulence, lb/ft² (Pa)

The terms of Equations 6.1 and 6.2 are mass times acceleration or force terms in Newton's second law of motion, ΣF = ma, as shown in Figure 6.2. Figure 6.2 is the x direction (Equation 6.2) and there are equivalent terms in the y direction equation. The equation shown in Figure 6.2 has been rearranged and multiplied by mass (ρ) to clearly indicate mass times acceleration terms versus force terms. The acceleration terms are local acceleration (time) and convective acceleration (converging or diverging stream lines). The force terms include changing hydrostatic pressure due to changing flow depth, the component of water weight acting on the sloping bed, atmospheric pressure gradient, bed shear stress, water surface shear stress due to wind, shear stresses caused by turbulence, and the pseudo force term due to the Coriolis Effect. The bed shear stress is evaluated using the Manning or Chezy relationships, though Manning is more commonly used. Surface shear is related to wind speed. Turbulence shear relates to turbulence exchange and horizontal diffusion of momentum. The Coriolis Effect is due to the fact that the model represents an area on the rotating earth. There is an apparent force which is caused by the resulting angular acceleration. Some of these forces are negligible in many river applications, including Coriolis, surface shear, and atmospheric pressure gradient, which apply most often to large bodies of water in tidal applications. The solution techniques for Equations 6.1 through 6.3 include the application of the finite difference method and finite element method.

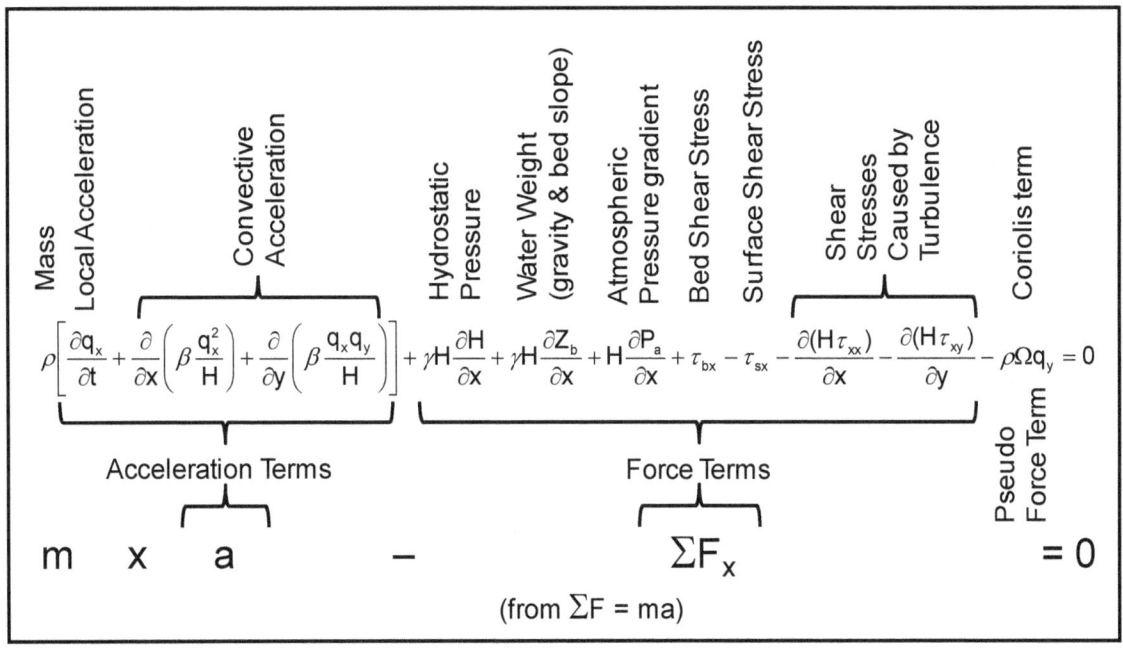

Figure 6.2. Terms in two-dimensional momentum transport equations, x direction.

6.3 TYPES OF TWO-DIMENSIONAL MODELS

6.3.1 Finite Element Method

The finite element method is well suited for solving differential equations over complex domains, which is why this manual recommends it for use in bridge hydraulics. The finite element method uses an unstructured mesh or grid and solves Equations 6.1 through 6.3 through numerical integration techniques for each element. Each element consists of nodes located at the corners and mid-sides where velocity and depth are computed. Elements are typically triangular or quadrilateral in shape, but curved sides are also possible by placing the mid-side nodes out of alignment with the corner nodes. The FST2DH model (FHWA 2003) also includes a quadrilateral element shape with a center node in addition to the mid-side and corner nodes. Figure 6.3 illustrates a variety of element types and shapes. The element sides do not need to align with the x- or y- directions, they do not need to have consistent size or orientation, and a mixture of triangular and quadrilateral elements are allowed. The unstructured mesh forms the geometric framework for the hydraulic computations.

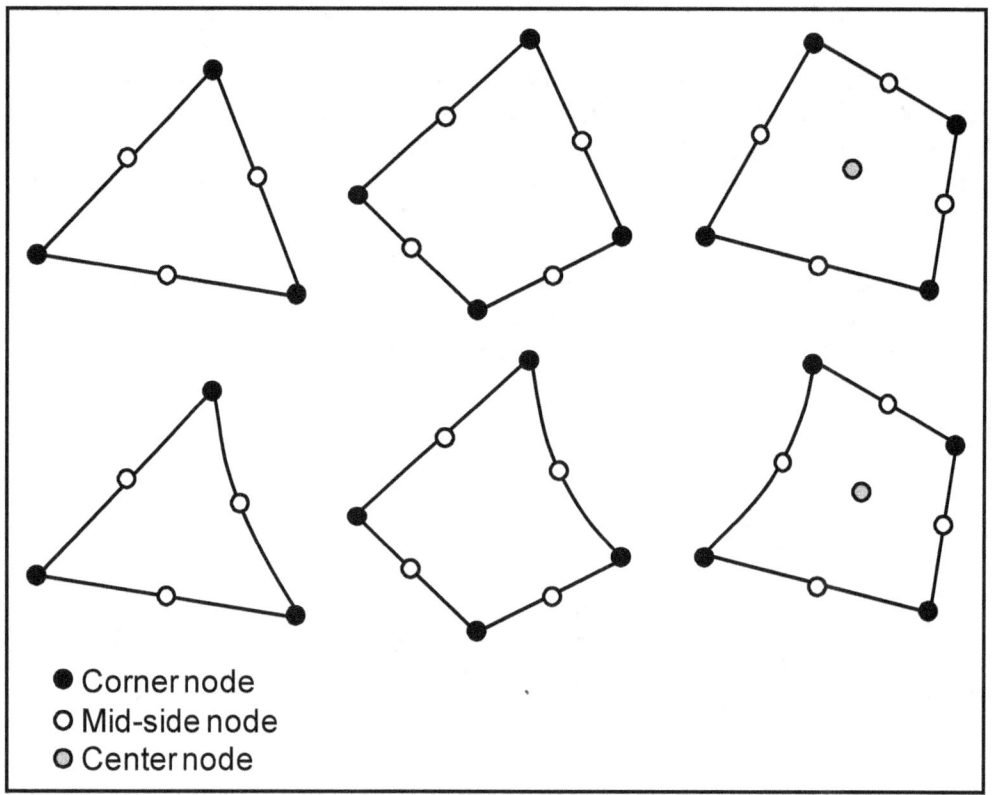

Figure 6.3. Element types and shapes.

Figure 6.4 is an example of a mesh layout using triangular and quadrilateral elements. The elements are arranged based on several criteria, which include:

- Accurate representation of the topography and bathymetry
- Accurate representation of land use and roughness variability
- Increased detail in areas of high velocity gradients (change in magnitude or direction)

Topography and bathymetry are represented by assigning elevations to the nodes. Land use is represented by varying roughness conditions (Manning n) by assigning material types to the elements. Figure 6.4 illustrates how triangular elements can be used to transition from large to small elements and to represent curved features. This allows for areas with greater topographic, land use and velocity variability to have greater detail. The velocity vector (x- and y- components) and flow depth are computed at each corner, mid-side, and center node.

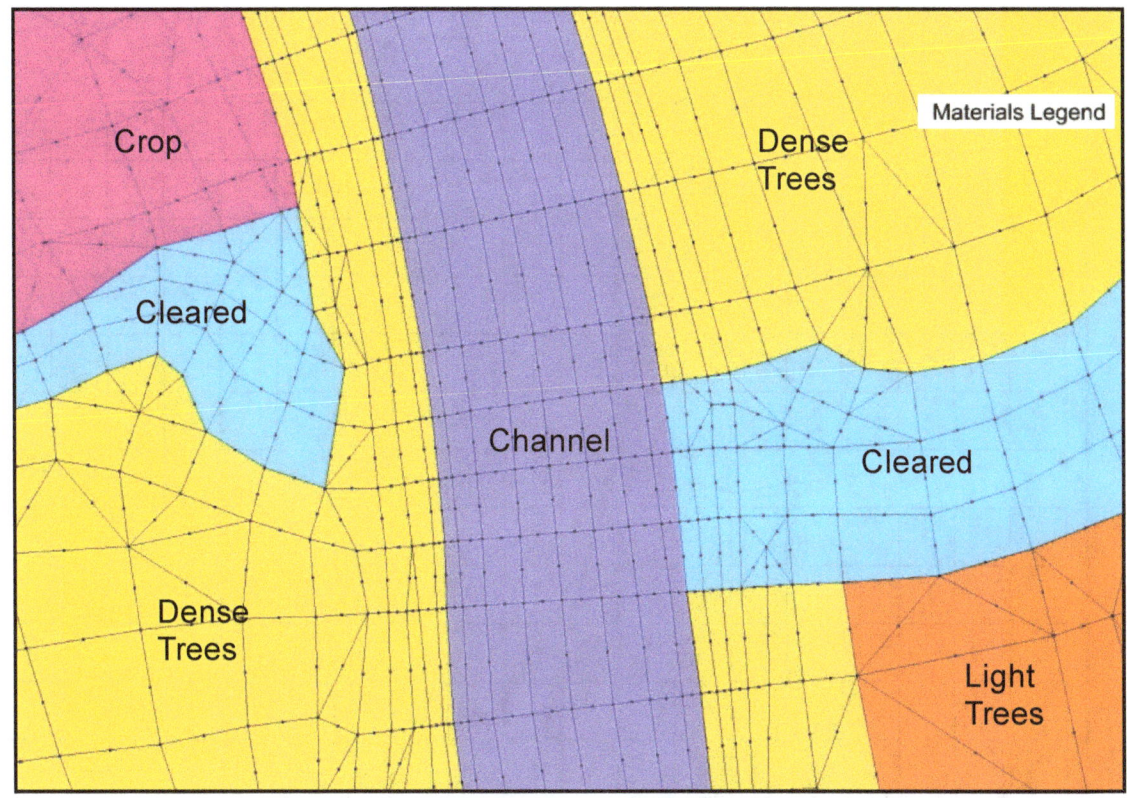

Figure 6.4. Example finite element network layout.

6.3.2 Finite Difference Method

The finite difference method is a numerical solution technique for differential equations that approximates derivatives with finite differences. The finite difference method for a two-dimensional model uses a structured grid of Δx and Δy increments over the domain. The values of velocity, depth and other variables are computed for each grid cell and the differences are used to approximate the derivatives ($\partial V/\partial x \approx \Delta V/\Delta x$). Because the grid is structured, solutions tend to be more rapid, though a large number of differential elements are often required in areas where there is little change in either input data or results. Figure 6.5 shows an illustration of an enhanced finite difference network. The network includes the finite difference cells, cells that have been disabled to represent a road embankment, a one-dimensional channel network that has replaced a number of the cells, and a one-dimensional culvert connection across the embankment. With suitably sized grid cells, even highly variable terrain can be well represented in a finite difference model, and a channel can be represented directly in the network without needing to use a one-dimensional representation. It is possible in some finite difference models to use nested grids where a finer grid is inserted into the coarse grid that covers the entire domain. Grid cell distances may need to be very small to accurately represent the flow field within a bridge opening. These small distances may be difficult to accommodate in a finite difference model when consistent size over the entire domain is required.

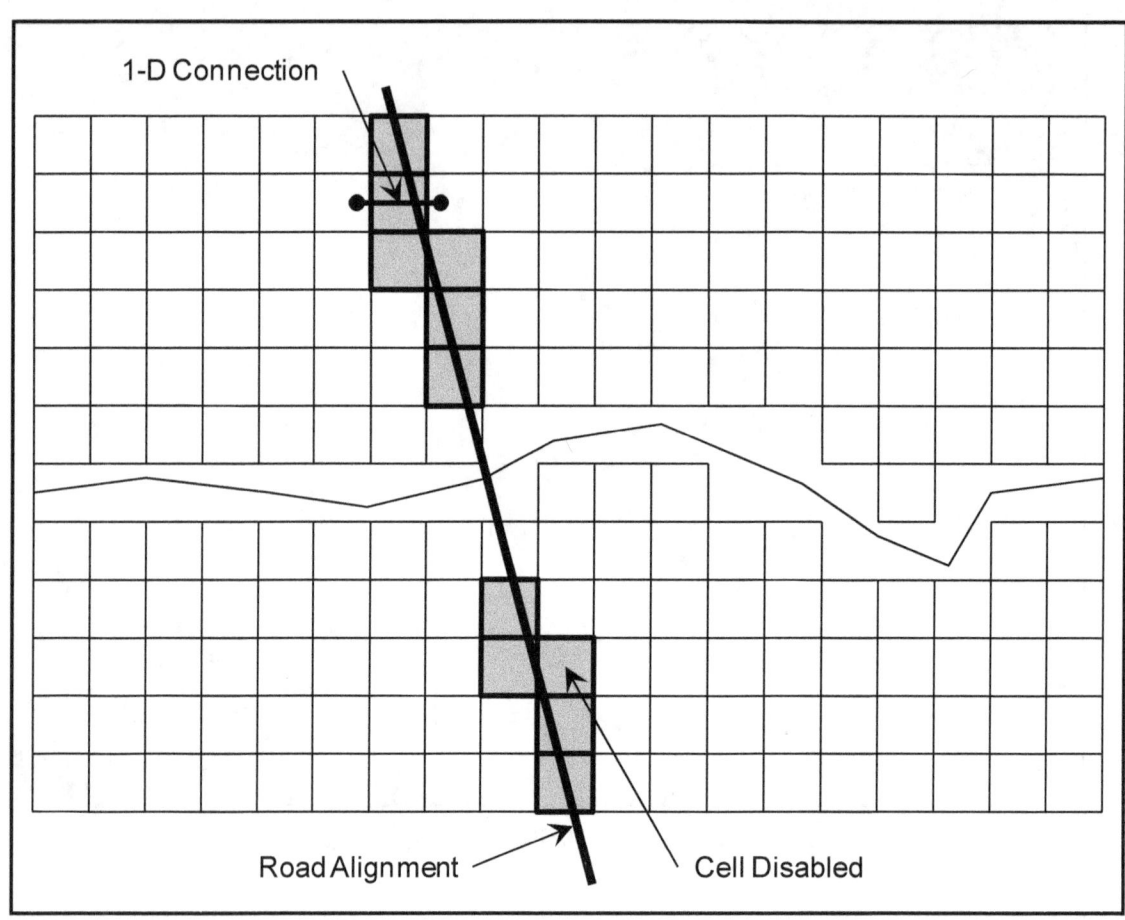

Figure 6.5. Example finite difference network.

6.4 GEOMETRIC REQUIREMENTS AND MESH QUALITY

6.4.1 Geometric Requirements

RMA2 (USACE 2009) a two-dimensional finite element model is very similar to FST2DH, but does not include some features that are desirable for bridge hydraulics. The RMA2 manual includes discussion of the importance of various aspects of two-dimensional models as they relate to the accuracy of simulation results. The discussion indicates that the most important aspects (60%) of model development are geometry and study design, which includes model extent. Boundary conditions are considered the second most important aspect (20%) followed by roughness (10%), eddy viscosity (6%), and "other" (4%). The amounts are intended only as a gauge of importance, but intuitively geometry must be accurate in order to develop an accurate result. If the ground elevation is incorrect, then one or more of the dependent variables (velocity, depth, and water surface elevation) must also be incorrect. Therefore, developing accurate geometry is fundamental to good modeling practice.

Figure 6.6 is a perspective plot showing a finite element mesh (black lines) with ground surface contours (white lines) and the ground surface elevation shaded with dark shading representing the lower elevation. The areas without elements are embankments that are above the expected water surface elevation and form a closed boundary where water flows along the element sides. The surface created by the elements should accurately simulate the three-dimensional surface of the channel bed and floodplain surfaces.

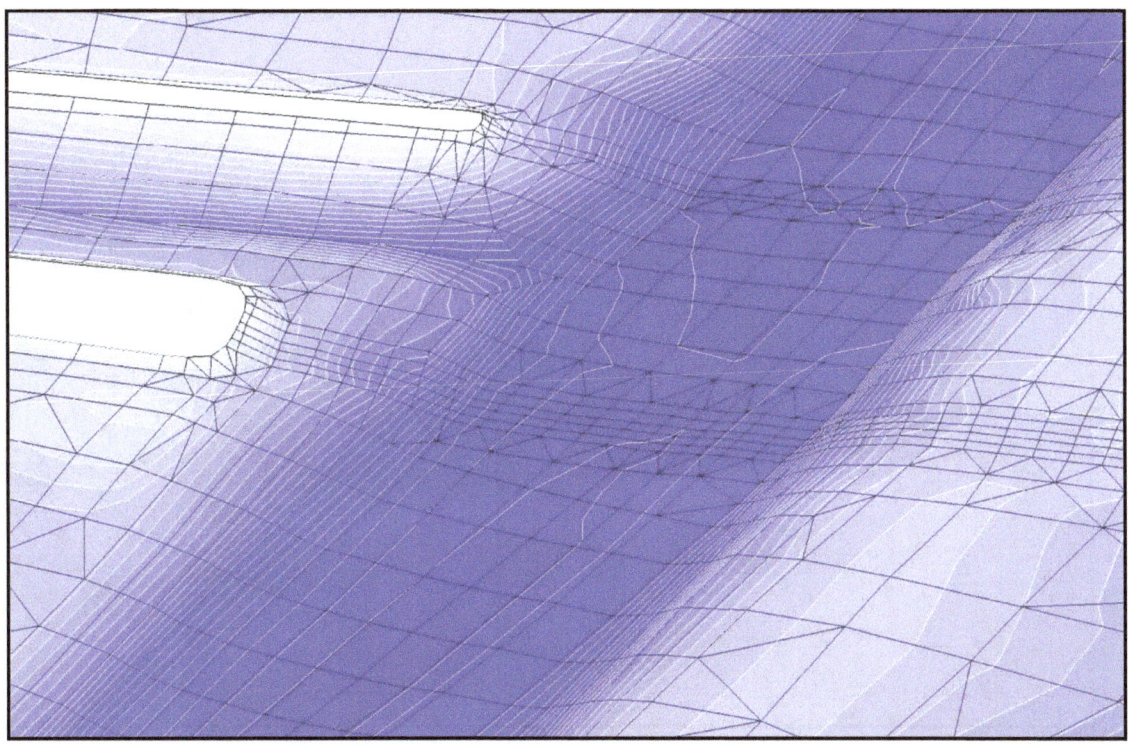

Figure 6.6. Example surface of finite element network.

Figure 6.7 shows a portion of a model domain that illustrates how topographic and bathymetric geometry, as well as model boundary conditions and the closed boundary combine to geometrically define the model domain. The contours of the model should reasonably replicate the contours of a detailed topographic map of the area. The closed boundary should enclose the area that can be inundated by the simulated flows. This model includes a navigation lock and dam as the upstream boundary (see Figure 4.3 for reference). The primary and auxiliary lock chambers at the bottom of the domain are defined by long walls and the three gates are separated by short walls. As indicated by the flow rates (arrows at the left end of the domain), during an extreme flood the three gates would convey the greatest flow and the two lock chambers would be used to discharge flow from upstream. A small amount of flow would bypass the structures and is shown as an inflow boundary at the top of the domain. The model could be used to evaluate potential impacts of other operational scenarios, such as more or less water being conveyed through the lock chambers.

Elements with curved sides are often used along closed boundaries. This improves the appearance of the mesh, but the main benefit is that a well-constructed curved boundary allows velocity vectors to be tangent to the mesh boundary. This is illustrated in Figure 6.8 where the boundary is formed with straight-sided elements on the left group and curved elements on the right group. The velocity vectors are shown at each of the nodes. The two vectors shown in heavy lines are not tangent to the mesh boundary nodes where straight elements sides are used. All the vectors are tangent to the mesh boundary nodes for the curved element side. Small "leaks" in the mesh occur for the left mesh and are avoided in the right mesh. The finite element mesh may not exactly maintain continuity due to these types of leaks. As a mesh quality review, continuity lines should be placed throughout the mesh to evaluate whether too much water is lost or gained in the system. Ideally, the total flow should be maintained, but a one or two percent discrepancy is acceptable. If there is excessive discrepancy in the flow rate, additional mesh refinement is needed.

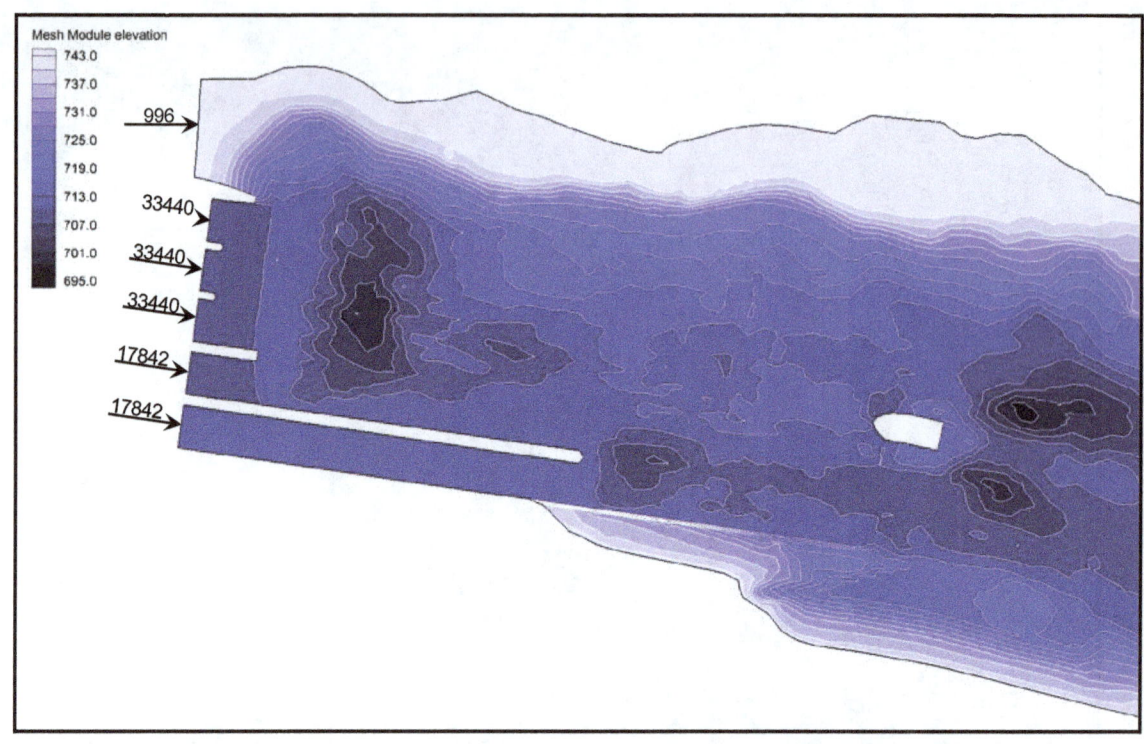

Figure 6.7. Topography and boundary of a finite element network.

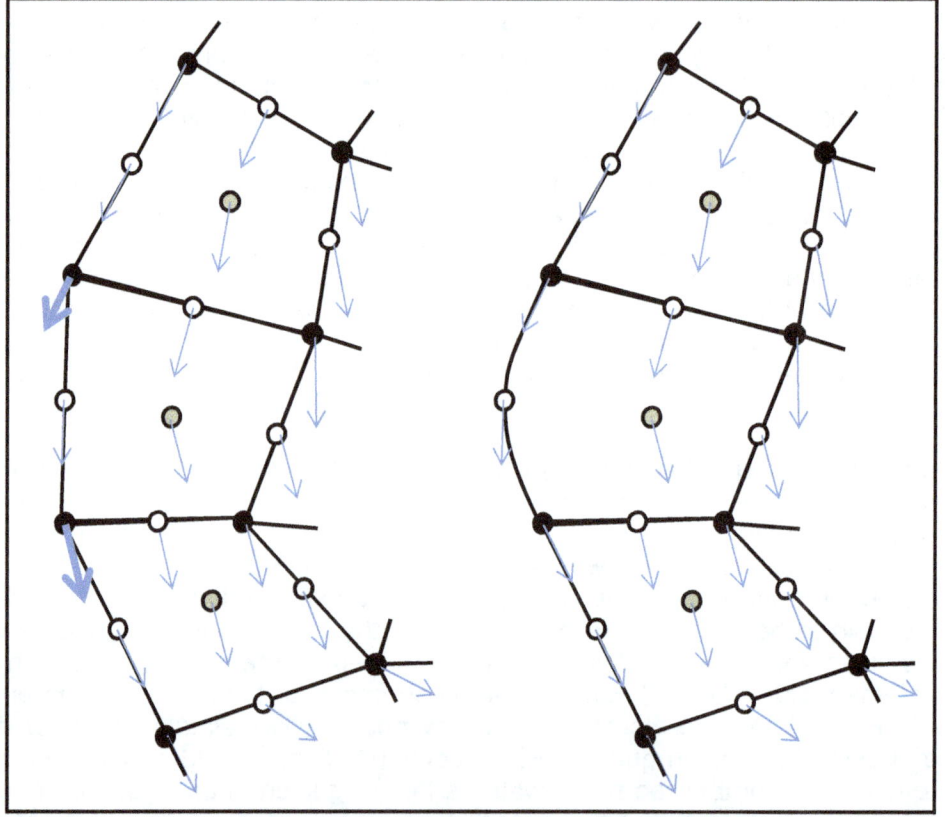

Figure 6.8. Illustration of curved mesh boundary.

6.4.2 Mesh Quality

In addition to accurate geometric representation, other aspects of mesh quality are important for accurate simulations. As discussed in Chapter 5, one-dimensional models should not have cross sections that are too widely spaced or have large changes in conveyance. In two-dimensional models, mesh refinement is required in areas of significant topographic or hydraulic variability. Hydraulic variability includes rapidly changing velocity magnitude, velocity direction, or flow depth. Figure 6.9 illustrates several other types of mesh quality considerations. These include oddly shaped elements, limits on interior angles, element aspect ratio, area change between elements, number of elements connected to a single node, and ambiguous gradients. The SMS (Surface-Water Modeling System, Aquaveo 2011) software for developing input and reviewing output for several two-dimensional models, including FST2DH and RMA2, includes tools for checking the mesh quality attributes shown in Figure 6.9.

Interior Angle. Figure 6.9 (a) shows triangular and quadrilateral elements that have interior angles that are too small or too great. Angles less than the 10 degree lower limit or more than the 130 degree upper limit should be avoided. The quadrilateral element also has an odd element shape with two long sides and two short sides.

Aspect Ratio. Figure 6.9 (b) shows elements with large aspect ratios, defined by element length divided by element width. These elements have aspect ratios of 15. The FST2DH manual recommends that aspect ratio be less than 12.5 and the RMA2 manual recommends 10 as an upper limit. However, a 10 degree interior angle on a triangle results in an aspect ratio of 5.7. Therefore, this manual recommends 5.7 as an upper limit of aspect ratio. It is easy to check the aspect ratio of a triangle or quadrilateral by measuring the length and width. The aspect ratios of triangles in an entire mesh can be checked by checking the interior angles with the angle computed as $\tan^{-1}(1/\text{Aspect Ratio})$. Using this approach, the aspect ratio of a quadrilateral can be checked using this approach by checking the interior angles of the quadrilateral split into triangles.

Maximum Area Change. Figure 6.9 (c) shows elements with large area changes compared to neighboring elements. The maximum recommended area change between elements is a 2:1 (0.5) ratio; twice or half as large. The velocity gradient can change rapidly over a small element but not for a large element. Therefore, values of derivatives may have a significant discontinuity between elements when the area change is too large and this can result in numerical instability and a poor representation of the flow field.

Number of Elements Connected to a Node. Figure 6.9 (d) illustrates too many elements connected to a single node. When too many elements connect to a node, too much weight is associated with the values of depth and velocity at that point in space. The maximum number of elements connected to a node should be no more than seven.

Ambiguous Gradients. Elements should be generally planar in elevation. When a quadrilateral is not relatively planar, it should be divided into two triangular elements. As illustrated in Figure 6.9 (e), when a quadrilateral has opposite corners that are higher or lower than both of the other corners, the topography of the element is ambiguous. For the example shown, the corner nodes and contours indicate that there is a saddle shape, but there is probably either a trough or ridge between the corners. The surrounding topography would clearly indicate which topographic feature is present. If the saddle is the true shape, then the quadrilateral should be divided into eight triangular elements.

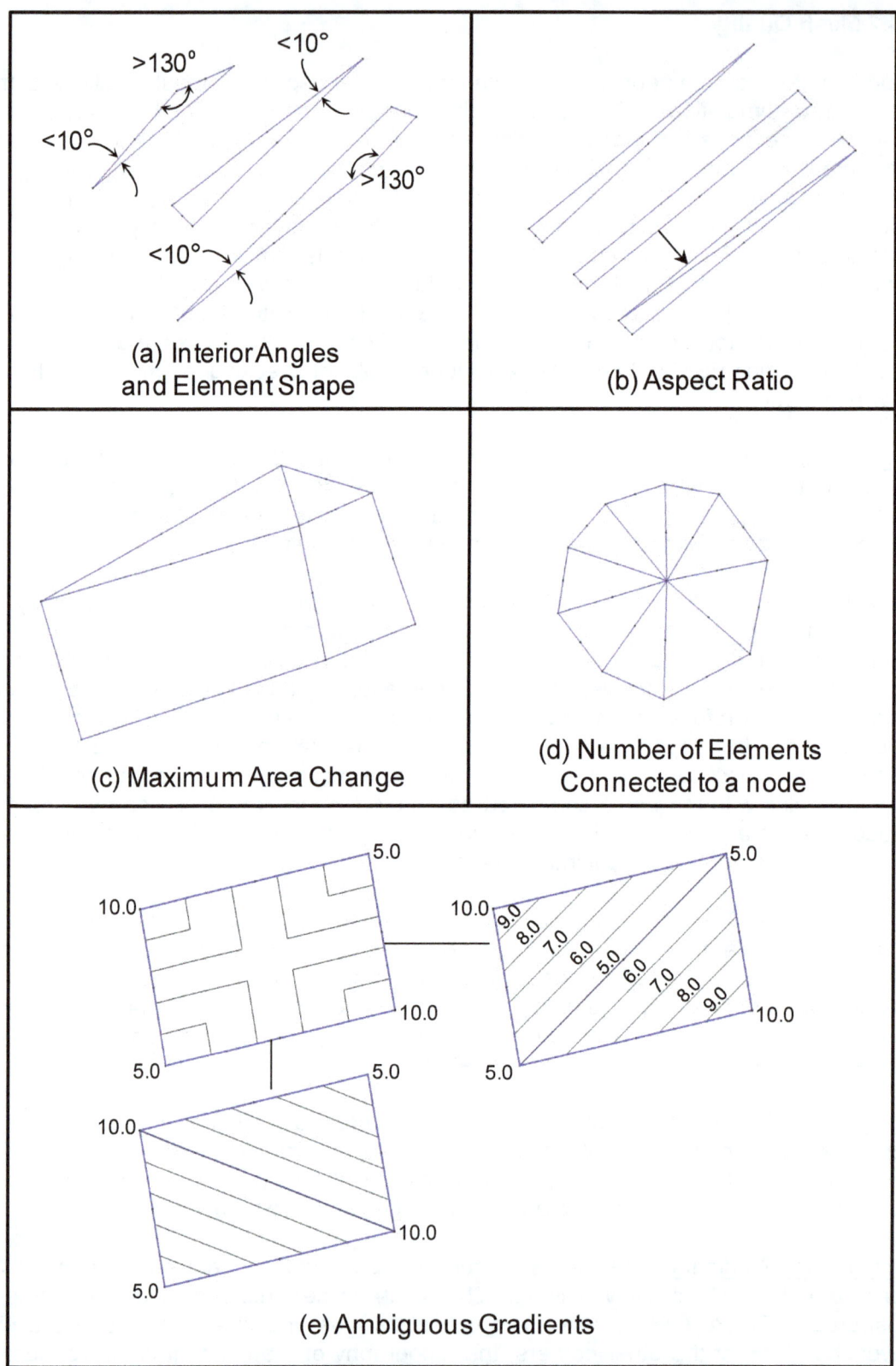

Figure 6.9. Finite element mesh quality considerations.

6.5 BRIDGE MODELING APPROACHES

For many bridges the primary hydraulic effect is the contraction and expansion of flow due to the constriction of flow by the roadway approach embankments. Two-dimensional models are ideally suited to compute the flow conditions and backwater associated with the expansion and contraction. Bridge hydraulics may also require analysis of pier and abutment obstruction, debris obstruction, submerged deck flow, road overtopping, bridge overtopping, and relief structures. Methods for analyzing these additional components are discussed in this section.

6.5.1 Pier Losses

There are three methods that can be used to include pier losses in two-dimensional models. For large piers, the pier layout can be incorporated directly into the mesh. For small piers, the drag force of the pier can be included as an additional force in the equations of motion for an element. The third approach is to make an adjustment to the Manning n of the bridge elements to account for the additional forces caused by pier drag.

<u>Disabled Element Approach.</u> For very large piers, it is often desirable to include the pier directly into the finite element network. This is illustrated in Figure 6.10. One pier is included by leaving a void in the mesh and the other pier is included by disabling the elements. In either case, the model will exclude flow from these areas by treating them as blocked. Flow that is approaching either of these piers must accelerate around the piers as it would in the prototype condition. Because velocity is depth averaged, vertical velocities are not included, and hydrostatic pressure is assumed, the complete flow field is not reproduced. A three-dimensional model or CFD model would be required to simulate the complete flow field.

The mesh is greatly refined in the areas adjacent to the piers because there are large velocity gradients around these obstructions. The disabled elements were used for the upper pier because it was part of a bridge design and the void was used because it is an existing pier. The disabled elements were assigned appropriate element types for the "no bridge" condition so the models with and without the new pier could be compared directly by maintaining the identical element pattern.

<u>Additional Force Method.</u> Because the equations of motion include the forces acting on a control volume defined by the element, the pier drag force can be included as an additional force in Equations 6.2 and 6.3. As illustrated in Figure 6.11, the FST2DH model (FHWA 2003) includes this option. The force is applied to the element as a whole so there is no obstruction in the element, nor is the force applied to a specific location in the element. The hydraulic engineer specifies the pier location (model x, y coordinate), dimensions (length and width), orientation, and drag coefficient. Multiple piers can be included in a single element. Because the pier dimensions and orientation are included, the projected area in the direction of flow is calculated. The flow is not deflected around the obstruction as with the disabled element approach, though for large piers in small elements there will be some flow redistribution around the element. If the computed drag force is large and the element size is small, the element may have a much reduced velocity and significant flow redistribution will occur. This condition may actually indicate that the drag force is underestimated for that pier because the velocity in the element is used for the calculation rather than the approach velocity upstream of the pier. The reasonableness of the result can be checked by computing the drag force using the true approach velocity and comparing the force with the force reported in the FST2DH output. If there is a large discrepancy, the model force can be increased by using an artificially large drag coefficient.

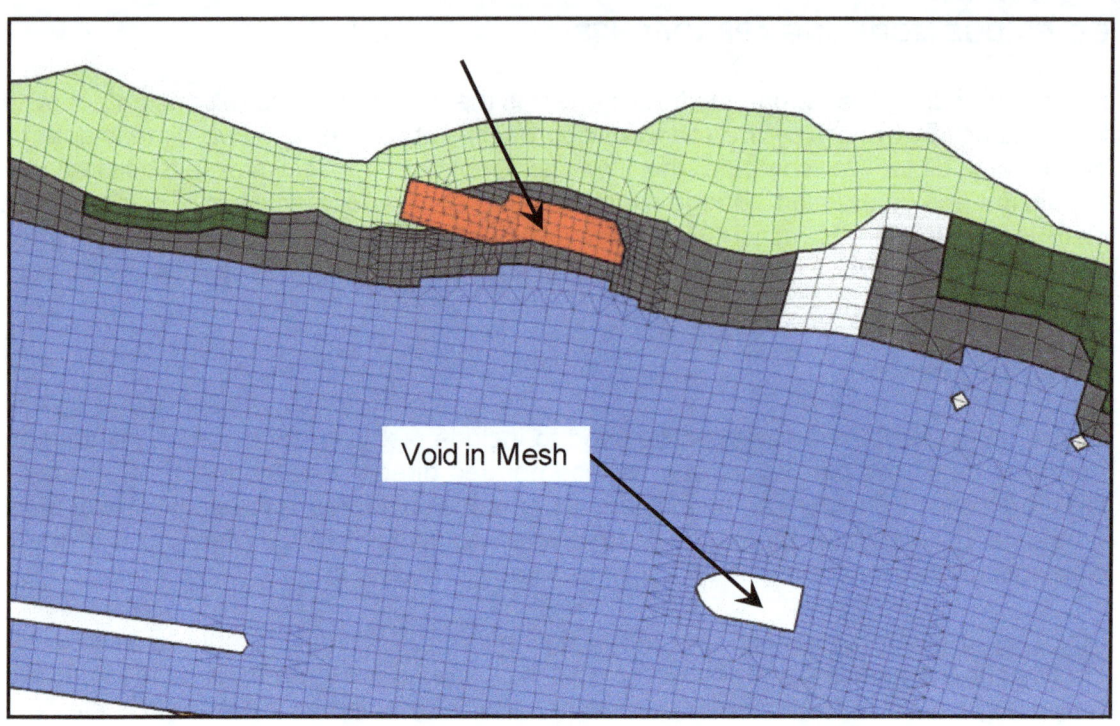

Figure 6.10. Disabled element approach for large piers.

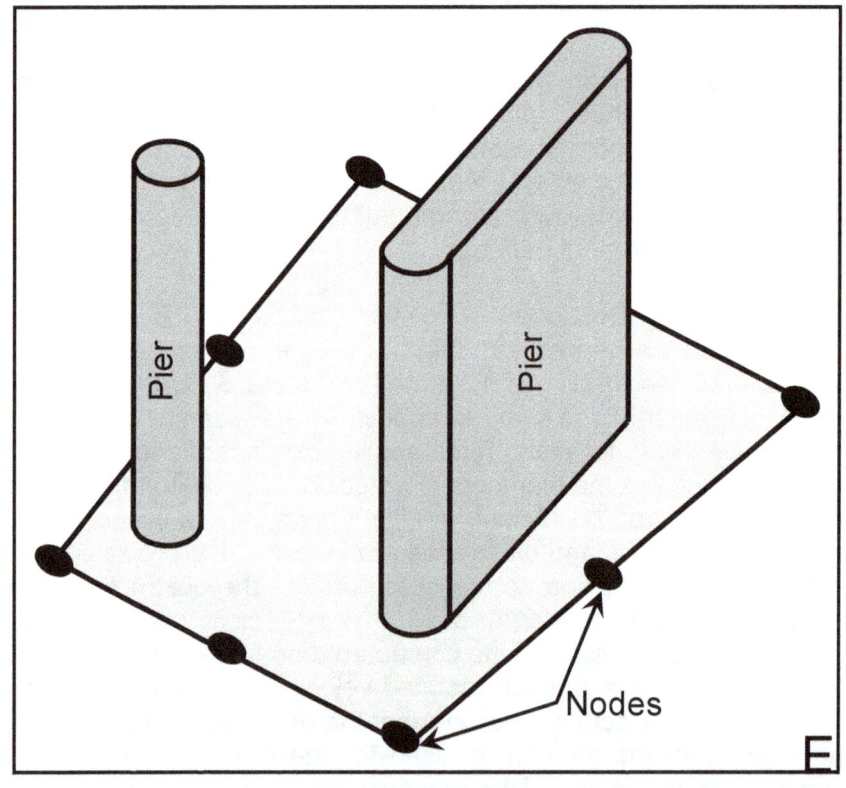

Figure 6.11. Pier drag force.

Increased Flow Resistance Method. Another approach to including the pier drag force is to increase the Manning n of the element. The force caused by pier drag is equated to an increase of shear stress so the total force is equivalent. The hydraulic engineer needs to compute the area of piers in an element or group of elements and apply the following equation to estimate the effective Manning n (n_e) that accounts for the additional force. One disadvantage of this approach is that angle of attack is not directly accounted for in the model but needs to be included when determining the pier width. The advantage of this approach is that the drag from a large number of piers can be incorporated into the model by adjusting the Manning n of some or all of the bridge elements. As with the additional force method, an obstruction is not modeled and flow redistribution may not occur. The effective Manning n is computed from:

$$n_e = \sqrt{n^2 + \frac{K_u^2 y^{1/3} \sum C_d A_p}{2g \sum A_E}} \tag{6.4}$$

where:

n_e = Effective Manning n of the element or group of elements
n = Manning n of the bed if no pier were present
C_d = Pier drag coefficient
A_p = Projected area of the pier in the direction of flow, ft² (m²)
A_E = Area of an element or group of elements in plan, ft² (m²)
y = Average flow depth, ft (m)
g = Gravitational constant, ft/s² (m/s²)
K_u = 1.486 for U.S. customary units and 1.0 for SI

6.5.2 Pressure Flow

Pressure flow occurs when the deck of the bridge is submerged. Flow depth becomes constrained by the deck so velocity must increase through the bridge opening. The two methods that can be used in FST2DH to simulate pressure flow are described below.

Capped Element Method. FST2DH is able to compute pressure flow for bridges directly within the model. The hydraulic engineer assigns low-chord elevations to the nodes that comprise the elements at a bridge. The continuity and momentum equations are adjusted to account for the constraint that the bridge deck imposes on flow depth, pressure head is used in place of flow depth for terms not associated with velocity, and an additional shear stress is used for flow in contact with the deck. Although this method accounts for many of the processes associated with pressure flow, it does not account for flow separation at the leading edge of the deck as would be associated with orifice and submerged orifice flow.

Increased Flow Resistance Method. The results of the capped element method can be compared with hand calculations of the orifice or submerged-orifice flow equations presented in Chapter 5. If the results differ significantly and the hydraulic engineer is more comfortable with the results of the equations, then the Manning n for the bridge elements can be manually adjusted until the results are consistent.

6.5.3 Weir Flow and Road Overtopping

Road overtopping is a common condition at highways during major floods. Vertical flow acceleration is often large for road overtopping conditions, so modeling the road geometrically in the finite element network is usually not recommended. Only when the embankment is low in comparison to the flow depth and vertical accelerations are negligible can the embankment be simulated geometrically.

In the FST2DH model, the weir equation is applied between two nodes and the computed flow is removed from the model at one node and entered back into the model at another node. This is illustrated in Figure 6.12. The road embankment is not included geometrically in the network and the roadway can either be a void in the network or included as disabled elements. The weir connection includes the elevation of the weir so the affects of the embankment are included. The nodes should be placed at the toe of embankment, but flow will not occur between the nodes until the weir elevation is exceeded. The weir coefficient and segment length along the roadway are also assigned to the node. This length is 2/3rd of the element side length for mid-side nodes and 1/6th of the adjoining element side lengths for corner nodes. The model should be set up with a one-to-one correspondence, including element side lengths, of nodes and elements across the embankment. Free flow and tail-water submerged flow can be simulated.

This approach to roadway overtopping is more accurate than in one-dimensional models because the water surface on both side of the roadway will vary along the roadway. Therefore, not only will weir flow be more accurately represented, but the flow frequency for initiation of overtopping will be more accurately predicted.

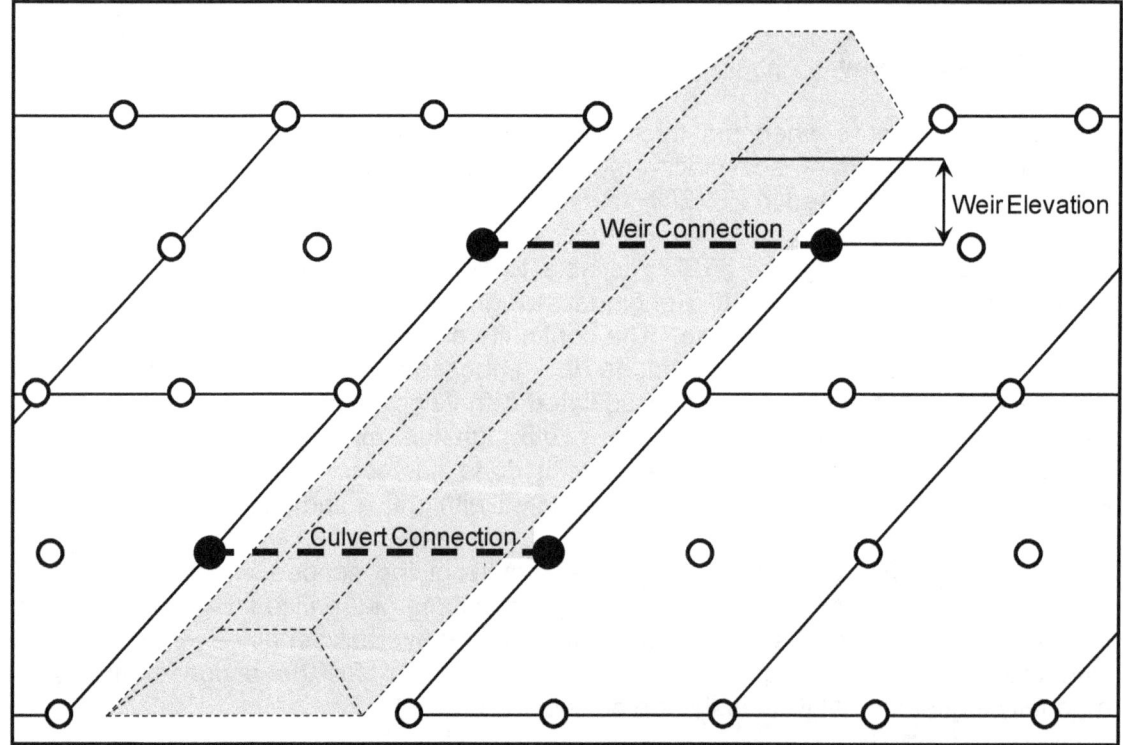

Figure 6.12. Weir and culvert connections.

6.6 MULTIPLE OPENINGS IN AN EMBANKMENT

Figure 6.13 shows velocity output from a two-dimensional model that includes multiple openings along the roadway embankment. Two-dimensional models are ideally suited for this condition because flow is distributed accurately throughout the model domain. The model shows high velocity flow along the entire length of the main channel and high velocities in the two relief bridges located on the floodplain. The model also shows flow concentration at the abutments.

There are no additional requirements for modeling multiple openings in a two-dimensional model than those required for modeling single openings. As with any two-dimensional model, the geometry, land use, roughness, boundary conditions, and model limits must be accurately represented.

As compared with one-dimensional modeling, there are no requirements for assigning stagnation zones or ineffective flow areas, so the need for judgment is reduced (see Section 5.8.4). There is also no assumption built into the computer code regarding energy balance at the openings or at a particular location upstream. Therefore, a more accurate distribution of flow among the bridge openings is computed.

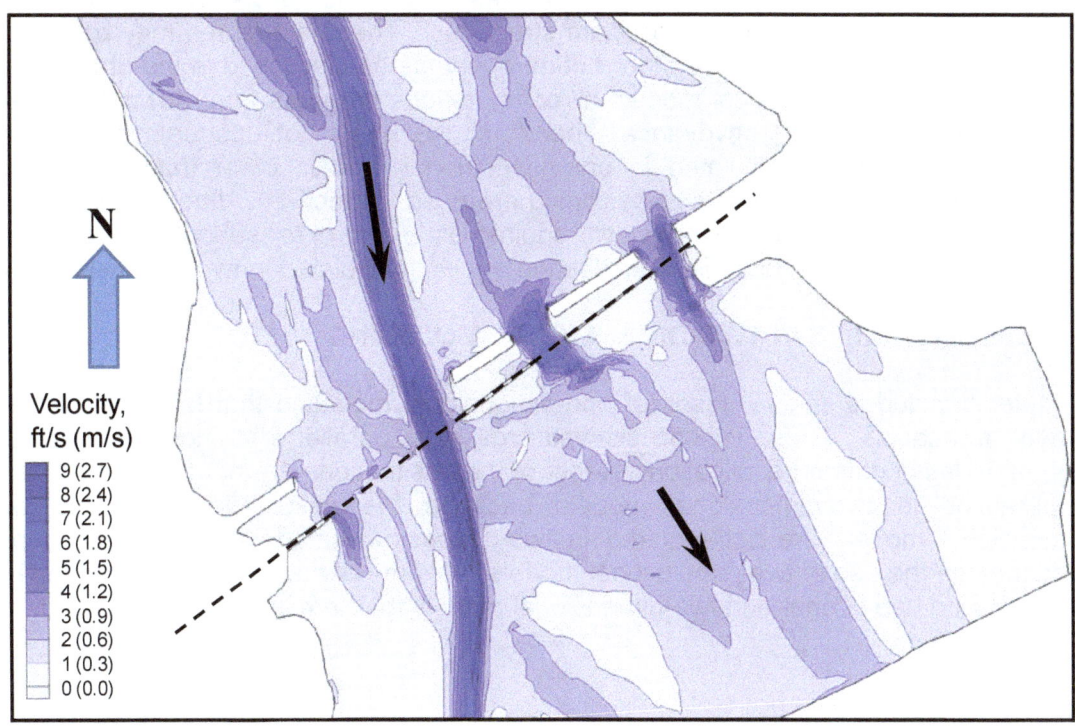

Figure 6.13. Two-dimensional model velocities with multiple bridge openings.

6.7 MULTIPLE OPENINGS IN SERIES

Parallel roadways have openings that align in series. In Figure 6.13 there is a railroad embankment that has similar main-channel and relief bridges to the upstream highway embankment. These openings are not identical in length, number of piers, or pier size. Also, the distance between the embankments is not consistent across the floodplain. As with multiple openings along an embankment, two-dimensional modeling is ideally suited for analyzing this type of situation. Prior to the availability of two-dimensional modeling it was difficult for hydraulic engineers to assess whether an upstream bridge or pier creates adverse conditions for a downstream structure. Therefore, when structural elements are not well aligned, or if questions exist regarding pier placement at an adjacent structure, two-dimensional models should be used. Depending on flood elevations, embankment heights and the amount of backwater created by the crossing, road overtopping may occur for one embankment and not another. These complex hydraulic conditions can be analyzed directly in two-dimensional models.

As with modeling multiple openings along an embankment, there are no additional requirements for modeling multiple openings in series in a two-dimensional model than those required for modeling single openings. Geometry, land use, roughness, boundary conditions, and model limits must be accurately represented. As compared with one-dimensional modeling, there are no requirements for estimating flow expansion and contraction between bridges or assigning ineffective flow areas, so the need for judgment is reduced. Therefore, a more accurate distribution of flow within each of the bridge openings is computed.

6.8 UPSTREAM FLOW DISTRIBUTION

As illustrated in the model in Figure 6.7, the results of which are shown in Figure 4.3, flow distribution may be affected by upstream structures. These structures may be river control structures, countermeasures, or other bridge openings as discussed in the previous section. In one-dimensional subcritical models all computations progress from downstream and flow is distributed based on conveyance. Therefore, the effects of upstream controls on flow distribution cannot be simulated in one-dimensional models other than by manipulating conveyance through Manning n or assigning areas as ineffective. Although the downstream water surface boundary condition is still required as a control for subcritical two-dimensional models, upstream impacts on flow distribution are well-simulated in two-dimensional models.

6.9 SPECIAL CASES IN TWO-DIMENSIONAL MODELING

Chapter 5 includes special cases of one-dimensional modeling that fall outside the typical model application. These include skewed crossings, parallel crossings, multiple openings and other less common applications. Most of these situations are not considered as special applications in two-dimensional models because the assumptions required by one-dimensional models are not required in two-dimensional models. This section provides guidance on the use of two-dimensional models for some of these cases, and compares and contrasts the use of one- and two-dimensional models for them.

6.9.1 Split Flow

Figure 6.14 is a two-dimensional model representation of the one-dimensional split flow model shown in Figure 5.23. In a one-dimensional model, separate reaches are required for the two split flow reaches and two more reaches are required upstream and downstream of the split flow reaches. In the one-dimensional representation, a trial and error process is used to apportion flow between the two reaches until an energy balance at the upstream combined-flow cross section is achieved. Two-dimensional modeling provides a better depiction of split flow hydraulics because the assumption of energy balance at a particular location is not made. In essence, the two-dimensional model is always using an iterative process to apportion flow throughout the network until the equations of motion are satisfied. Therefore, solution to the split flow problem is intrinsic to two-dimensional hydraulic analysis. The assignment of a specific location for energy balance is not a requirement in two-dimensional modeling, nor is it even possible to make that assignment. Another advantage of two-dimensional modeling is that if flow over the island from one "reach" to the other occurs, the two-dimensional model would not need a special lateral connection that is required by the one-dimensional model.

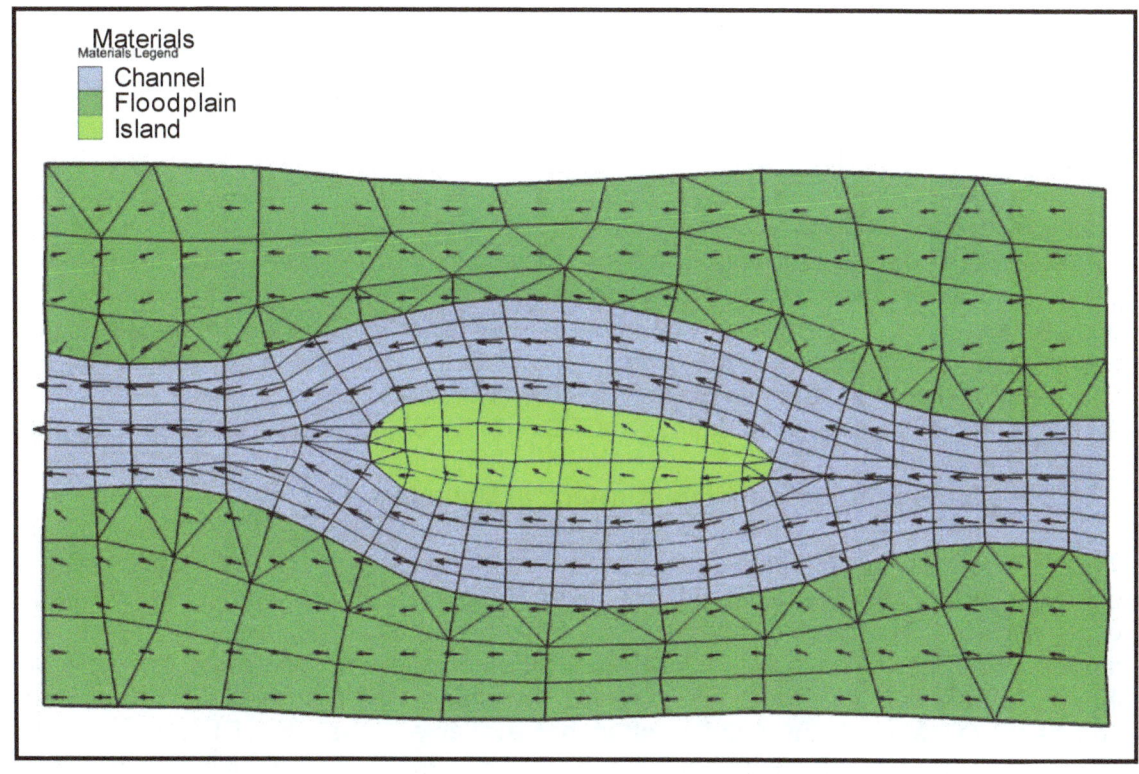

Figure 6.14. Illustration of split flow two-dimensional model.

6.9.2 Lateral Weirs

In one- and two-dimensional models, lateral weirs are locations where flow is removed from the model domain laterally relative to the dominant flow direction. This is illustrated for a one-dimensional model in Figure 5.26. In the two-dimensional model a weir connection as described in Section 6.5.3 can be used. Flow would be removed from nodes along one side of the road embankment and entered at nodes along the other side. As with one-dimensional models, lateral outflows that leave the model domain can also be achieved in two-dimensional models using weirs, culverts, or discharge boundary conditions.

6.9.3 Culverts

Culverts are rarely significant components of a two-dimensional model because culverts are much smaller than bridges and flow through the culvert is analyzed very well using specific relationships that were developed for that purpose. Flow approaching or downstream of the culvert may be two-dimensional, such as water moving along an embankment that turns to flow through the culvert. Figure 6.12 illustrates a culvert connection in a FST2DH two-dimensional model. A culvert connection is handled similarly to a weir connection in that flow is removed from one boundary node and reentered at a corresponding boundary node on the opposite side of the embankment. The culvert dimensions, entrance and exit invert elevations, entrance conditions, and Manning n are required input.

6.9.4 Debris

There is no automated method for including debris on piers in an FST2DH model as there is in HEC-RAS. However, the area of the debris blockage can be included with the pier dimensions and an increased force will be computed using the additional drag force method or increased flow resistance method.

6.10 TWO-DIMENSIONAL MODELING ASSUMPTIONS AND REQUIREMENTS

6.10.1 Gradually Varied Flow

Although two-dimensional modeling provides a much more complete analysis of bridge hydraulics than one-dimensional modeling, especially as it relates to flow distributions and lateral velocity components, two-dimensional models do not account for vertical velocities and accelerations. Therefore, the flow is assumed to have a hydrostatic pressure distribution, vertical velocities are neglected, and flow circulation is not simulated. If these flow features are an important aspect of the flow hydraulics, such as detailed analysis of flow at a pier, then three-dimensional models, CFD models, or physical models are required.

6.10.2 Flow Distribution and Water Surface at Boundaries

Two-dimensional models make various assumptions about flow distribution and water surface elevation at boundaries. Water surface boundaries are usually treated as a level water surface. This may not be an accurate representation, so, as with one-dimensional models, it is best to have the downstream boundary located well away from the point of interest. Although it is possible to enter a varying water surface boundary, the data necessary to establish the input is often not available. The FST2DH model will usually distribute water at the upstream boundary very reasonably. Other models may or may not do as well. In any case, the model results should be evaluated to make sure that the upstream boundary is not unduly influencing the solution. The upstream boundary should also be located well away from the location of interest. As a rule-of-thumb, the upstream and downstream boundaries should be at least one floodplain width upstream and downstream of the bridge crossing. If flow is not fully expanded at the boundaries and largely one-dimensional, then the model extent should be increased.

6.10.3 Model Step-Down and Convergence

Unlike one-dimensional models, which have a control either at the downstream or upstream boundary depending on flow regime, two-dimensional models do not compute flow by progressing from one boundary to another. The starting condition for most two-dimensional models is a uniform pool of water that inundates the entire model domain. Once a solution is achieved for this condition, the model head boundary is stepped down by small increments until the desired water surface boundary is achieved. The model must achieve a reasonably stable and converged solution for each intermediate run. This process can be tedious and at times difficult if the model becomes unstable. There are many approaches to achieving an efficient step-down process and a stable, yet accurate target condition. These include stepping down water surface elevation at a low discharge and then stepping up discharge, and using high Manning n and viscosity terms during the step-down process and then stepping down these coefficients. The SMS software has automated run-control that includes various step-down procedures (Aguaveo 2011).

Once the model water surface and discharge boundary conditions, and other input variables including Manning n and viscosity terms, have reached the target values, the model must be run for a sufficient number of iterations to converge on a numerically valid solution. Convergence criteria related to water surface, velocity, or unit discharge can be set to halt program execution once the criteria are met.

6.10.4 Wetting and Drying

Most two-dimensional models require all the nodes of an element to be wetted for that element to be included in the flow computations. If a single node is dry, then the entire element is removed from the computational network. Dry elements along a discharge or water surface boundary should be avoided as it may cause model termination. The wetting and drying process can create numerical instabilities but also may exclude areas with significant conveyance when the elements are large. It is desirable to have a relatively smooth boundary along the wet/dry boundary.

There are several techniques for maintaining model stability when areas of the network are dry. These include incorporating additional network refinement and using depth-variable Manning n and increasing its value for shallow depths. Some models, including FST2DH and RMA2, allow the engineer to assign porosity to the ground. By assigning a low porosity value, this approach allows for a very small amount of flow even at nodes where the water surface is below the node elevation.

(page intentionally left blank)

CHAPTER 7

UNSTEADY FLOW ANALYSIS

7.1 INTRODUCTION

Almost all flow in rivers and streams is to some extent unsteady, i.e., it changes with time. Also, the rate of flow and the depth usually vary along the river. In many applications, flow may be assumed to be uniform along a short reach of the channel. Among the most important causes of unsteady flow are the following:

1. Runoff from precipitation (rainfall event and/or snowmelt); when depth and velocity of flow in a river change rapidly with time
2. Unsteady or transient flows released from reservoirs during operations for flood control, hydropower generation, recreation, and wildlife management, etc.
3. Tidal-generated waves (astronomical tides)
4. Dam-break floods
5. Wind-generated storm surges or seiches
6. Landslide-generated waves
7. Earthquake-generated tsunami waves
8. Irrigation flows affected by gates, pumps, diversions, etc.

There are several computer models that have been developed for simulating one-dimensional flow. Fread of the National Oceanic Atmospheric Administration's National Weather Service (NOAA's NWS) developed two unsteady flow models having the capability to simulate flows through a single stream or a system of interconnected waterways. DWOPPER was the original model and was later replaced by FLDWAV (Jin and Fread 1997). The HEC-RAS model (USACE 2010c) incorporates the UNET (USACE 2001) unsteady flow algorithms for a full network of natural and constructed channels. The HEC-RAS model is the most widely used 1-D model in the US.

Figure 7.1 illustrates the basic properties of flow hydrographs. The downstream flow hydrograph exhibits a delay (travel time) and is reduced (attenuation) when there is no additional flow contributions between the upstream and downstream locations. Hydrologic routing focuses on the discharge hydrograph. For bridge hydraulics, hydrodynamic simulation is preferred because hydraulic variables (velocity, water surface elevation, depth, etc.) are computed throughout the channel reaches represented in the model domain.

7.1.1 Unsteady Flow Equations – Saint-Venant Equations

The discussion in this section emphasizes the one-dimensional equations and applications; however the two-dimensional equations also will be presented. Solutions to one-dimensional flow problems are conveniently viewed in a three-dimensional coordinate space, in which two of the axes are distance along the channel and time, x and t, respectively, as shown in Figure 7.2. The third coordinate axis corresponds to the solution, such as discharge $Q(x,t)$. Similar three-dimensional surfaces can be used to represent the variation in depth $Y(x,t)$, water surface elevation $h(x,t)$, or velocity $V(x,t)$. From Figure 7.2, the initial hydrograph is given through time at the upstream cross section at $x = x_0$. As the flood wave travels downstream, the hydrograph is attenuated and lagged in time. Figure 7.2 also illustrates that the condition at a location $x > 0$ will not change for some amount of time (lag time) before the initial upstream change in flow propagates downstream.

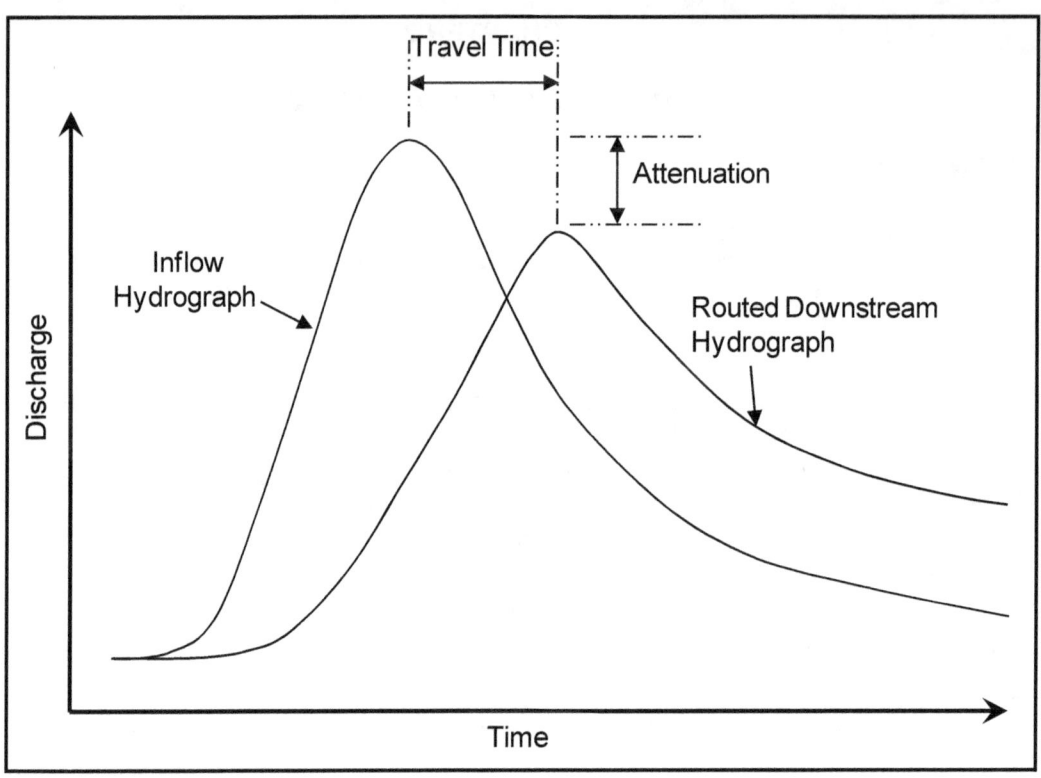

Figure 7.1 Unsteady flow hydrographs.

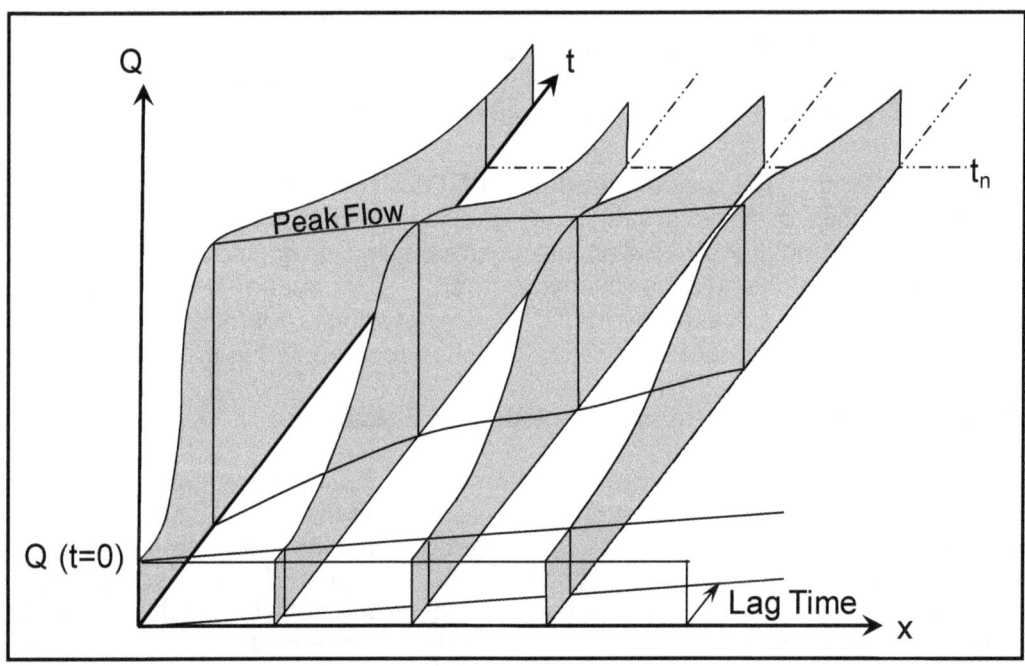

Figure 7.2. Unsteady solution of discharge versus x and t.

The initial conditions for an unsteady flow model are first a solution of the steady state flow, Q at time $t = t_0$ (i.e., steady state water surface profile computation), and the inflow hydrograph at the most upstream channel cross section. In Figure 7.2, this indicates that flow and depth would be known along the x axis for time $t = t_0$ and the flow or depth would be known at the upstream cross section $x = x_0$.

Looking at only the x-t plane in Figure 7.3, known water surface elevations and flow along the x-axis are indicated by the squares at time $t = t_0$ and the known inflow hydrograph ordinates at cross section x_0 are illustrated by the circles. The unknowns are discharge and depth along the channel at successive time lines until the entire surface is known. In Figure 7.2 the solution has reached time t_n.

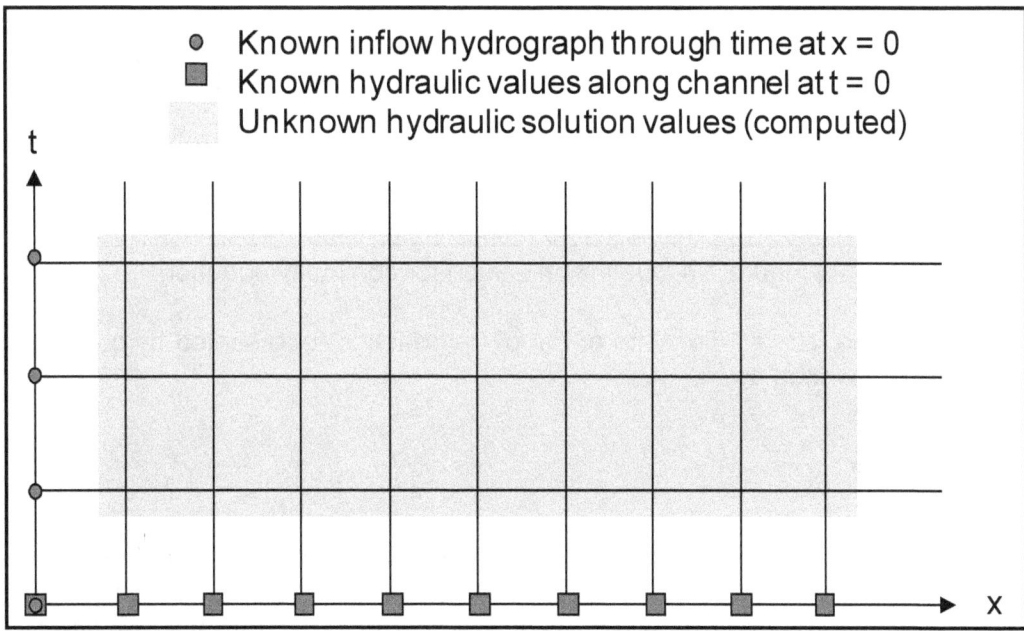

Figure 7.3. x-t plane illustrating locations for known and computed values.

The equations governing the flow of water in general are based on known physical principles, conservation of mass and momentum. In unsteady flow analysis the continuity and the momentum equations must be solved explicitly because the flow and the elevation of the water surface are both unknown.

7.1.2 Unsteady Continuity Equation

The continuity equation for one-dimensional unsteady flow considers water that accumulates or depletes in a control volume (storage). In steady flow the conservation of mass can be written as $Q = AV$ where the discharge is constant and the only unknown is the water surface. Consider a short length (Δx) of channel as shown in Figure 7.4. The derivations make the following assumptions (Chaudhry 2008):

1. The pressure distribution is hydrostatic
2. The channel bottom slope is small so that the flow depth measured vertically is almost the same as the flow depth normal to the channel bottom (i.e., $\sin \theta \approx \tan \theta \approx \theta$, where θ is the angle between the channel bottom and the horizontal datum)
3. The velocity distribution at the channel cross section is uniform
4. The channel is prismatic; that is, the channel shape remains unchanged with distance

5. The friction losses in unsteady flow may be computed using the empirical formulas (i.e., the Manning equation) for steady-state flows

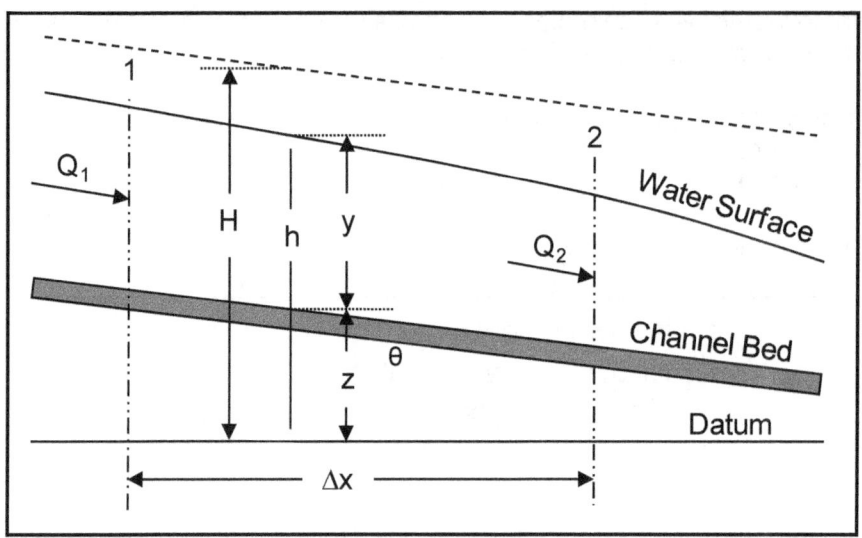

Figure 7.4. Definition sketch for continuity equation.

Assuming that the lateral inflow into or out of the reach is represented by q∆x, the continuity equation can be written as:

$$Q_2 - Q_1 = \frac{\partial Q}{\partial x} \Delta x \tag{7.1}$$

The partial derivative is necessary since Q is changing with both time and distance along the channel. Given that h represents the water surface elevation above the datum (h = z + y), the volume of water between sections 1 and 2 is increasing at the rate $T(\partial h/\partial t)\Delta t$. From the conservation of mass, the change in the flow must be equal to the change in the channel storage, and for a short length of channel ∆x where T is the top width of the water surface as shown in Figure 7.5, $(\partial A/\partial t)\Delta x \approx T(\partial y/\partial t)\Delta x$.

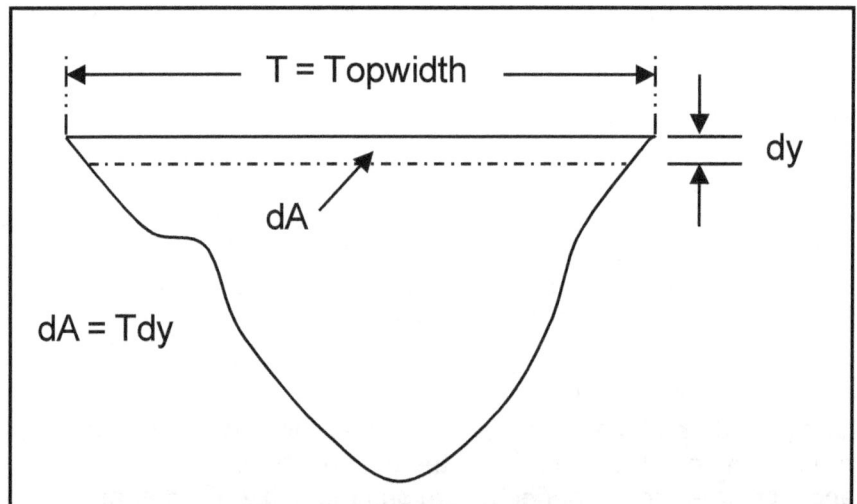

Figure 7.5. Channel cross section relating topwidth to area.

Substituting for Q = AV, the continuity equation can be written as:

$$\frac{\partial AV}{\partial x}\Delta x + T\frac{\partial y}{\partial t}\Delta x = q\Delta x \tag{7.2a}$$

$$A\frac{\partial V}{\partial x} + VT\frac{\partial y}{\partial x} + T\frac{\partial y}{\partial t} = q \tag{7.2b}$$

In order to account for off-channel storage, Fread (1981) wrote the continuity equation as:

$$\frac{\partial Q}{\partial x} + \frac{\partial (A + A_o)}{\partial t} = q \tag{7.2c}$$

where A is the active cross-sectional area of flow and A_o is the inactive (off-channel storage) cross-sectional area.

7.1.3 Dynamic Momentum Equation

Applying Newton's second law to the elemental length between Sections 1 and 2 of Figure 7.4 yields:

$$\sum F_x = ma = \rho A \Delta x \left(\frac{dV}{dt}\right) = \rho A \Delta x \left(V\frac{\partial V}{\partial x} + \frac{\partial V}{\partial t}\right) \tag{7.3}$$

The net forces causing flow down slope in Figure 7.4 are (1) the force resisting the shear force (i.e., action versus reaction) and (2) the difference in the hydrostatic forces acting on the element. Other forces may be present for special cases. These could include pier drag forces and wind surface shear. The shear force can be represented by $\tau_0 W P \Delta x$ and the difference in the hydrostatic forces in the downstream direction is given by $\gamma A \Delta h$ assuming the water surface slope and channel bed slope are small.

Substituting these forces into Equation 7.3 gives the equation of motion:

$$-\gamma A \Delta h - \tau_0 W P \Delta x = \rho A \Delta x \left(V\frac{\partial V}{\partial x} + \frac{\partial V}{\partial t}\right) \tag{7.4}$$

Equation 7.4 can be solved for τ_0 to give the following expression.

$$\tau_0 = -\gamma R\left(\frac{\partial h}{\partial x} + \frac{V}{g}\frac{\partial V}{\partial x} + \frac{1}{g}\frac{\partial V}{\partial t}\right) \tag{7.5}$$

Noting that h = y + z and $\tau_0/\gamma R = S_f$ then Equation 7.5 can be written as

$$S_f = -\frac{\partial z}{\partial x} - \frac{\partial y}{\partial x} - \frac{V}{g}\frac{\partial V}{\partial x} - \frac{1}{g}\frac{\partial V}{\partial t} \tag{7.6}$$

Substituting $S_0 = -\partial z/\partial x$ for the bed slope resulting in

$$S_f = S_0 - \frac{\partial y}{\partial x} - \frac{V}{g}\frac{\partial V}{\partial x} - \frac{1}{g}\frac{\partial V}{\partial t} \qquad (7.7)$$

Solved together with boundary conditions, Equations 7.2 and 7.7 are the complete dynamic wave equations for one-dimensional unsteady flow. Chapter 6 provides the complete set of dynamic equations for two-dimensional flow analysis and includes additional force terms that are difficult to represent in one-dimensional models.

<u>Dynamic Wave Equation Terms</u>. Meanings of the various terms in the dynamic wave equations are as follows (Henderson 1966):

Continuity equation:

- $A(\partial V/\partial x)$ = prism storage

- $VT(\partial y/\partial x)$ = wedge storage

- $T(\partial y/\partial t)$ = rate of rise

- q = lateral inflow per unit length

Momentum equation:

- S_f = friction slope (frictional forces)

- $\partial z/\partial x = S_0$ = bed slope (gravitational effects)

- $\partial y/\partial x$ = pressure differential

- $(V/g)(\partial V/\partial x)$ = convective acceleration

- $(1/g)(\partial v/\partial t)$ = local acceleration

Depending upon the relative importance of the various terms of the momentum equation, the equation can be simplified for various applications. Approximations to the full dynamic wave equations are accomplished by combining the continuity equation with the various simplification of the momentum equation. The most common approximations of the momentum equation are shown in Figure 7.6 (Henderson 1966). Although the time derivative in Equation 7.7 is only included in the full dynamic wave equations, each of the approximations can be used in unsteady flow analysis by coupling with the unsteady continuity equation (Equation 7.2).

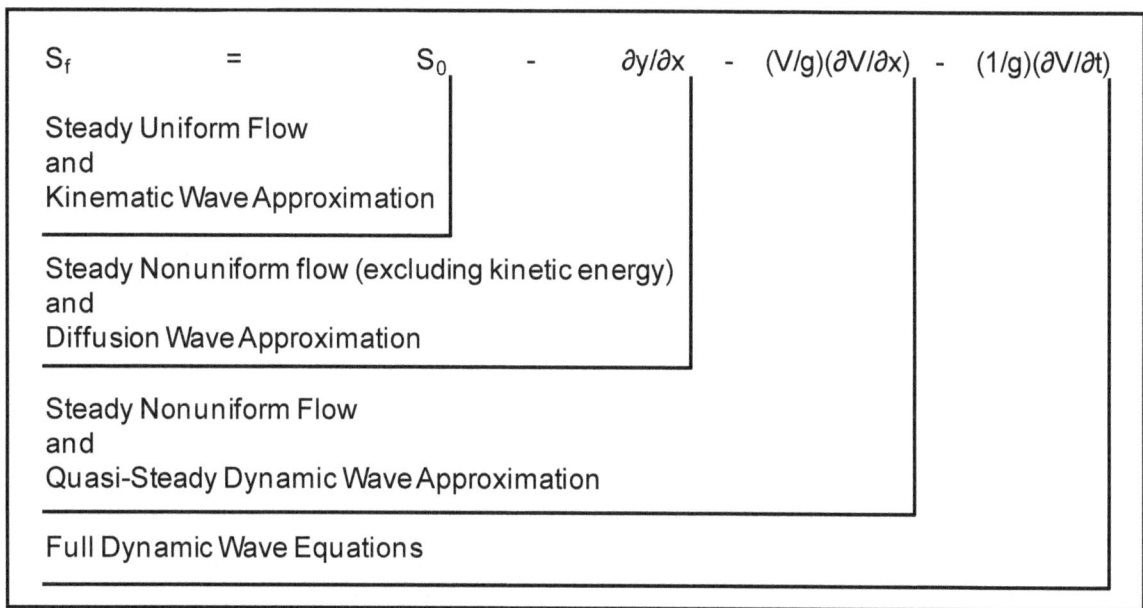

Figure 7.6. Approximations of the Momentum Equation.

7.2 MODEL UPSTREAM AND DOWNSTREAM EXTENTS

The minimum number of cross sections in a steady state riverine bridge analysis includes upstream (Approach), downstream (Exit) and bridge cross sections. The water surface elevation is specified for the Exit cross section and a discharge is specified for the simulation. As discussed in Section 4.2, additional upstream and downstream cross sections are often warranted to assess hydraulic impacts of the bridge design and to decrease uncertainties associated with inexact boundary conditions. All of the factors discussed in Section 4.2 apply to unsteady flow models.

Study limits, cross sections, and reach lengths comprise the geometry of the hydraulic model (Figure 7.7). Once the geometry is created, all other information required to complete a hydraulic model can be added.

The model limits of an unsteady model must be carefully chosen to include all potential storage upstream and downstream of the location of interest. If a bridge hydraulic model including only the minimum number of cross sections were used as the unsteady model geometry, the simulation would be inaccurate because storage and routing effects would be significantly under represented. Therefore, unsteady models almost always require much longer upstream and downstream limits than steady models.

In many cases the water surface elevation at the downstream boundary of a hydraulic model is not known, or a control such as a hydraulic structure is not present. In these situations, a downstream boundary condition is assumed, which establishes the starting water surface elevation. Several assumptions can be made to estimate the downstream water surface elevation including normal and critical depth.

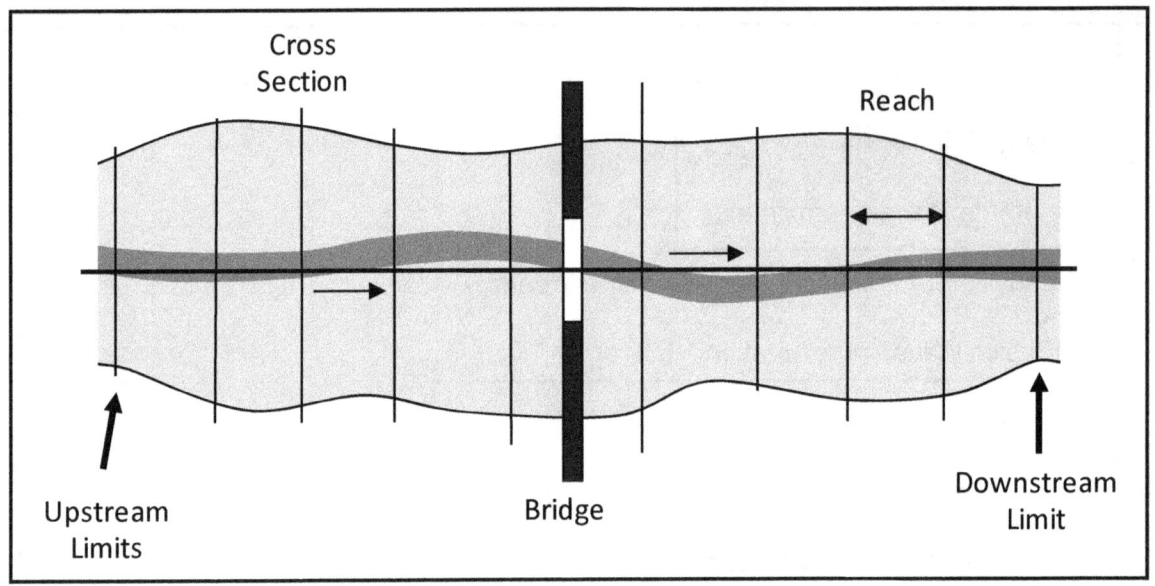

Figure 7.7. Study limits upstream and downstream.

7.3 DIFFERENCES BETWEEN STEADY AND UNSTEADY FLOW ANALYSIS

In addition to model limits, a major difference between unsteady-flow analysis and steady-flow analysis is the information needed at the boundaries of the stream system. In steady-flow analysis, knowledge of one elevation at the downstream boundary is needed to start the computations for subcritical flow, or at the upstream boundary for supercritical flow.

The typical approach to riverine modeling has been to use hydrologic routing to determine discharges and steady state analysis to compute water surface elevations. This practice is a simplification of true river hydraulics, which is more completely represented by unsteady flow. However, many practical problems in bridge and culvert hydraulics can be adequately solved using these traditional methodologies. This section identifies some of the differences between steady and unsteady flow analysis.

1. Steady flow analysis utilizes both hydrologic analysis for the determination of the flow and hydraulic analysis to compute the water surface profile. Steady flow analysis assumes that although the flow is steady, $\partial Q/\partial t = 0$, it can vary with respect to space $\partial Q/\partial x \neq 0$. As seen from the continuity equation for unsteady flow, Equation 7.2, both the flow and depth can vary with time and space.

   The steady state water surface profiles compute the water surface for a specified flood event (i.e., 100-year flood). The primary assumption of this computation is that:

   - Peak flow coincides with the peak stage
   - Flow can be adequately estimated at all locations along the channel reach, and
   - Peak stages occur simultaneously over a short reach of channel

2. For small bed slopes (i.e., slopes less than 0.0004) or highly transient flows, such as tidal influences or dam breach flood waves, the peak stages do not necessarily coincide with the peak discharges, and the rating curves of stage versus discharge are not single valued. Actual rating curves are looped due to the changing of the energy slope throughout the flood event. This means that two discharges are possible at the one stage depending on whether the stage occurs on the rising limb or falling limb of the

hydrograph. The magnitude of this loop can be affected by any of several hydraulic parameters, the most significant of which may be backwater effects from downstream. Each flood will follow a different loop. Figure 7.8 illustrates looped rating curves for two floods with the arrows designating the rise and recession limbs of the discharge hydrographs. The inner loop is for a slower rate of rise and fall which creates a narrower loop.

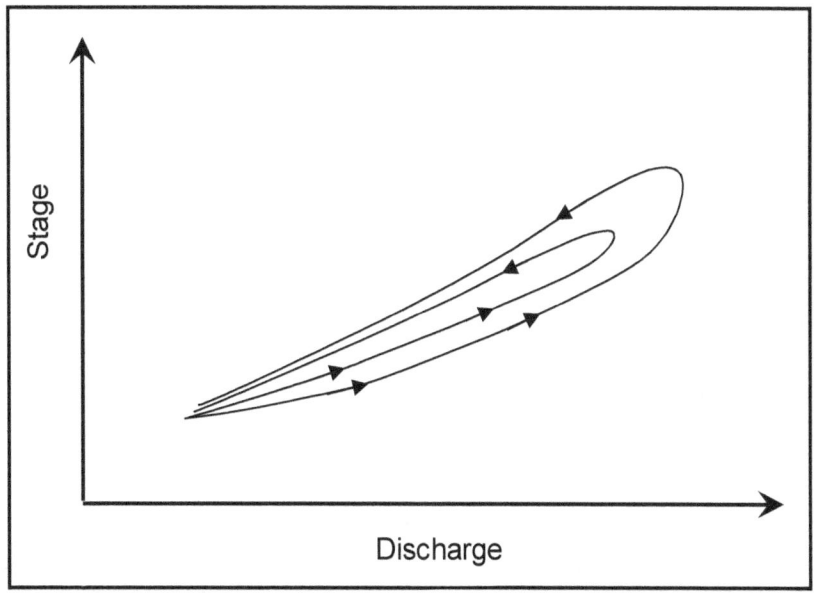

Figure 7.8. Looped rating curves showing the rising and recession limbs of a hydrograph.

3. The total flow downstream from a junction of a two tributaries is not necessarily the combination of the two flows. Backwater from the flow at the junction can cause water to be stored in upstream areas, reducing the flow combinations.

4. Tributary flows entering a main stream channel may experience a flow reversal caused by flow in the main stem backing up into the tributary or vice versa e.g., when a large tributary flood enters the main channel during a period of low flow.

5. If the inflow or stage at a boundary is changing rapidly, the acceleration terms in the momentum equation are important and thus unsteady flow is a more robust and complete computation. Examples of the phenomena are dam break analysis, rapid gate openings and closures. Regardless of the slope of the channel, unsteady flow analysis should be used for all rapidly changing hydrographs.

6. For full networks, where the flow divides and recombines, unsteady flow analysis should be considered for subcritical flow. Unless the problem is simple, steady flow analyses cannot accurately compute the flow distribution. When flow divides and recombines in the split-flow reaches the length of the channels, the resistance to flow, and channel geometry will differ. This causes the flood wave to travel through the reaches at different speeds, which in turn affects the flow distribution in the reaches. To accurately determine the flow distribution, unsteady flow modeling is preferred over steady state modeling.

7.4 SOLUTION SCHEMES FOR THE SAINT-VENANT EQUATIONS

Several numerical methods have been used to solve the Saint-Venant Equations.

1. The Method of Characteristics – The method of characteristics is a technique for solving partial differential equations. The method is valid for any hyperbolic partial differential equation such as the Saint-Venant Equations. The method reduces the partial differential equation to a family of ordinary differential equations for which the solution can be integrated from some initial data. Typically the problems are reduced to simplified conditions such as wide rectangular approximation for the channel geometry. Henderson (1966) has an excellent overview of the method applied to open channel hydraulics.

2. Finite Difference Numerical Methods – The finite difference methods are numerical methods based upon the approximations that permit replacing differential equations by finite difference equations. These finite difference approximations are algebraic in form, and the solutions are related to grid points for approximating the solutions to differential equations using finite difference equations to approximate derivatives. The finite difference solution involves three steps: (1) dividing the solution into grids or nodes, (2) approximating the given differential equation by finite difference equivalence that relates the solutions to grid points, and (3) solving the difference equations subject to the prescribed boundary conditions and/or initial conditions.

The error in the solution is defined as the difference between its approximation and the exact analytical solution. The two sources of error in the finite difference methods are round-off error, (loss of precision due to computer rounding of decimal quantities) and truncation error or discretization error, (difference between the exact solution of the finite difference equation and the exact quantity assuming perfect arithmetic). In order to solve a problem, the problem's domain must be discretized. This is usually done by dividing the domain into a uniform grid (see Figure 7.9). In Figure 7.8 and the following finite difference equations, the solution domain shows positions in space as i, i+1, i+2, etc. and positions in time as j, j+1, j+2, etc. Note that this means that finite-difference methods produce sets of discrete numerical approximations to the derivative, often in a "time-stepping" manner.

A common and accepted procedure for solving the one-dimensional unsteady flow equations is the four-point implicit scheme, also known as the box scheme. More information is needed for unsteady-flow analysis than for steady flow analysis. For the example shown in Figure 7.7, a single channel with no special features (excluding the bridge) is divided into 9 computational elements (reaches) yielding 10 nodes (cross sections). With two unknowns (depth and flow) at each node, there are 20 unknowns but only 18 equations (2 per computational element). Therefore, the unknowns cannot be determined without some additional information at the boundaries of the system.

The time derivatives are approximated by a forward difference quotient centered between i and i + 1 points along the x-axis, i.e.,

$$\frac{\partial K}{\partial t} = \frac{K_i^{j+1} + K_{i+1}^{j+1} - K_i^{j} - K_{i+1}^{j}}{2\Delta t_j} \tag{7.8}$$

where K represents any solution variable (e.g., velocity, discharge, flow depth, etc.).

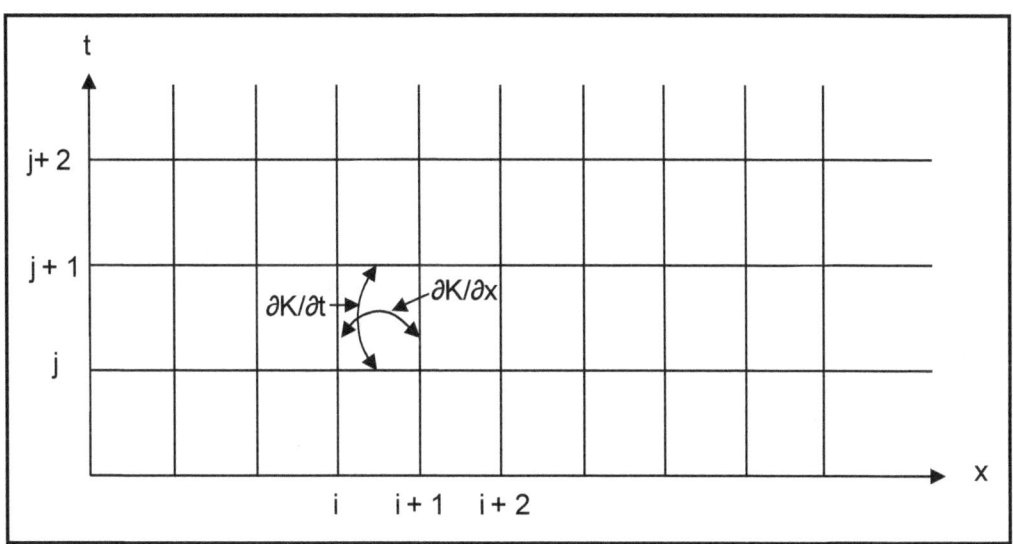

Figure 7.9. Discrete x-t solution domain for any variable.

The spatial derivatives are approximated by a forward difference quotient positioned between two adjacent time lines according to weighting factors of θ and 1-θ, i.e.,

$$\frac{\partial K}{\partial x} = \theta \left(\frac{K_{i+1}^{j+1} - K_i^{j+1}}{\Delta x_i} \right) + (1-\theta) \left(\frac{K_{i+1}^{j} - K_i^{j}}{\Delta x_i} \right) \tag{7.9}$$

Variables other than derivatives are approximated at the time level where the spatial derivatives are evaluated by using the same weighting factors, i.e.,

$$K = \theta \left(\frac{K_{i+1}^{j+1} + K_i^{j+1}}{2} \right) + (1-\theta) \left(\frac{K_{i+1}^{j} + K_i^{j}}{2} \right) \tag{7.10}$$

The influence of the θ weighting factor on the accuracy of the computations was examined by Fread (1974), who concluded that the accuracy decreases as θ departs from 0.5 and approaches 1.0. This effect becomes more pronounced as the magnitude of the computational time step increases. Usually, a weighting factor of 0.60 is recommended to minimize the loss of accuracy while avoiding instability. When the finite difference operators as defined in the three equations above are used to replace the derivatives and other variables, the four-point implicit difference equations are obtained.

The terms associated with the jth time line are known from either the initial condition or previous computations. The initial conditions are values of h and Q at each node along the x axis for the first time line (j=1).

Since there are four unknowns and only two equations, the algebraic approximation of the Saint-Venant Equations cannot be solved in an explicit or direct manner. However, if the equations are applied to each of the N-1 rectangular grids between the upstream and downstream boundaries (Figure 7.9), a total of 2N-2 equations with 2N unknowns can be formulated (where N denotes the number of nodes). Then, prescribed boundary conditions, one at the upstream boundary and one at the downstream boundary, provide the necessary two additional equations required for the system to be determinate.

In order to solve the unsteady flow equations, the state of the initial conditions of water surface elevation and discharge (h and Q) must be known at all cross sections at the beginning ($t = t_0$) of the simulation (represented by the squares in Figure 7.3). This is the initial condition of the flow, which is typically steady nonuniform flow and would be solved using HEC-RAS (steady) or WSPRO computer models.

When the flow is subcritical, information at both the upstream and the downstream boundary of the system is also needed. The information supplied at a boundary is called a boundary condition. This information can be in one of three forms: flow known as a function of time (flow hydrograph), water-surface elevation known as a function of time (stage hydrograph), or a relation between flow and water-surface elevation (rating curve or energy slope for river conditions). The upstream boundary is typically a flow hydrograph (represented by the circles in Figure 7.3) and the downstream boundary is typically a known relation between flow and water-surface elevation (a rating curve or energy slope for river conditions). If the riverine system is influenced by a tidal condition, then the downstream boundary condition is almost always modeled using a stage hydrograph of tides.

The information supplied at a special feature internal to the stream system is often called an internal boundary condition. In unsteady-flow analysis, internal boundary conditions are approximated as steady-flow relations because the special features generally are short enough that the changes in momentum and volume of water within the special features are small. The isolation and description of the special features is a major component of unsteady-flow analysis.

The same computational problems can arise for unsteady-flow analysis as for steady-flow analysis because both analyses use algebraic approximations to the differential and integral terms. These approximations are developed for a computational element of finite length. If the computational element is too long, an incorrect solution results. The difference between the analyses is that in unsteady-flow analysis the computational problems are more complex and more frequent than in steady-flow analysis. The increased frequency is primarily because unsteady-flow analysis involves computations over a wide range of water-surface elevations, whereas most steady-flow analysis involves computations over a narrow range of water-surface elevations. Furthermore, the time dimension results in additional complications. Generally, the closer cross-sections are spaced, the shorter time-step is required. Therefore the need to reduce cross-section spacing must be balanced with the length of the simulation.

7.5 CHANNEL AND FLOODPLAIN CROSS SECTION

Figure 7.10 illustrates the interaction between the channel and the floodplain flows. When the river is rising, flow moves laterally away from the channel, inundating the floodplain and filling available storage areas. As the depth increases, the floodplain begins to convey water downstream. When the river stage is falling, water moves back toward the channel from the floodplains supplementing the main flow in the channel.

Even though the flow is two-dimensional, because the primary direction of the flow is along the main channel, it can be approximated by a one-dimensional representation. Off-channel ponding areas can be modeled as storage areas that exchange water to and from the channel or to other storage areas within the floodplain. For this case, modeling the flow using a steady state approximation will produce very different results than if modeled using an unsteady approximation due to differences caused by storage.

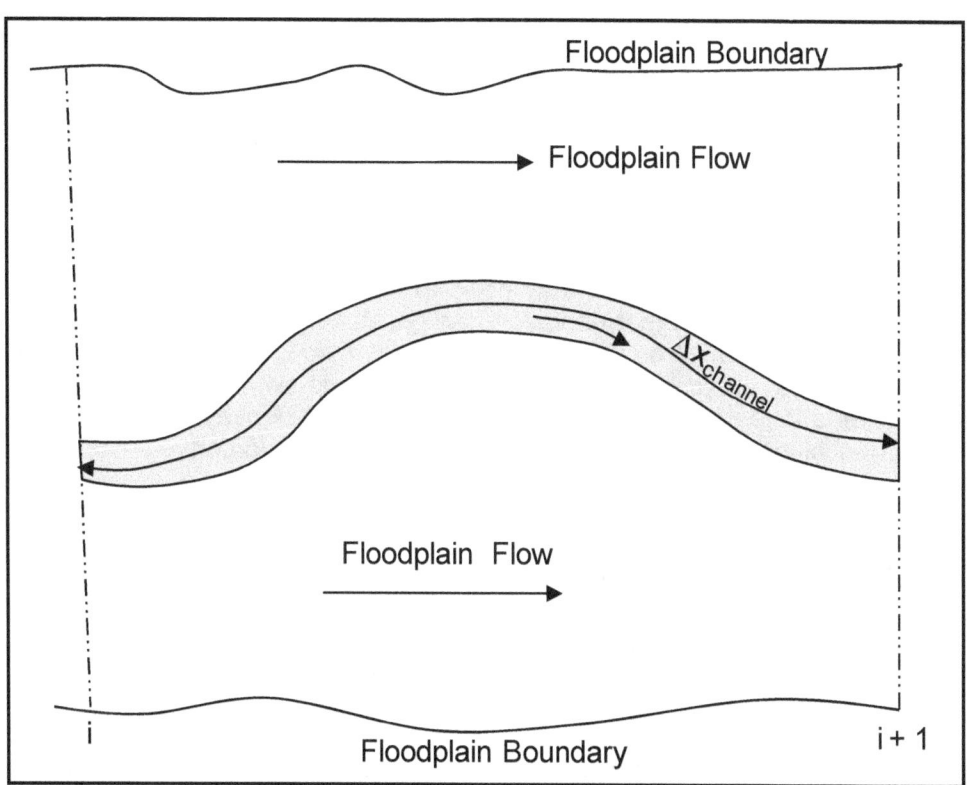

Figure 7.10. Channel and floodplain flows.

Applying steady flow to model the channel and overbank flows illustrated in Figure 7.10 can be addressed in many different ways. One approach is to ignore overbank conveyance entirely, assuming that the overbank is used only as storage. The HEC-RAS model has several geometry features that can be used to modify cross section data to better simulate the actual hydraulics. Under optional cross section properties a series of program options are available to restrict flow to the effective flow areas of cross sections. Among these capabilities are options for: ineffective flow areas, levees, and blocked obstructions.

The ineffective flow option allows the user to define areas of the cross section that will contain water that is not actively being conveyed. Ineffective flow areas are often used to describe portions of a cross section in which water will pond, but the velocity of that water in the downstream direction is close to zero. This water should be included in the storage calculations and other wetted cross section parameters, but it is not included as part of the active flow area. The volume of water that is already in storage does not attenuate the flow although the change in volume does.

In unsteady flow modeling it is always important to account for channel storage as well as conveyance. This is because the hydrograph volume must be simulated and portions of the flow may store, but not convey flow. It is useful to think of storage-only portions of a cross section as having extremely high n-values, which effectively eliminates the conveyance.

For unsteady flow, when the water surface exceeds the trigger elevation, the ineffective flow area will either begin to convey flow or remain ineffective depending upon its type (permanent on non-permanent). For non-permanent ineffective flow areas, once the water surface is higher than the trigger elevation, the entire ineffective flow area becomes effective. Water is assumed to be able to move freely in that area based on the roughness, wetted perimeter, and area of each subdivision. The left and right overbanks are no longer considered storage but are now active flow areas.

Occasionally, ineffective flow areas may need to remain ineffective permanently. When the water surface is below the trigger elevation, the permanent ineffective flow area behaves like the non-permanent area. For permanent ineffective flow areas, when the water surface elevation surpasses the trigger elevation, the area below the trigger elevation remains effective. Water above the trigger elevation is assumed to convey flow. This option is useful to avoid numerical instability associated with sudden change in conveyance.

7.6 STORAGE AREAS AND CONNECTIONS

Unsteady flow analyses are used to predict the temporal and spatial variations of a flood hydrograph as it moves through a river reach. The effects of storage and flow resistance within a river reach are reflected by changes in hydrograph shape and timing as the floodwave moves downstream. Figure 7.11 shows the major changes that occur to a discharge hydrograph as a floodwave moves downstream.

The HEC-RAS computer program provides an option to enter off-channel storage areas as ponding areas that are either in-line or off-line. Storage areas are treated as simple reservoirs. The use of storage areas allow for more stable and faster computations than representing a region with cross sections. The continuity equation is used to account for volume of the storage area and the flow into and out of the storage is accomplished with the storage indication method (i.e., as reservoir routing also sometimes called level pool routing). The momentum equation is not computed and the storage is computed by volume/elevation methods of either an area time relationship to account for depth or inputting an elevation/volume curve. Figure 7.12 is an example of an off-line storage area used in the HEC-RAS computer program.

Storage areas can be connected to a cross section(s) using a lateral connection, placed at the top or bottom of a reach, or connected to another storage area. The only data needed to describe storage areas are storage versus elevation. Two methods are available for this in the program: surface area times depth, or interpolation from an entered rating curve of elevation versus volume.

The data for storage area connections are a combination of procedures available in the computer model. The storage area can be connected to river reach with a lateral connection that can include gated structures and culverts, or entered as a stage-discharge rating curve. An initial water surface elevation or a storage area is also required for simulation.

7.7 HYDRAULIC PROPERTY TABLES

The HEC-RAS computer program has several features that aid in the computation and trouble shooting of the unsteady flow program for problems that may be encountered during a computer run. The Geometric Preprocessor is one such feature. It is used to process the geometric data into a series of hydraulic properties tables, rating curves, and a family of rating curves. This is done in order to speed up the unsteady flow calculations. Instead of calculating hydraulic variables for each cross section during each iteration the program interpolates the hydraulic variables from the tables.

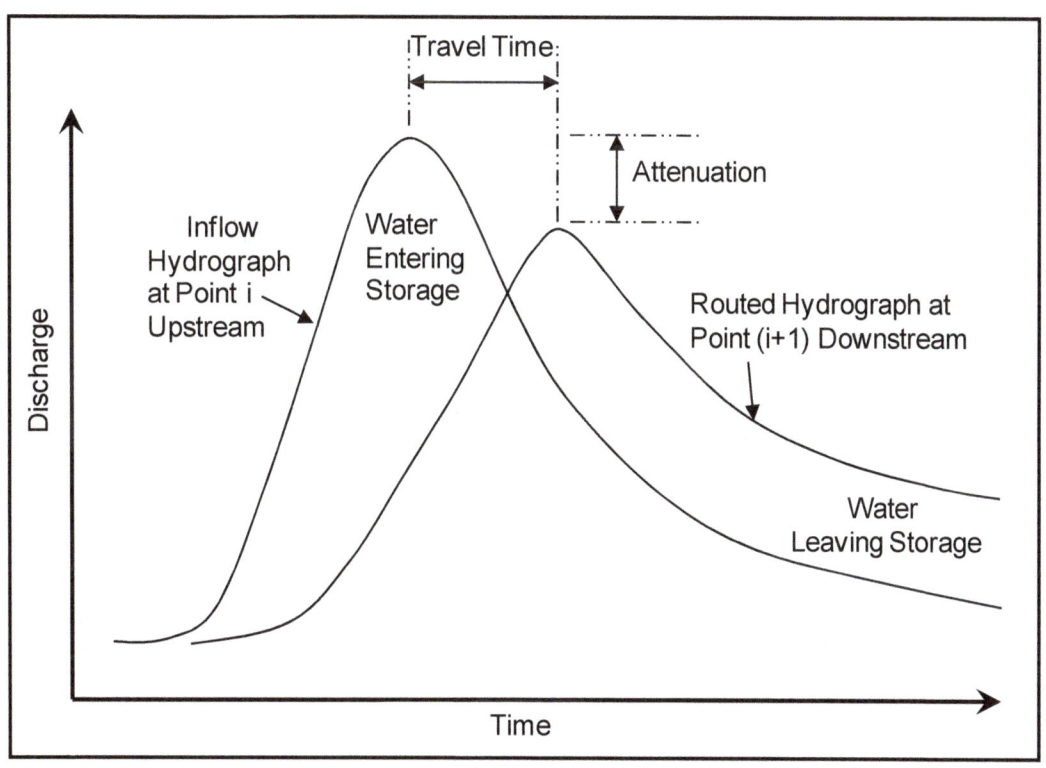

Figure 7.11. Routed hydrograph characteristics.

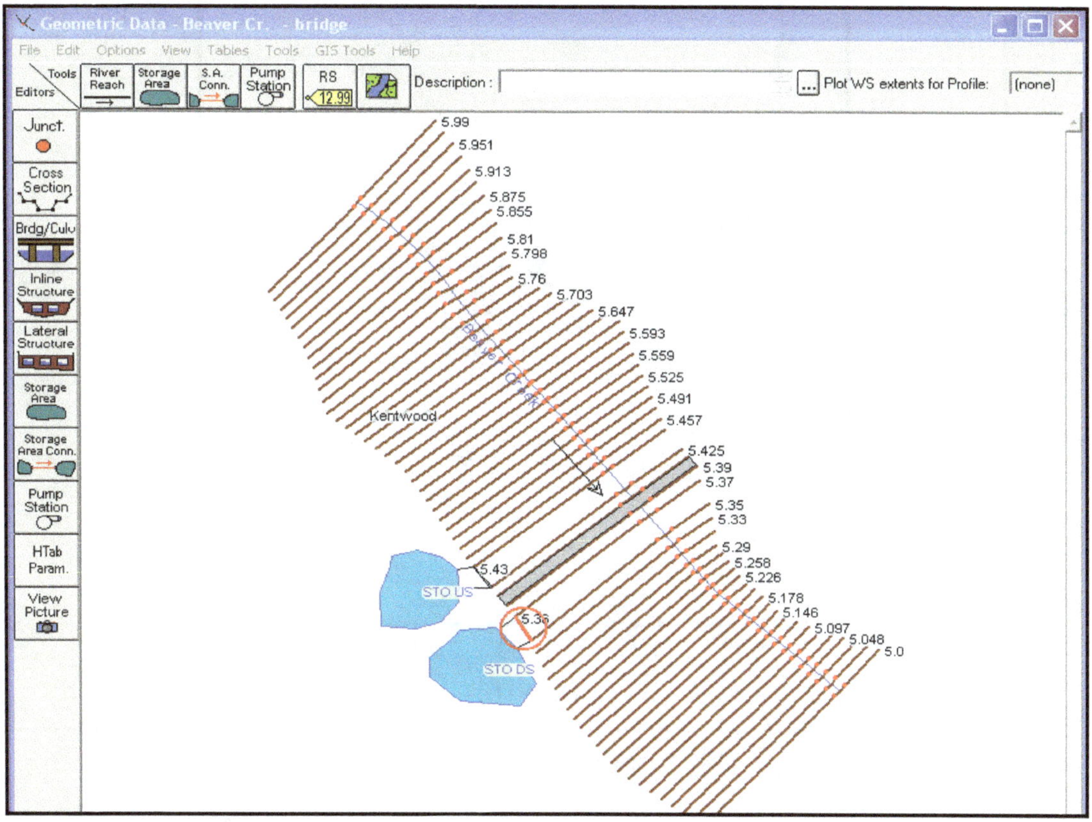

Figure 7.12. Illustration of an off-line storage area using the HEC-RAS computed model.

Cross sections are processed into tables of elevation versus hydraulic properties of area, conveyances, and storage (see Figure 7.13 for how the channel is subdivided). The user is required to set the interval to be used for spacing the point in the cross section tables. This interval is very important, in that it will define the limits of the table that is built for each cross section. The interval must be large enough to encompass the full range of stages that may be incurred during the unsteady flow simulations. On the other hand, if the interval is too large, the tables will not have enough resolution to accurately depict the changes in area, conveyance, and storage with respect to elevation. Another benefit of the hydraulic tables is that they can be used to troubleshoot geometric problems that arise when running unsteady flow models. The output from the geometric processor can be viewed either as hydraulic property tables or plots of the rating curves. Viewing the graphical output is a useful diagnostic tool for examining cross section geometry. The relationship of area, storage, and conveyance should be examined for abrupt changes with elevation. Any abrupt change should be reviewed to determine the overall significance for that particular run.

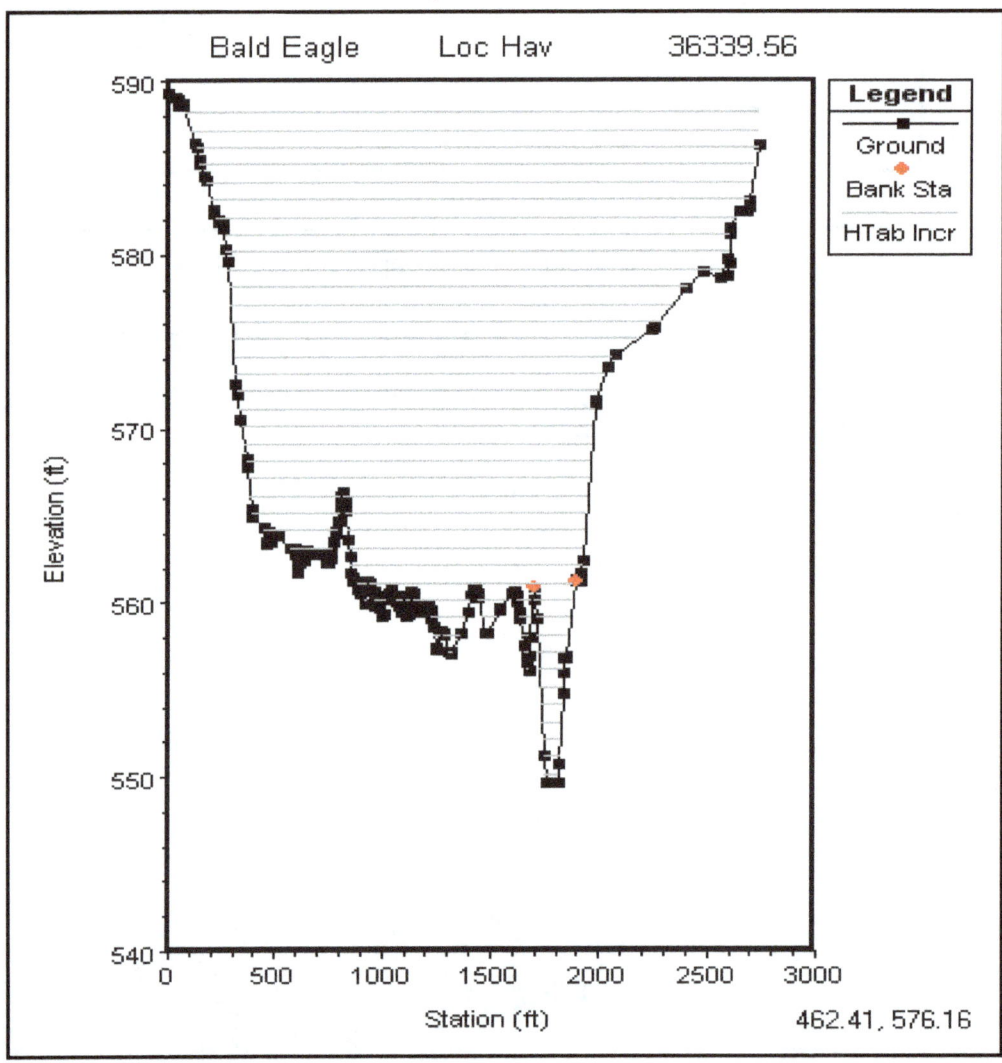

Figure 7.13. Cross section hydraulic table increments.

Figure 7.14 shows an anomaly in the conveyance for the right overbank flow at elevation 214.2. The cross-sectional plot shown in Figure 7.15 shows that the overbank Manning n value is a constant for the right overbank having a value of 0.06. Due to the computational scheme in HEC-RAS which subdivides the cross section by the number of horizontal n values, a problem could exist in determining the proper conveyance. By adding another Manning n value beginning at station 1024, the program will subdivide the cross section by the two n values and recompute the conveyance. Note that when the program was rerun with two n (but the same) values for the right overbank, the conveyance curve was much smoother as shown in Figure 7.16. The feature of viewing these characteristic curves can help the modeler troubleshoot many of the geometry properties that will cause the unsteady flow computation to be unstable.

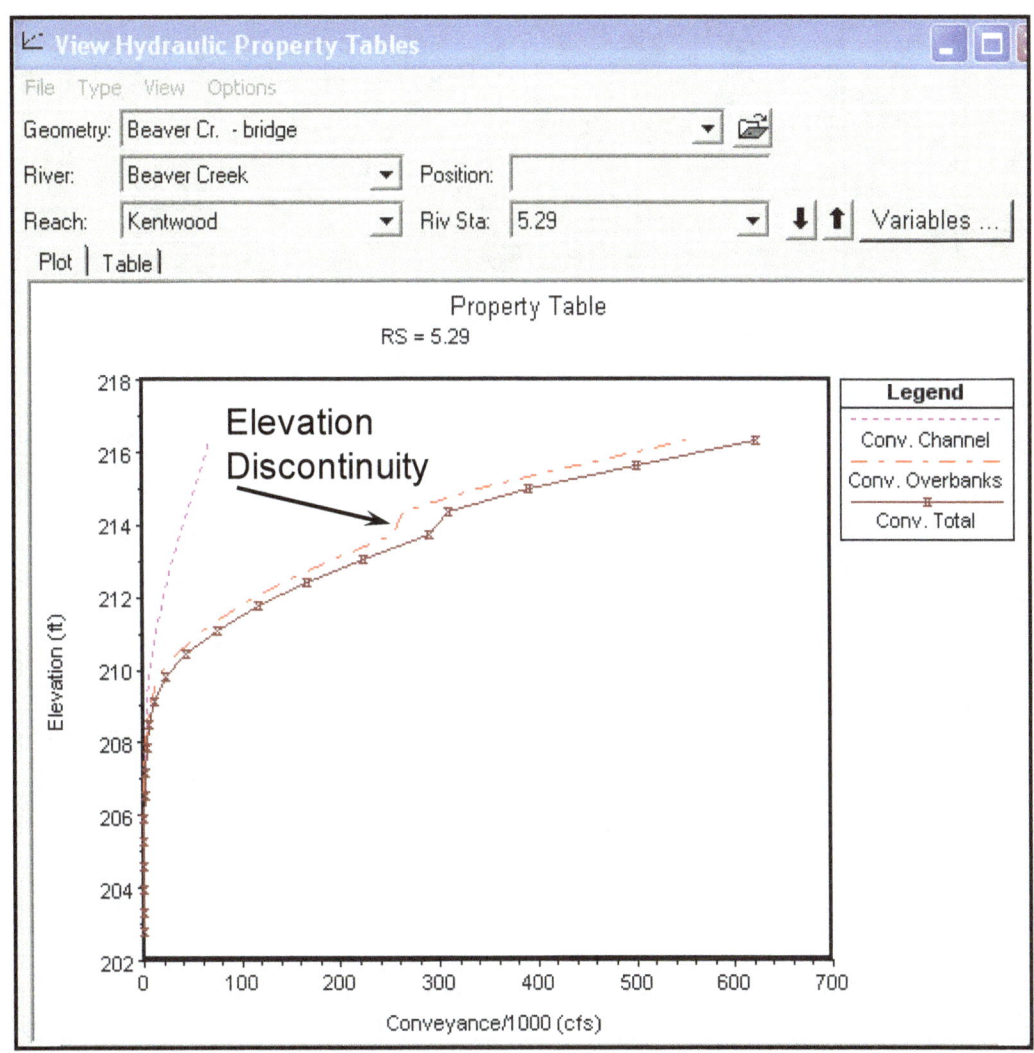

Figure 7.14. Conveyance properties versus elevation for a single cross section.

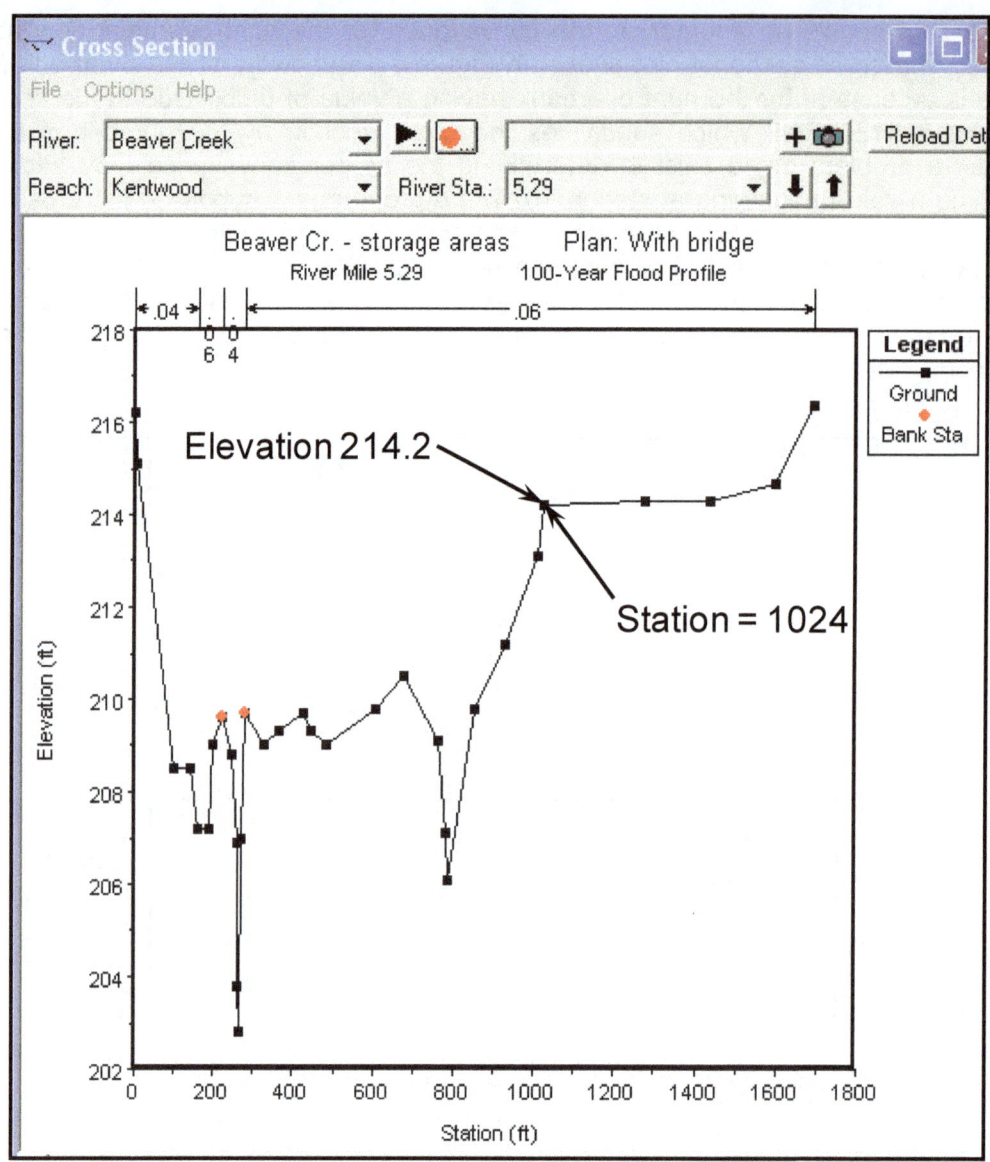

Figure 7.15. Channel cross section for an unsteady flow model.

7.8 NUMERICAL STABILITY

Model accuracy can be defined as how well the numerical solution matches the true solution. Accuracy depends upon the assumptions and limitations of the model (i.e., one-dimensional model with a single water surface verses a two-dimensional model). It also depends upon the accuracy of the (1) geometric data, cross section data, Manning n values, bridges and culverts; (2) flow data and boundary conditions; and (3) the numerical accuracy of the solution scheme.

If the one-dimensional unsteady flow equations are assumed to represent the flow conditions through a river system, then only an analytical solution of these equations will yield an exact solution. Therefore, any finite difference solutions are going to be approximate.

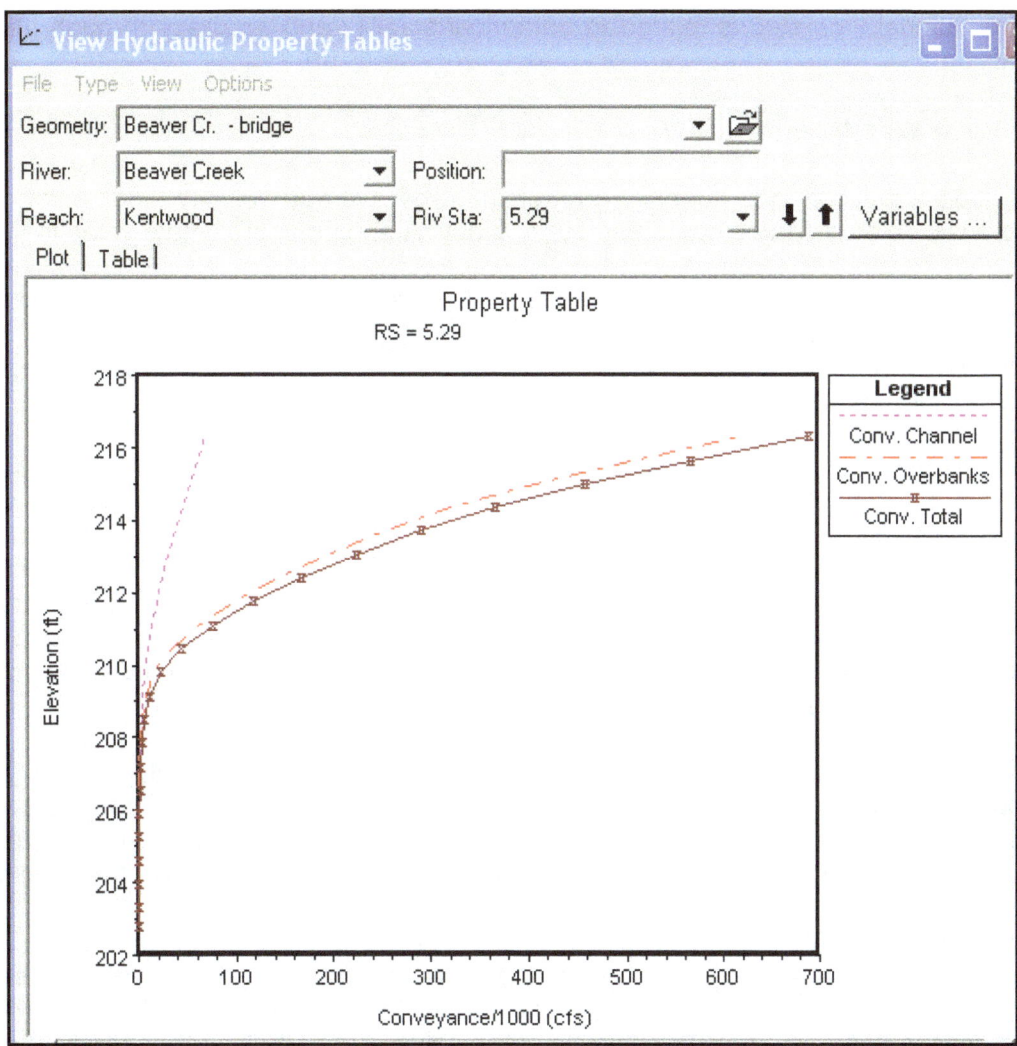

Figure 7.16. Conveyance properties versus elevation for a single cross section with right overbank subdivided.

A numerical model is unstable when numerical errors grow to the extent at which the solution begins to oscillate, or the errors become so large that the computations cannot continue. Factors affecting model stability include: cross section spacing, computational time step, theta weighting factor, solution iteration, and solution tolerances.

Rapidly rising hydrographs can cause computational problems, instability and non-convergence, when applied to numerical approximations of the unsteady flow equations. This is the case when an implicit, non-linear finite difference solution technique is used, which is the case for most of the numerical solutions of the Saint-Venant equations. However, many computational problems can be overcome with proper selection of the time step Δt and the distance step Δx.

Cross section spacing should be placed at representative locations to describe changes in geometry. Additional cross sections should be added at locations where changes occur in discharge, slope, velocity, and roughness. Cross sections also should be added at structures located along the river reach. Bed slope plays an important role in cross section spacing. Steeper slopes require more sections, and streams flowing at high velocities also will require more cross sections.

Computational time step is related to numerical stability and accuracy through the Courant Condition.

$$C_r = V_w(\Delta t/\Delta x) \leq 1.0 \text{ or } \Delta t \leq \Delta x/V_w \tag{7.12}$$

The flood wave speed is normally greater than the average velocity. For most rivers, the flood wave velocity can be calculated by $V_w = dQ/dA$ and an approximate value is $V_w = 1.5V$.

The Courant condition may yield time steps that are too restrictive (i.e., a larger time step could be used and still maintain accuracy and stability. Fread (1981) found for many practical unsteady flow problems that the Courant conditions can be relaxed and values greater than 1 yield satisfactory results.

The theta weighting factor is applied to the finite difference approximations when solving the unsteady flow equations. Theoretically, theta can vary from 0.5 to 1.0. However a practical limit is from 0.6 to 1.0. A theta of 1.0 provides the most stability, while a value of 0.6 provides the most accuracy. When choosing theta, there is a balance between accuracy and computational robustness. Larger values of theta produce solutions that are more robust and less prone to blowing up. Small values of theta, while more accurate, tend to cause oscillations in the solution, which are amplified if there are large numbers of internal boundary conditions.

At each time step derivatives are estimated, the equations are solved, and all of the computation nodes then are checked for numerical error. If the error is greater than the allowable tolerances, the program will iterate. The default maximum number of iterations in the HEC-RAS program is set at 20. More iterations will generally improve the solution.

Within the HEC-RAS program two solution tolerances can be set or changed. The water surface calculation is set to 0.02 feet and the storage area elevation solution is set at 0.05 feet. These default values should be acceptable for many river simulations. Making the tolerances larger can reduce the stability of the solution, and making them smaller can cause the program to go to the maximum number of iterations.

7.9 TWO-DIMENSIONAL UNSTEADY FLOW MODELS

The governing equations for two-dimensional unsteady flow (Saint-Venant equations) are presented in Chapter 6. The equations in two dimensions include additional terms, such as wind stress, that are difficult to represent in one-dimensional models.

Just as the one-dimensional unsteady flow solution is more complex than the one-dimensional steady flow solution, the two-dimensional solution is much more complex than the one-dimensional solution. The two-dimensional modeling approach is most appropriate to calculate:

- Water levels and flow distributions around islands
- Flow at bridges having one or more relief openings
- In extremely contracting and expanding reaches
- Into and out of off-channel storage or flow situations such as overtopping of a levee
- Flow at river junctions
- Circulation and transport in water bodies with wetlands
- Water surface elevations and flow patterns in large rivers, reservoirs, and estuaries

Boundary conditions are required throughout the simulation just as in the one-dimensional modeling. They are applied along the flow boundaries of the solution, and are required to eliminate the constants of integration when the governing equations are numerically integrated to solve for U, V, and h in the interior domain. External boundary nodes along the downstream end of the network are typically assigned a water surface elevation and boundary nodes along the upstream end of the network are typically assigned flow or discharge. The use of boundary condition specification removes either the depth, or one or both of the velocity components from the computations.

Dynamic simulations are used to model situations where water levels, flow rates, and velocities can change over time, such as an estuary where ocean tides influence the flow. For tidal flow situations, starting a model at low tide usually will more quickly attain realistic flow conditions throughout the model domain. This is somewhat similar to setting critical depth as a downstream boundary condition for a one-dimensional model; the M_2 curve will converge more quickly than an M_1 backwater curve. If good prototype tidal data is not accessible, then one alternative is to access or generate synthesized harmonic tidal data. Several software packages which will generate harmonic tidal data at most USGS station locations are available.

7.10 TIDAL WATERWAYS

The HEC-25 (FHWA 2004, 2008) manuals provide guidance on tidal hydrology, hydraulics, and coastal issues related to highways. Tidal waterways are defined as any waterway either dominated or influenced by tides and hurricane storm surges. The first step in evaluation of highway crossings is to determine whether the bridge crosses a river which is influenced by tidal fluctuations (tidally affected river crossing) or whether the bridge crosses a tidal inlet, bay or estuary (tidally controlled). The flow in tidal inlets, bays and estuaries is predominantly driven by tidal fluctuations (with flow reversal), whereas the flow in tidally affected river crossings is driven by a combination of river flow and tidal fluctuations. Therefore, tidally affected river crossings are not subject to flow reversal, but the downstream tidal fluctuation acts as a cyclic downstream control. Tidally controlled river crossings will exhibit flow reversal.

Tidally affected river crossings are characterized by both river flow and tidal fluctuations. From a hydraulic stand point, the flow in the river is influenced by tidal fluctuations which result in a cyclic variation in the downstream control of the tail water in the river estuary. The degree to which tidal fluctuations influence the discharge at the river crossing depends on such factors as the relative distance from the ocean to the crossing, riverbed slope, cross-sectional area, storage volume, and hydraulic resistance. Although other factors are involved, the relative distance of the river crossing from the ocean can be used as a qualitative indicator of tidal influence. At one extreme, where the crossing is located far upstream, the flow in the river may only be affected to a minor degree by changes in tail water control due to tidal fluctuations. As such, the tidal fluctuation downstream will result in only minor fluctuations in the depth, velocity, and discharge through the bridge.

As the distance from the crossing to the ocean is reduced, again assuming all other factors as equal, the influence of the tidal fluctuations increases. Consequently, the degree of tail water influence on flow hydraulics at the crossing increases. A limiting case occurs when the magnitude of the tidal fluctuations is large enough to reduce the discharge through the bridge crossing. Because of the storage of the river flow at high tide, the ebb tide will have a larger discharge and velocities than the flood tide.

Wind is a significant component of surge at a coastline. The U.S. Army Corps of Engineers Storm Surge Analysis manual (USACE 1986b) indicates that wind is the greatest component of storm surge and that the peak surge occurs in the area of maximum winds. Use of a wind field as a two-dimensional boundary condition may be necessary to model some tidal waterway conditions. In determining the forces on bridges, the properties of the flow that have the greatest impact are the height of the water and its velocity (FHWA 2009c). Wind fields can create waves that significantly affect the bridge structure and therefore need to be analyzed in all river crossings that are tidally influenced.

The damage to highway bridges in recent hurricanes was due primarily to wave attack on storm surge (see Chapter 10). The damage was caused as the storm surge raised the water level to an elevation where larger waves could strike the bridge superstructure. Individual waves produce both an uplift force and a horizontal force on the bridge decks. The magnitudes of these forces depend on wave characteristics and on the inundation of the deck. The magnitude of wave uplift force from individual waves can exceed the weight of the simple span bridge decks. The total resultant force is able to overcome any resistance provided by the typically small connections between the pile caps and bridge decks. The decks begin to progressively slide, "bump," or "hop" across the pile caps in the direction of wave propagation.

The buoyancy of the bridge decks caused by air pockets trapped under the bridge decks contribute to the total force on the individual bridge decks when the deck is submerged, i.e., when the storm surge elevation exceeds the bridge deck elevation. However, bridge decks that were elevated above the storm surge still-water elevation were still damaged in both Hurricanes Ivan and Katrina by waves.

This conclusion, that wave loads were the primary cause of damage, is based on post-storm inspections of the damaged bridges in combination with numerical model hindcasts of the wave and surge conditions during the storms, some exploratory laboratory tests, and a review of the related coastal engineering literature.

CHAPTER 8

BRIDGE SCOUR CONSIDERATIONS AND
SCOUR COUNTERMEASURE HYDRAULIC ANALYSIS

8.1 INTRODUCTION

The most common cause of bridge failures is from floods eroding bed material from around bridge foundations. Scour is the engineering term for the erosion of soil, alluvium or other materials surrounding bridge foundations (piers and abutments) by flowing water. The HEC-18 and HEC-20 manuals (FHWA 2012b, 2012a) are the primary FHWA resources for guidance on evaluating scour and stream instability at bridge crossings. Safe bridge design must account for scour conditions that may occur over the life of the bridge. Scour is greatest during flood events when flow velocity and depth is highest, but the event-related scour is in addition to the long-term stream instability components of channel shifting, aggradation, and degradation.

Each of the scour components discussed in this chapter should be considered during bridge design. It is important for bridge engineers to recognize that these scour and stream instability components be considered over the life of the bridge. No equations for predicting scour are provided in this chapter because updated equations and procedures may be incorporated into future versions of HEC-18 and HEC-20, and because every type of scour is not discussed in this chapter.

The FHWA HEC-18 and HEC-20 manuals are the primary source of guidance and procedures for incorporating scour and stream instability into safe bridge design. The American Association of State Highway and Transportation Officials (AASHTO 2010) LRFD Design Specifications includes the following statements, regarding the factors related to scour and stream instability that should be considered in bridge design:

- Evaluation of bridge design alternatives shall consider stream instability, backwater, flow distribution, stream velocities, scour potential, flood hazards, tidal dynamics (where appropriate) and consistency with established criteria for the National Flood Insurance Program.
- Studies shall be carried out to evaluate the stability of the waterway and to assess the impact of construction on the waterway.
- (Consider) whether the stream reach is degrading, aggrading, or in equilibrium.
- (Consider) the effect of natural geomorphic stream pattern changes on the proposed structure.
- For unstable streams or flow conditions, special studies shall be carried out to assess the probable future changes to the plan form and profile of the stream and to determine countermeasures to be incorporated in the design, or at a future time, for the safety of the bridge and approach roadways.
- For the design flood for scour, the streambed material in the scour prism above the total scour line shall be assumed to have been removed for design conditions.
- Locate abutments back from the channel banks where significant problems with ice/debris buildup, scour, or channel stability are anticipated.
- Design piers on floodplains as river piers. Locate their foundations at the appropriate depth if there is a likelihood that the stream channel will shift during the life of the structure or that channel cutoffs are likely to occur.

- The added cost of making a bridge less vulnerable to damage from scour is small in comparison to the total cost of a bridge failure.

This is only a partial list of AASHTO's specifications and commentary related to scour and stream instability. These topics are a significant aspect of safe bridge design, and are a complex combination of hydrologic, hydraulic, fluvial-geomorphic, erosion, scour, sediment transport, geotechnical, and structural considerations. The following sections describe scour and stream instability processes, how to obtain data from hydraulic models for computing scour, and numerical modeling of scour countermeasures.

8.2 SCOUR CONCEPTS FOR BRIDGE DESIGN

During flood flows, water is conveyed in the river channel and in the floodplains adjacent to the channel. Figure 8.1 illustrates a representative bridge crossing and the flow characteristics for a flood condition. The figure includes streamlines and velocity contours. The more widely spaced streamlines are in the floodplain and divide the flow into 5 percent increments. The closely spaced streamlines are in the channel and divide the flow into 10 percent increments of flow. The road embankments constrict the flow into the bridge opening where flow velocities are highest. Upstream of the bridge constriction, where flow is fully expanded in the floodplain, approximately 65 percent of the flow is in the channel and 35 percent is in the floodplains. In the bridge opening, approximately 90 percent of the flow is in the channel and 10 percent is in the floodplain area between the channel banks and abutments (setback area). Flow velocity is less than 5.7 ft/s (1.7 m/s) in the upstream channel and 1.2 ft/s (0.37 m/s) in the upstream floodplains. This compares with velocities in bridge opening as high as 8.8 ft/s (2.7 m/s) in the channel and 4.4 ft/s (1.34 m/s) in the setback areas. The higher velocities in the bridge opening generate much higher shear stresses and are much more erosive than the upstream flow velocities. In addition to the increased velocities, bridge structural elements (piers and abutments) locally obstruct flow and cause additional erosion at these locations.

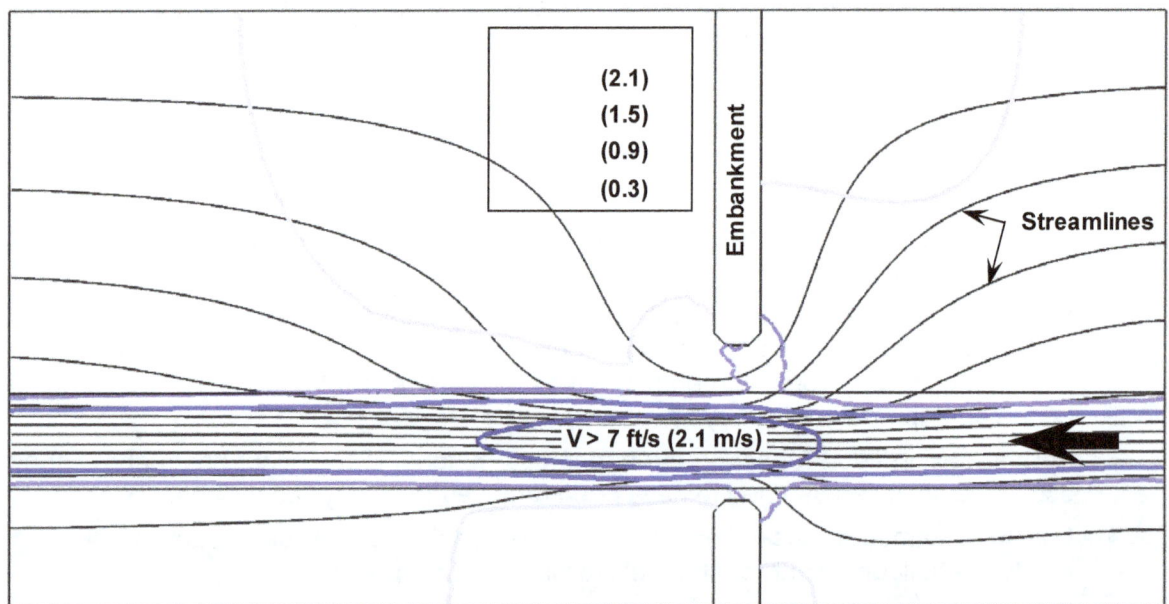

Figure 8.1. Velocity and streamlines at a bridge constriction.

Scour is a significant concern during extreme flood events and bridges should be designed to withstand the scour produced by these events. Channel geometry, which includes aggradation, degradation, channel shifting, and channel widening, also changes during the life of a bridge. Therefore, potential for stream instability should be a part of safe bridge design.

8.3 TYPES OF SCOUR

8.3.1 Contraction Scour

Contraction scour is a sediment imbalance process that occurs during floods when the sediment supply from upstream is less than the sediment transport capacity in the bridge opening. There are two sediment supply conditions for contraction scour; clear water and live bed. Clear-water contraction scour occurs when the upstream flow velocity is insufficient to transport bed material. The HEC-18 manual (FHWA 2012b) includes equations for determining the critical velocity when bed material movement is initiated, which depends on flow depth and particle size. Clear-water conditions occur for fine sediment sizes (sands and fine gravel) only when flow velocity is small and for coarse sediment sizes (coarse gravel and cobbles) even for relatively high velocity. Live bed conditions occur when there is sufficient flow velocity to transport bed material upstream of the bridge. Very fine sediment (clay and silt) is often not found in channel beds in significant amounts and does not generally play a role in either clear-water or live-bed contraction scour. The water may be turbid due to suspended transport of silt and clay, but is still considered as clear-water from the standpoint of bed material transport.

For clear-water contraction scour, the flow velocity in the bridge opening is sufficient to move bed material even though the upstream flow velocity is too low for bed material movement. For live-bed contraction scour, the higher flow velocity in the bridge opening has a greater capacity for transporting sediment that is the upstream flow velocity. In either case, there is an imbalance between sediment supply and sediment transport capacity, and contraction scour occurs. The channel bed erodes and lowers, thereby increasing the flow depth and decreasing the flow velocity until the bed material transport capacity equals the supply from upstream. The erosion process takes time so depending on the duration of the flood, the ultimate scour may not be achieved. Accurate contraction scour calculations depend on having accurate estimates of flow distribution at the approach and bridge cross sections. Flow is divided into channel, left floodplain and right floodplain in the fully expanded flow upstream of the bridge, and divided into channel, left setback (floodplain) and right setback areas under the bridge. These subarea discharges control the contraction scour process.

<u>Live-Bed Contraction Scour.</u> Live-bed scour almost always occurs in river channels during flood events. Exceptions to this expectation are boulder-bed and bedrock channels that are not alluvial. Channels that have significant levels of diversion and/or flood control may also not be live-bed because the channel forming flows no longer occur. Figure 8.2 includes a plan and profile sketch to illustrate the flow variables for live-bed contraction scour. At the approach section (cross section 1), the flow velocity in the river channel is high enough to transport bed material. The total sediment transport in the approach channel depends on the flow depth (y_1), velocity (V_1), discharge (Q_1), width (W_1), and sediment size (represented by the median bed material particle size, D_{50}). At the bridge section (cross section 2), floodplain flow has entered the channel so the channel discharge (Q_2), velocity, and sediment transport capacity are greater than in the channel at the approach section. A hydraulic model includes a surveyed cross section at the bridge so the flow depth in the model is a pre-scour depth (y_0). The channel width at the bridge section (W_2) is often similar to the upstream width, but

may be wider or narrower. The bridge section channel may also be partially blocked by piers or by abutments that encroach into the channel, which results in W_2 less than W_1. The live-bed scour equation is presented in HEC-18 (FHWA 2012b). The equation yields the total flow depth including scour (y_2), and the scour is the difference between this depth and the pre-scour flow depth ($y_{s-c} = y_2 - y_0$). Because it is assumed that bed material size is consistent along the channel reach, bed material size is only used to determine whether or not live-bed conditions exist and does not actually appear in the live-bed contraction scour equation. For live-bed conditions, a functional relationship for contraction scour is:

$$y_{s-c} = fn\ (y_1, Q_1, W_1, Q_2, W_2, y_0) \tag{8.1}$$

Larger amounts of contraction scour occur for greater differences between channel discharge at the approach and bridge sections. Also, scour increases for narrower channel widths at the bridge section. The worst case live-bed contraction scour occurs then the bridge abutments and road embankments encroach into the channel and the entire floodplain flow is conveyed in the constricted channel at the bridge opening. Live-bed contraction scour decreases as the abutments are set back farther from the channel banks and when fewer or narrower piers are located in the channel.

Clear-Water Contraction Scour. Clear-water contraction scour is expected in the setback areas under a bridge. The fully expanded floodplain flow upstream of the bridge usually has a low velocity and would not be expected to mobilize the granular floodplain materials. Floodplains are also often comprised of cohesive materials and vegetated. Therefore, although very fine particles (silts and clays) may be transported in suspension, there is little potential for bed material transport or live-bed scour in floodplains. Flow velocity in the setback area under the bridge is, however, often high enough to cause erosion. Clear water scour is, therefore, an erosion process based on flow velocity and shear stress. Figure 8.3 includes a plan and profile sketch of the clear-water contraction scour variables. The important variable at the approach section (section 1) is velocity (V_1), but is only used to determine whether the velocity is less than the critical velocity for bed material transport. This comparison should be made if there is any uncertainty about whether the upstream flow is transporting bed material. The channel as well as the setback areas could have clear-water contraction scour, but most often only the setback areas will. If there is a relief bridge through the embankment on the floodplain, this opening will also typically have clear-water contraction scour.

The clear-water contraction scour equation is a function of only the hydraulic conditions in a particular subarea, not upstream conditions. These variables include discharge (Q), width (W), and flow depth before scour (y_0). Clear-water contraction scour occurs until the lowering of the ground, which increases depth and decreases flow velocity, produces a non-eroding velocity. The non-eroding velocity is a function of grain size (D_{50}) for non-cohesive soils and is a function of critical shear stress (τ_c) for cohesive soils. The HEC-18 manual (FHWA 2012b) includes equations for clear-water contraction scour. As with the live-bed contraction scour equation, the clear-water contraction scour equation yields a total depth including scour (y_2) and the predicted scour is the difference between this depth and the pre-scour depth ($y_{s-c} = y_2 - y_0$). For clear-water conditions, a functional relationship for contraction scour is:

$$y_{s-c} = fn(Q, W, D_{50}\ (\text{or}\ \tau_c), y_0) \tag{8.2}$$

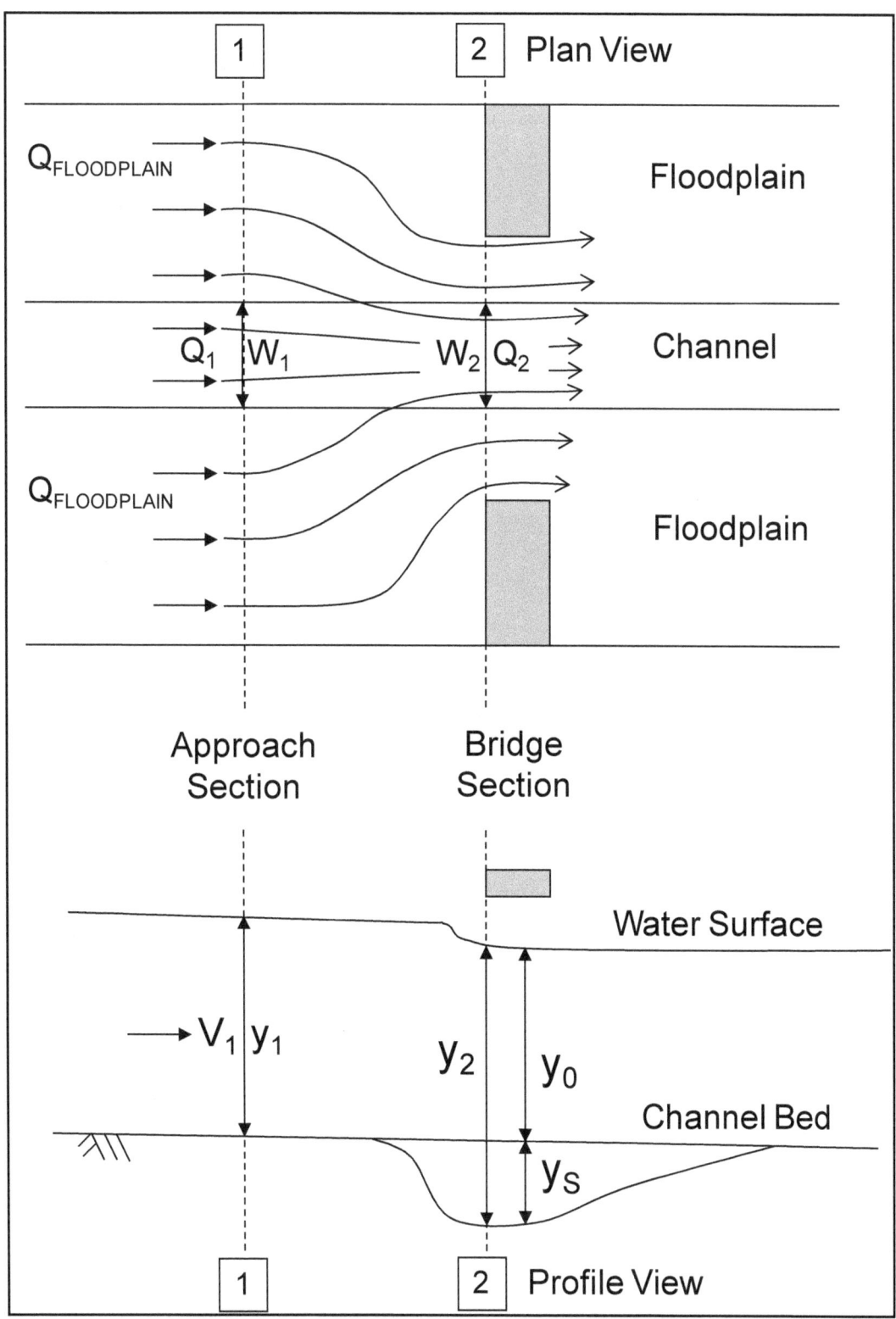

Figure 8.2. Live-bed contraction scour variables.

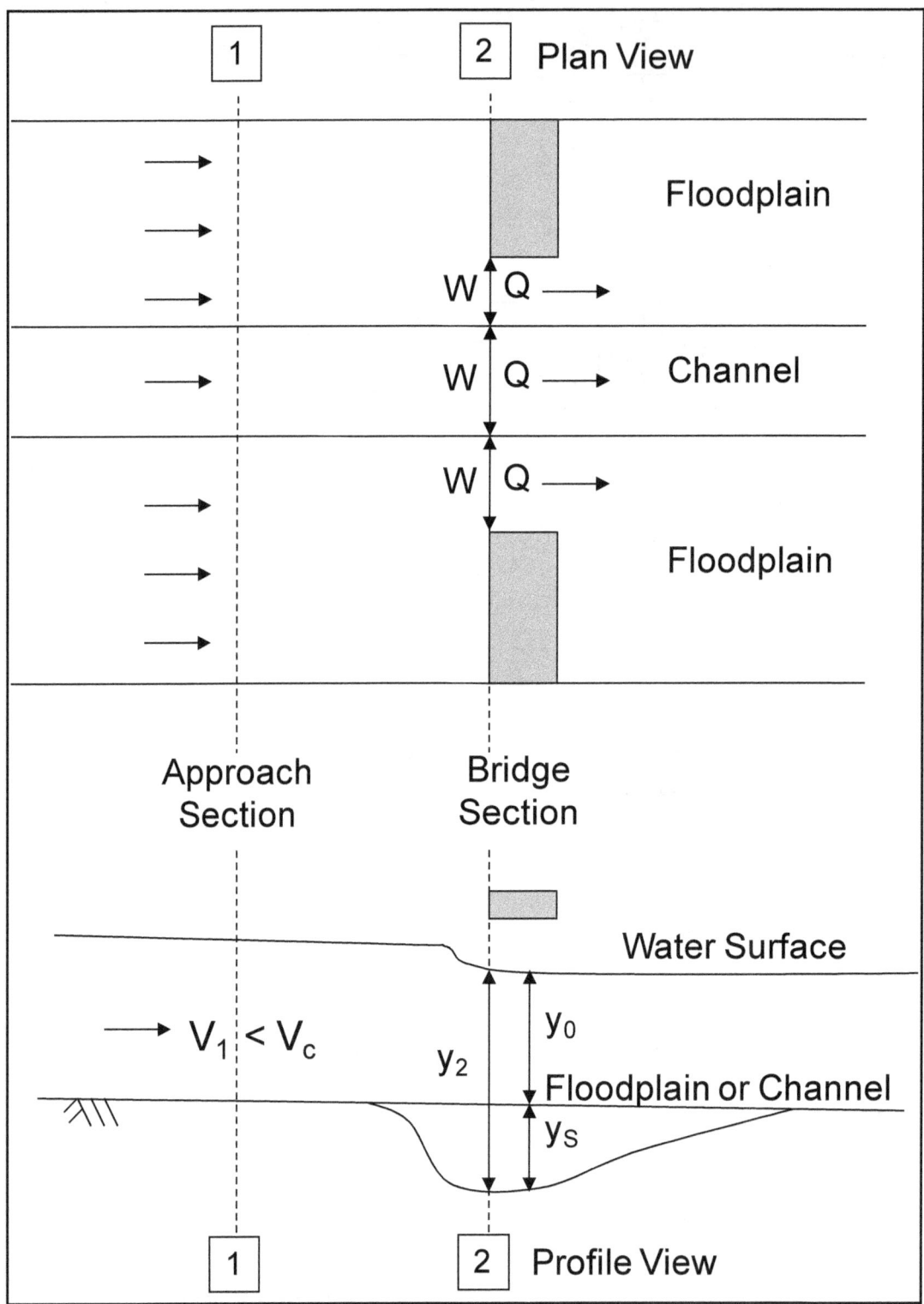

Figure 8.3. Clear-water contraction scour variables.

Vertical Contraction Scour (Pressure Scour). Pressure scour is another type of contraction scour, but created by a vertical constriction rather than a horizontal constriction. It can occur even if no horizontal constriction is present. Pressure scour may be either live-bed or clear-water depending on the upstream flow and sediment characteristics. Prediction of pressure flow scour at an inundated bridge deck may be important for safe bridge design and for evaluation of scour at existing bridges. An experimentally and numerically calibrated-formula reference was developed by FHWA (2012c) to calculate pressure scour depth under various deck inundation conditions. This formula is included in HEC-18 (FHWA 2012b). Figure 8.4 illustrates the flow characteristics at a fully submerged bridge deck. The depth in the area of maximum scour is comprised of three components, which are h_c (the vertically contracted depth not including scour), y_s (the scour depth), and t (the boundary layer thickness). Pressure conditions can significantly increase total scour at a bridge because flow depth is referenced to the bottom of the bridge deck rather than the water surface, and because the boundary layer thickness is an additional scour component. Using the contraction scour, y_{s-c}, computed from the relationships represented in Equations 8.1 and 8.2, and referencing flow depth to the bottom of the deck the functional relationship for vertical contraction scour is:

$$y_{s-vc} = fn\ (y_{s-c},\ t) \tag{8.3}$$

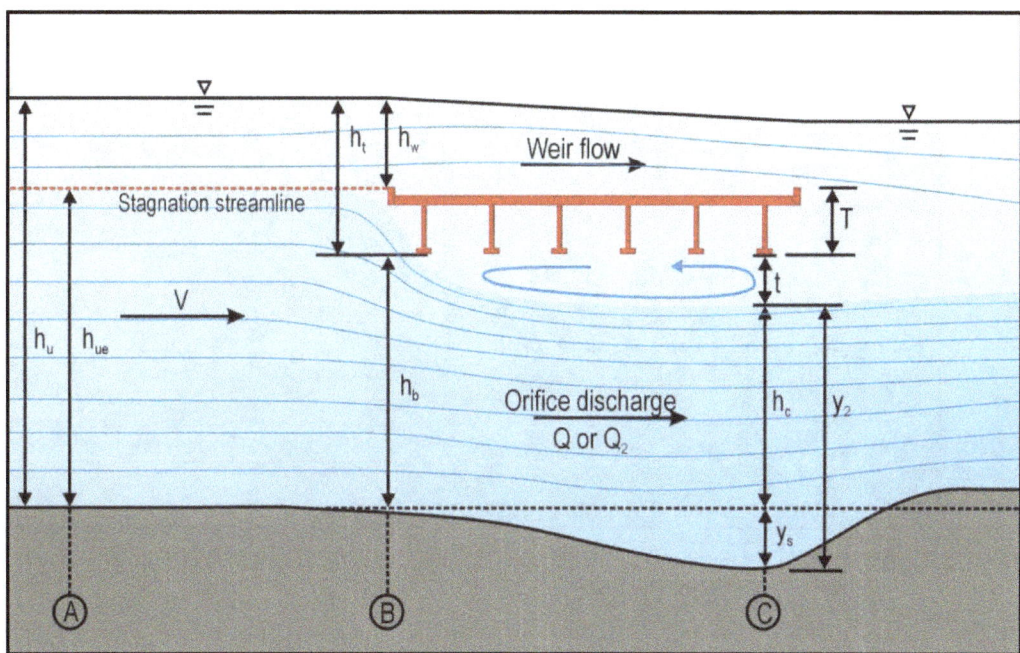

Figure 8.4. Vertical contraction scour.

8.3.2 Local Scour

Local scour occurs where the flow field is disrupted by an obstruction. The term "local" is used because scour is in the vicinity of the obstruction, not across the entire channel or bridge section. The flow is redirected and accelerates, vortexes form, and there is increased turbulence. The two most common types of local scour at bridges are pier scour and abutment scour. Ice and debris can also impact local scour.

Pier Scour. Pier scour is illustrated in Figure 8.5. The velocity upstream of the pier accelerates around the pier and flow is directed downward along the front face of pier. A "horseshoe" vortex forms where the downward flow reaches the bed and the size of the vortex increases as the scour hole enlarges. The flow around the pier sheds vortexes on the sides of the pier. Sediment deposition occurs in the wake area downstream of the pier. Pier scour can occur in clear-water or live-bed conditions. There are many factors that influence the magnitude of pier scour. Hydraulic factors are velocity (V), depth (y), and angle of attack (θ) of the flow approaching the pier, but outside the influence of the pier. Pier shape (ξ, including circular, square, sharp, rounded-, or rectangular-nosed), pier width (a), and pier length (L) also contribute significantly to the amount of pier scour. Complex pier geometries that include pile groups, pile caps, and footings must also be considered when computing pier scour. Sediment size (D_{50}), density ($\Delta = \rho_s/\rho - 1$) and gradation (σ) are included in some pier scour equations. Although pier scour may appear to be a relatively simple process, the calculations are often cumbersome for all but the simplest cases. HEC-18 (FHWA 2012b) includes several pier scour equations for various conditions. A functional relationship for pier scour is:

$$y_{s-p} = fn\ (V,\ y,\ \theta,\ a,\ L,\ \xi,\ D_{50},\ \Delta,\ \alpha) \tag{8.4}$$

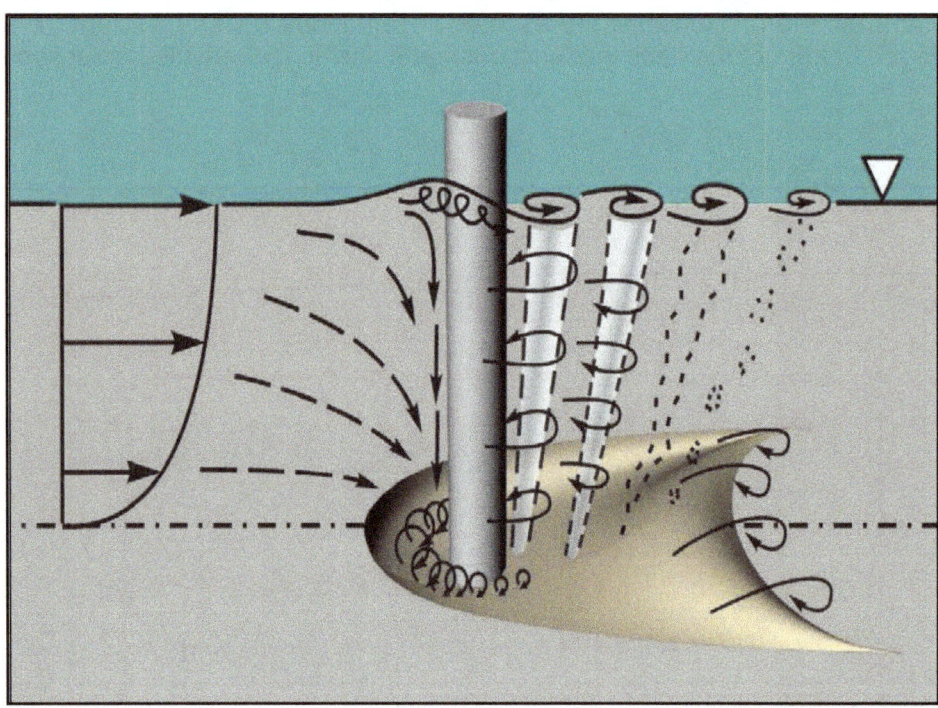

Figure 8.5. The main flow features forming the flow field at a cylindrical pier (NCHRP 2011a).

Abutment Scour. Scour occurs at abutments when the roadway embankment and abutment obstruct the flow. Abutment scour is a type of local scour, but is related to contraction scour because the embankment is the primary cause of flow constriction.

NCHRP (2011b) conducted an evaluation of abutment scour processes and prediction methods. The conclusions and recommendations that pertain to abutment scour evaluation and safe bridge design include:

- Contraction scour should be viewed as the reference scour depth for calculating abutment scour. Abutment scour should be taken as the product of the contraction scour caused by flow acceleration through the constricted opening multiplied by a factor accounting for large-scale turbulence. This approach would replace the current approach for adding contraction scour to a separately computed abutment scour.

- Abutments should be designed to have a minimum setback distance from the channel bank of the main channel with riprap protection of the embankment and a riprap apron to protect against scour. The setback distance should accommodate the apron width recommended in HEC-23 (FHWA 2009a).
- Two-dimensional models should be used on all but the simplest bridge crossings as a matter of course.

Abutment foundations should be designed to be safe from long-term degradation, lateral migration, and contraction scour; and protected from abutment scour with riprap and/or guide banks, dikes, or revetments. NCHRP (2010a) developed abutment scour equations that account for a range of abutment types, abutment locations, flow conditions and sediment transport conditions. HEC-18 (FHWA 2012b) includes these relationships. Figure 8.6 illustrates abutment scour processes. Where abutments are set well back from the channel, abutment scour is located entirely on the floodplain. Where abutments are set in or close to the channel, abutment scour can occur entirely in the channel or in the floodplain and channel. When the abutment is set close to the channel the channel sediment and floodplain soil characteristics, including grain size and cohesion, factor into the proportion of scour that will occur in the floodplain versus channel. Because abutment scour is related to contraction scour, the scour relationship is similar in form to contraction scour, $y_{s-a} = y_{max} - y_0$. y_{s-a} is the abutment scour depth, y_{max} is the maximum flow depth resulting from the combination of contraction and abutment scour, and y_0 is the flow depth in the vicinity of the abutment prior to scour. The value of y_{max} is related to y_2 in the live-bed and clear-water contraction scour equations and to an amplification factor, α_A, related to large-scale turbulence structures. Abutment shape (ξ, including spill-through and wing wall abutment types) also affects abutment scour. An abutment scour functional relationship can be expressed as:

$$y_{s-a} = fn(y_2, \alpha_A, y_0) \tag{8.5}$$

Abutment scour can result in geotechnical failures of the embankment or channel bank materials. Once the geotechnical failure depth is reached, scour will not increase in depth but will progress laterally, potentially creating a free-standing abutment foundation that would act more as a pier from the standpoint of scour.

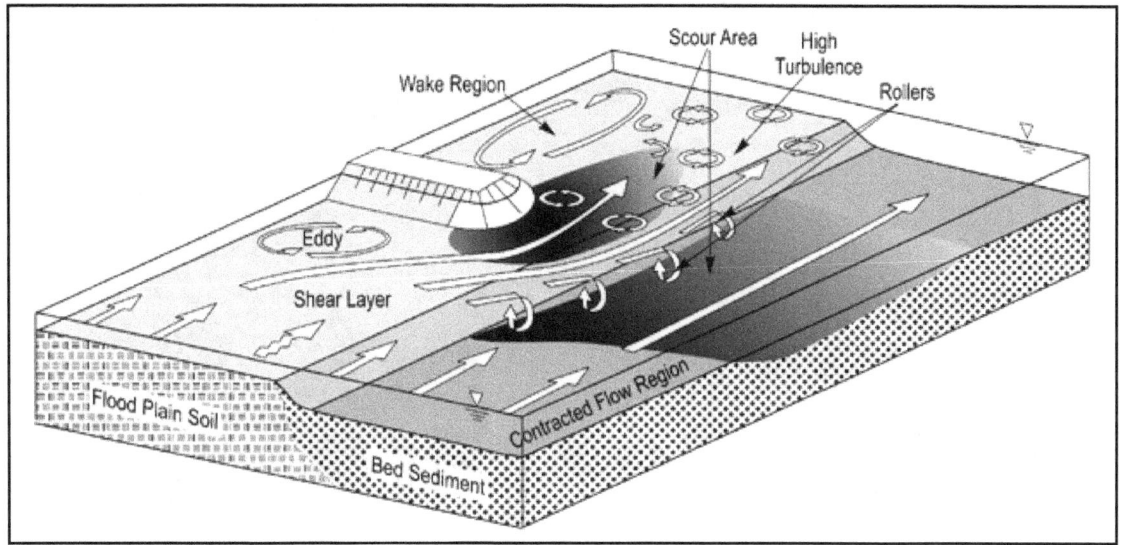

Figure 8.6. Flow structure in floodplain and main channel at a bridge opening (NCHRP 2011b).

8.3.3 Debris Scour

Debris is a common problem at bridges, especially during floods. Debris loading and impact forces can damage piers, decks, and girders and debris can reduce the waterway opening thereby increasing upstream flooding. All types of scour can be increased due to debris collection. Contraction scour is increased when debris blocks a portion of the bridge opening and pressure scour is increased when debris collects on the bridge deck and girders.

Increased pier scour, as shown in Figure 8.7, is the most common type of debris scour problem. Debris clusters are highly variable from one bridge location to another and at the same bridge from one flood to the next (NCHRP 2010b). HEC-20 (FHWA 2012a) provides guidance on identifying upstream debris production potential and debris collection potential depending on pier location in the channel. HEC-18 (FHWA 2012b) provides scour relationships for debris clusters on piers. Figure 8.8 illustrates that debris scour at a pier depends on the flow impacting the pier and the flow plunging below the debris blockage. The plunging flow creates a scour hole just downstream of the leading edge of the debris cluster and the pier obstruction creates a local scour hole. The maximum scour depth at a pier may occur when the debris cluster size and flow depth cause these two scour holes to coincide. This debris cluster size may not be the largest anticipated to collect at a pier.

HEC-RAS (USACE 2010c) can hydraulically simulate debris blockage at individual piers (Figure 8.9) by reducing flow area in the bridge opening. Simulating debris in two-dimensional models would require use of the methods discussed in Chapter 6 for including pier drag. For debris collected at the deck, the low chord of the bridge would need to be adjusted.

Figure 8.7. View down at debris and scour hole at upstream end of pier.

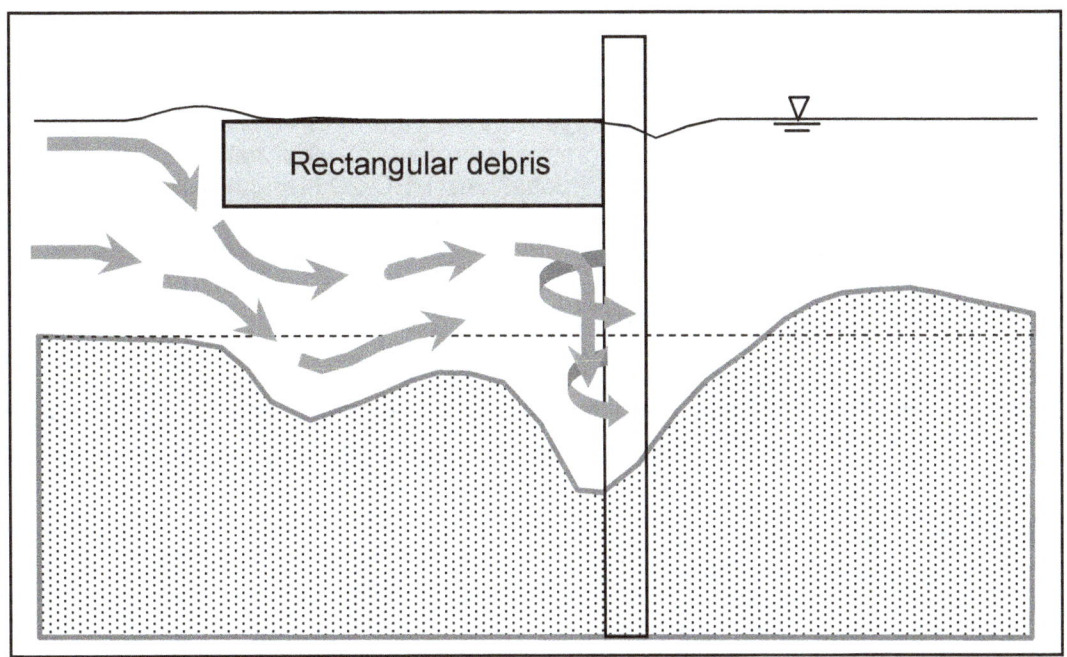

Figure 8.8. Idealized flow pattern and scour at pier with debris.

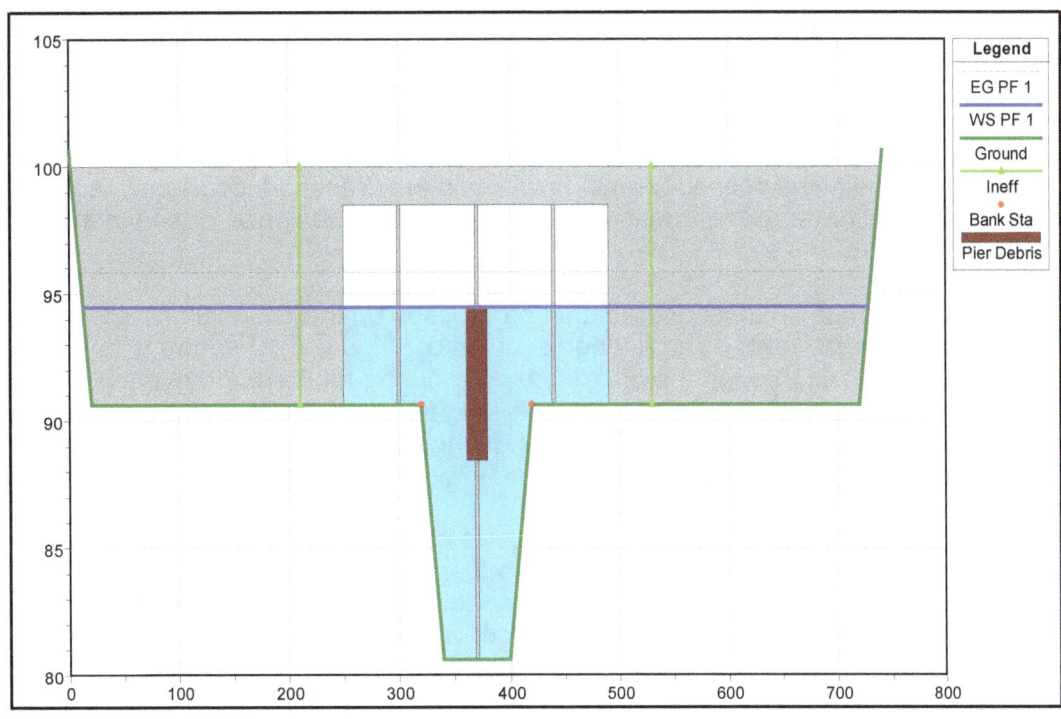

Figure 8.9. Debris in HEC-RAS hydraulic model.

8.3.4 Channel Instability

In the context of safe bridge design, channel instability includes any channel change that can threaten a bridge foundation. The change may be natural or result from a variety of human activities. Channel instability can create changes in channel geometry that expose foundations and increase scour during floods. The HEC-20 Manual (FHWA 2012a) provides guidance on evaluating channel instability at bridges. Even though these changes may be gradual or episodic, they are usually cumulative and are considered long-term because they alter the channel over the life of the bridge. Therefore, the potential for vertical and horizontal change must be considered in safe bridge design.

Channel instability not only considers the existing conditions, but also potential future conditions. Factors that may need to be considered when assessing potential channel instability include:

- Channel size and form
- Flow and flood history
- Valley and floodplain setting
- Geologic and other vertical or horizontal controls
- Channel and floodplain materials
- Vegetation and land-use
- Sediment sources and supply

Vertical Instability. Vertical change includes aggradation and degradation resulting from a long-term excess or deficit in sediment supply, and from degradation caused by headcutting. Long-term trends in discharge also impact channel geometry because channels that convey larger flows tend to be wider and deeper. If a channel consistently conveys more water than it has historically, the channel will enlarge. This can occur due to increased runoff from urbanization, from climate change, and many other causes. Bridge inspection files that include repeat cross section measurements are useful in identifying aggradation and degradation problems and trends. The sediment transport chapter (Chapter 9) includes the discussion of sediment continuity and how sediment transport concepts can be used to analyze aggradation and degradation when there is an imbalance of sediment supply and transport capacity.

Headcuts occur when channel degradation progresses up the channel and are caused when the downstream base level of a channel is lowered. Figure 8.10 shows a headcut that will migrate upstream and through the bridge crossing during future runoff events. Features of a headcut that can threaten a bridge include long-term degradation that persists after the headcut has migrated upstream of the bridge, plunge pool when headcut is under the bridge, and channel widening that occurs because bed lowering can destabilize channel banks.

Lateral Instability. Figure 8.11 shows progressive channel migration over a 72 year period at a highway crossing. The channel banks were identified and traced from historic and recent aerial photography. These banklines not only show trends of channel migration down valley and across valley, but also variability in channel width through time. The channel migration process includes erosion of the bank materials, bank geotechnical failures, transport of the eroded and failed materials, and sediment accretion on the insides of bends (point bars). Reviewing historic aerial photography is not only useful for identifying the potential for lateral instability problems at a bridge, but can be used to make predictions of channel location during the life of the bridge. These photo-comparison techniques are presented in HEC-20 (FHWA 2012a). As illustrated in Figure 8.12, a single flood can also cause extreme channel migration and widening, which for some regions can present significant challenges for bridge design.

Figure 8.10. Headcut downstream of a bridge.

Figure 8.11. Meander migration on Wapsipinicon River near De Witt, Iowa.

Figure 8.12. Channel widening and meander migration on Carson River near Weeks, Nevada.

8.3.5 Evaluating Channel Instability

Vertical and lateral instability is often identified during bridge inspections, through channel reconnaissance during bridge design, or through comparison of recent and historic aerial photography. Hydraulic modeling and sediment transport analysis can also be used to evaluate channel instability. As discussed in Chapter 9, sediment transport analysis can be used to evaluate channel aggradation and degradation trends over the life of a bridge. Even when sediment transport modeling is not performed, hydraulic models, especially two-dimensional models, can provide insight into vertical and lateral channel instability potential. Locations where channel flow velocity is much higher than up- or downstream may be prone to bed or bank erosion. Models can be used to predict a future condition, but they can also be used to evaluate potential future conditions by configuring the model for expected channel changes.

Model results should never be interpreted without considering the river characteristics. Geologic controls, sediment characteristics, vegetation characteristics, and manmade features may counteract erosion that may be expected from reviewing model results. It is important the channel reconnaissance be performed and that the hydraulic engineer develops an understanding of a wide range of fluvial geomorphic processes and potential channel response as discussed in HEC-20 (FHWA 2012a).

8.4 COMPUTING SCOUR

Each of the types of scour relies on hydraulic variables as input to the scour calculations. These variables include velocity, depth, discharge, flow width, unit discharge, and flow direction. The quality and accuracy of hydraulic modeling directly impact the accuracy of scour calculations. If model geometry is inaccurate, bank stations are not correctly or consistently defined, Manning n values are not accurate, or model assumptions are violated, then the poor quality of the hydraulic input data used in scour calculations can result in unreasonable and incorrect scour estimates.

The variables listed above all depend on the suitability of the hydraulic model to define flow distribution. For pier scour, the velocity and depth upstream of the pier are required input. For contraction scour, the amounts of flow in the channel relative to the floodplains both upstream and in the bridge opening are required. Abutment scour depends on the same flow distribution information as contraction scour, but also requires an estimate of flow concentration adjacent to the abutment.

The rest of this section provides discussion of extracting the necessary hydraulic information from one- and two-dimensional models. Recognizing that two-dimensional models provide more accurate representations of the flow field and flow distribution, FHWA encourages the use of two-dimensional modeling for all but the most straightforward bridge crossings.

8.4.1 One-Dimensional Models

Figure 8.13 shows the minimum number of cross sections for a one-dimensional bridge hydraulic model. The Exit cross section is required to establish the downstream boundary condition for the model. Contraction and abutment scour calculations require channel and floodplain discharges at the Approach section and bridge crossing. In a HEC-RAS model, the Crossing includes bridge and roadway geometry data placed between two cross sections that are adjacent to the bridge and roadway. One-dimensional models can provide estimates of hydraulic variables by computing incremental conveyance throughout the cross section and distributing flow in proportion to conveyance.

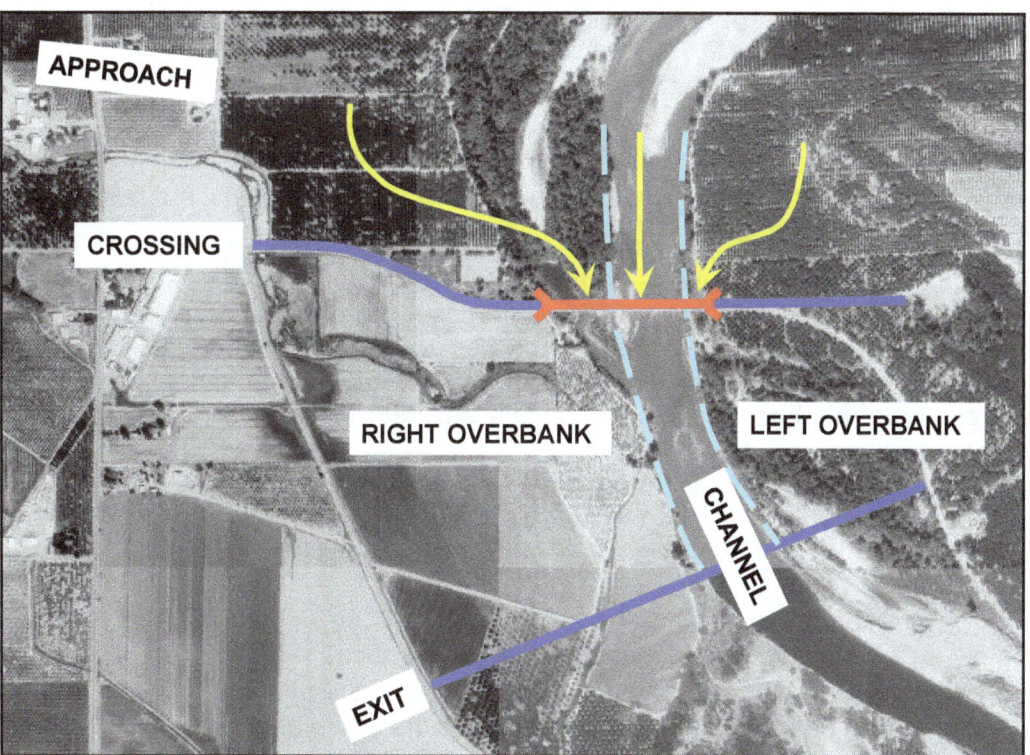

Figure 8.13. Cross section locations at bridge crossings in one-dimensional models.

Figure 8.14 is a graphical representation of flow distribution at the bridge crossing from a HEC-RAS model. The results are also available as tabular output from the HEC-RAS program. The cross section shown is adjacent to the bridge and roadway immediately upstream of the bridge. This cross section is used in scour calculations to avoid pier influence. The diagonally hatched areas are ineffective flow areas created by the embankment blockages. This figure shows low velocity in the overbank areas under the bridge where flow depth is low and Manning n is high, and high velocity in the channel where flow depth is high and Manning n is low.

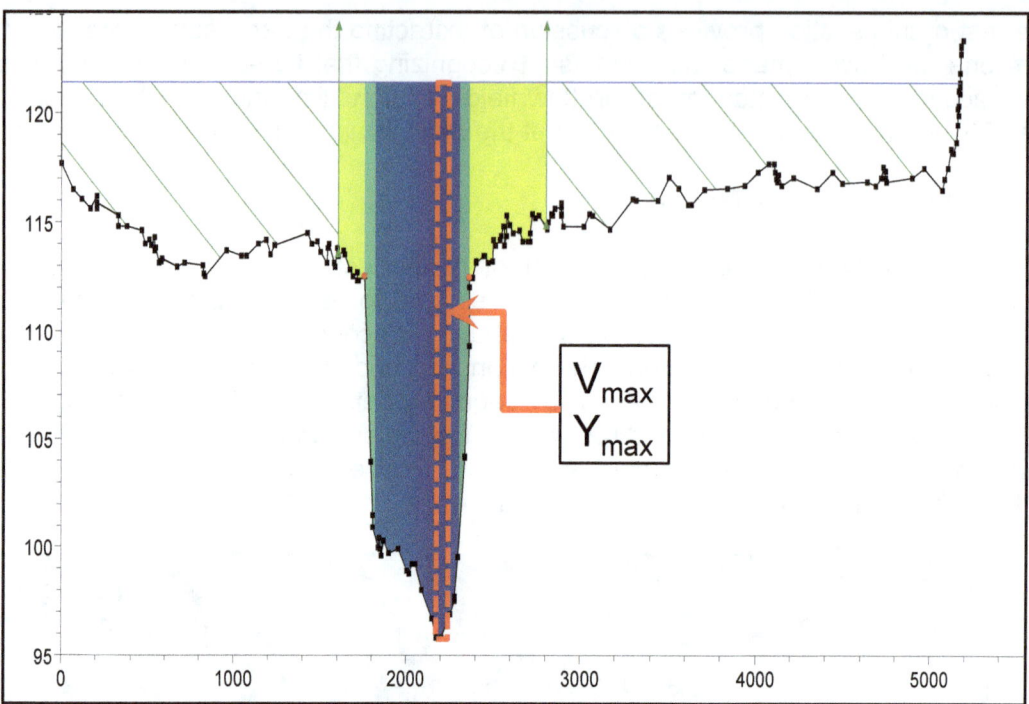

Figure 8.14. Flow distribution from one-dimensional models.

Pier scour calculations require velocity and depth upstream of the pier. Maximum velocity and depth, highlighted in Figure 8.14, are often used for pier scour calculations because they produce the most conservative results and because a shift in the thalweg could direct the highest velocity to any pier in the channel. The overbank and channel discharges used for contraction and abutment scour calculations are determined by proportioning channel and overbank conveyance.

It is important to remember that the flow distribution in HEC-RAS is approximate and based on several assumptions, including:

- Flow is gradually varied.
- Flow is distributing relative to incremental conveyance.
- There is a level water surface across the entire cross section.
- There is a single value of energy slope across the entire cross section.

Although these assumptions affect the entire flow distribution to some extent, the area where there is the greatest error is near the abutment where much higher velocity and flow concentration (unit discharge) are expected.

8.4.2 Two-Dimensional Models

Two-dimensional model results are shown graphically as velocity contours and vectors in Figure 8.15. Contours of depth are also available as graphical output. The figure depicts a complex flow situation where a highway crosses a channel and wide floodplain. There is a long, main channel bridge, a shorter relief bridge on the floodplain (upper right corner of the figure) and another relief bridge further along the embankment (not shown in the figure). There is also a narrow railroad embankment, which has a main channel bridge and two relief bridges, downstream of the wide highway embankment.

For pier scour calculations, point values of velocity and depth can be obtained at any location. Flow direction can also be determined from the model output to estimate angle of attack at a pier. Figure 8.15 also shows four flow lines. Flow lines, also called flux lines and continuity lines, are used in two-dimensional models to compute the discharge through an area. The flow lines in this figure are positioned to compute channel discharge at the bridge opening and approaching the bridge upstream, overbank flow in the wide floodplain under the main channel bridge, and total flow in the relief bridge. The area, length, average velocity, and average depth can also be determined from the flow line output. These variables provide the input data for contraction and abutment scour calculations.

Figure 8.15 also shows the flow concentration (high velocity) at the two abutments of the main channel bridge. This type of flow concentration is not available output from one-dimensional models. Unit discharge can be computed at any point in the two-dimensional model by multiplying velocity and depth, or at any flow line by dividing discharge by width (flow line length). Although this is a much more accurate representation of flow than a one-dimensional model, two-dimensional models also make simplifying assumptions, which include hydrostatic pressure and no vertical velocity components.

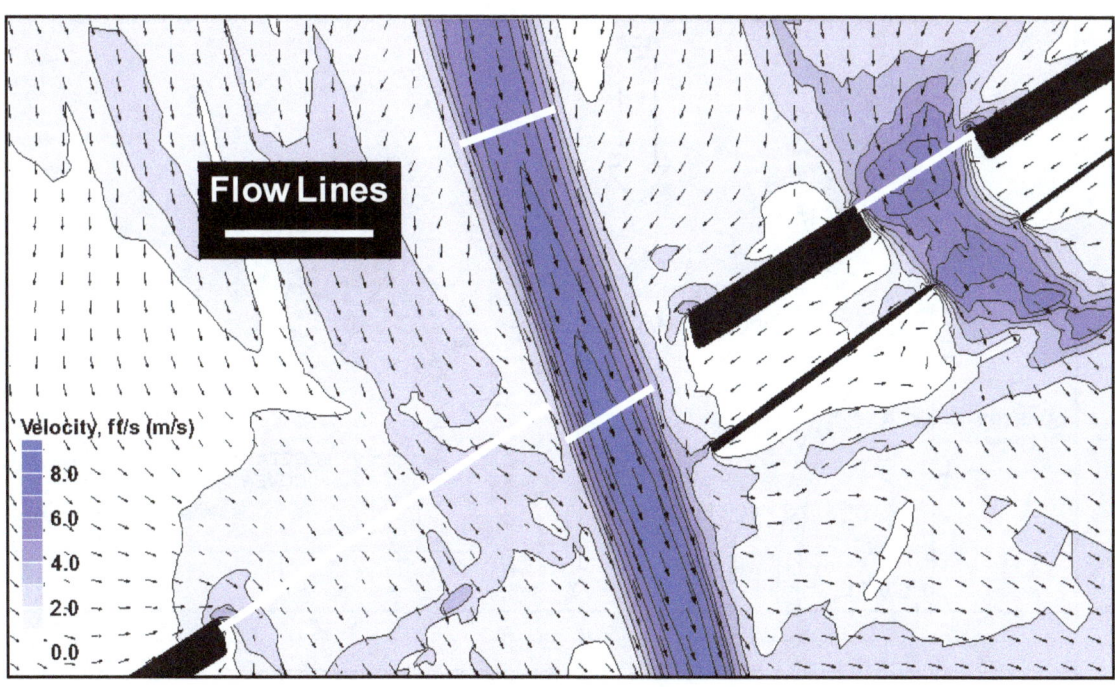

Figure 8.15. Velocity and flow lines in two-dimensional models.

8.5 HYDRAULIC ANALYSIS OF SCOUR COUNTERMEASURES

Scour and stream instability have always threatened the safety of bridges over water. Countermeasures are intended to control, inhibit, change, delay, or minimize these threats. HEC-23 (FHWA 2009a) provides guidance on selecting and designing countermeasures for various types of threats considering the range of river characteristics that are encountered. In addition to countering erosion and scour, with few exceptions countermeasures also alter flow and need to be included in hydraulic models. This section describes hydraulic modeling considerations for several countermeasures.

8.5.1 Revetments and Vegetation

Channel bank revetments and vegetation are the most common type of lateral stream instability and bank erosion countermeasure. Revetments are placed directly on the channel bank and include riprap, articulating concrete blocks, various types of mattresses, and may be used in combination with vegetation. Hydraulic modeling of revetments and vegetation includes adjusting geometry to represent earthwork and assigning representative values of Manning n for the countermeasure material.

8.5.2 Guide Banks

When embankments encroach on wide floodplains, significant amounts of flow may approach the bridge opening parallel to the roadway embankment. The resulting flow concentration and large scale turbulence can generate large amounts of scour at the abutment. Flow separation can also reduce the effective bridge opening. Guide banks (as shown in Figure 8.16) can be used to prevent severe abutment scour and reduce flow separation. Scour may still occur, but is expected only at the upstream end of the guide bank.

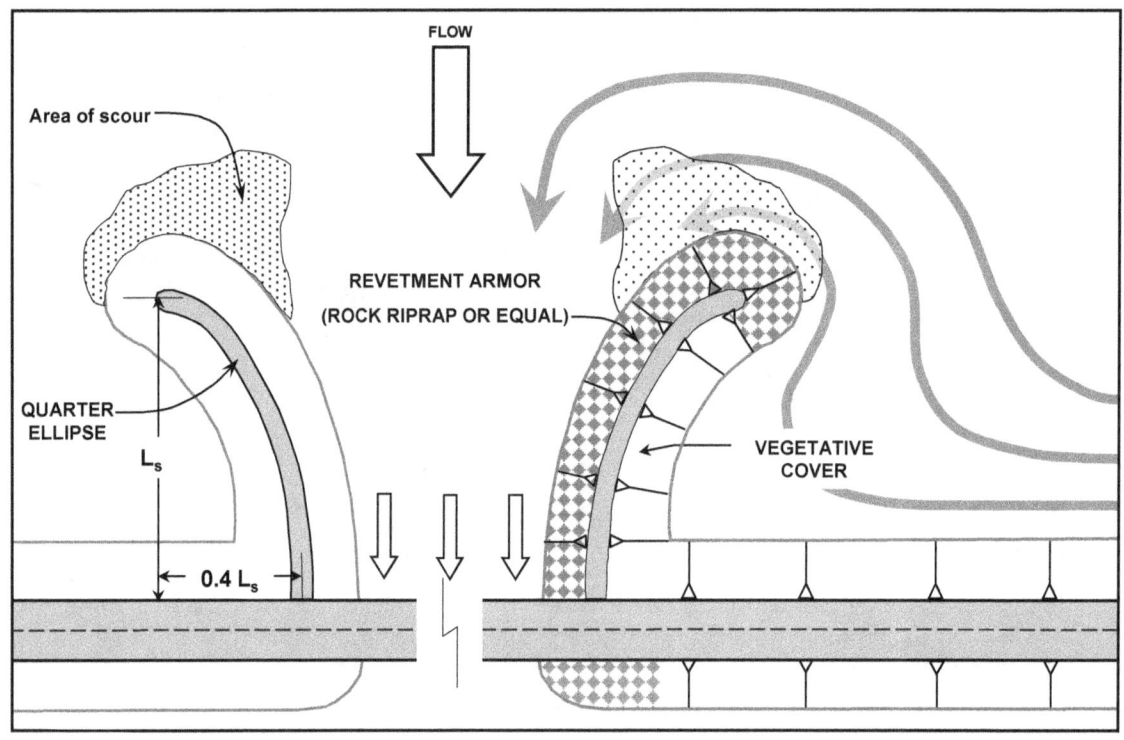

Figure 8.16. Typical guide bank (modified from FHWA 1978).

For one-dimensional modeling, one additional cross section should be located at the upstream ends of the guide bank. The cross section should only include active flow in the area between the two guide banks. It is reasonable to use lower values of the contraction coefficient as compared to the value used to represent the more abrupt flow transition resulting when no guide bank is used. The water surface elevation can differ greatly from the front side to the back side of guide bank. The energy grade elevation at the cross section at the upstream end of the guide banks is a reasonable estimate of the elevation of ponded water along the back side of the guide bank. The HEC-23 manual (FHWA 2009a) provides the SBR (Set Back Ratio) method to estimate the flow velocity at an abutment from one-dimensional model results. The SBR method can also be used to estimate the flow velocity at a guide bank.

As shown in Figure 8.17, the geometry of guide banks can be included directly in two-dimensional models. The finite element mesh shown in this figure demonstrates that areas of rapid change in velocity magnitude or direction require a more refined network of elements. The unstructured mesh of the finite element network also allows for detailed assignment of cover type, i.e. Manning n. Figure 8.18 shows the flow field around this guide bank and the flow around the abutment at the other end of the bridge. There is flow separation under the bridge right abutment (left side of figure) but not on the guide bank side. Flow velocities are also much lower at the guide bank protected side.

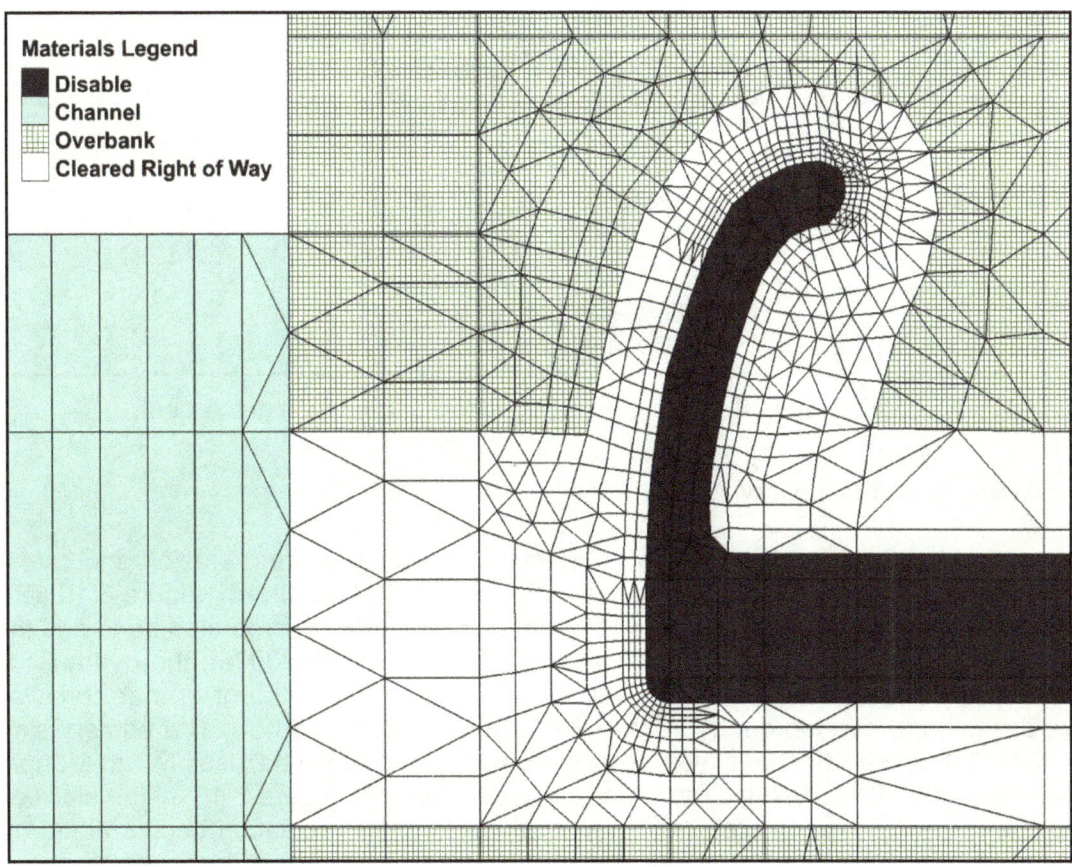

Figure 8.17. Guide bank in a two-dimensional network.

Figure 8.18 clearly illustrates some of the benefits of two-dimensional modeling for bridge scour analysis. The true flow field is much better simulated in two-dimensional models. Because flow direction is computed intrinsically by two-dimensional models the angle of attack used for pier scour becomes more deterministic, though potential for future change must always be considered. The two-dimensional model also shows that the right abutment, which does not include a guide bank, has flow separation and a portion of the bridge opening is not effective for conveying flow. The maximum velocity at the guide bank is also much lower than at the opposite abutment. Therefore, the required riprap size is much smaller for the guide bank.

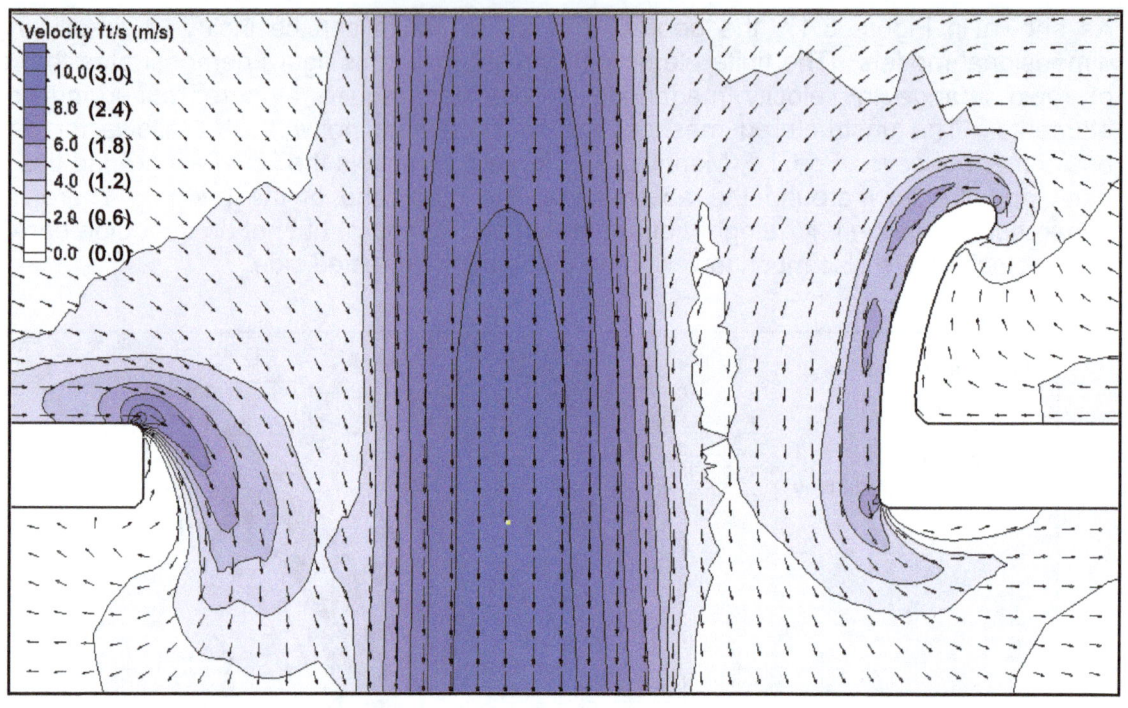

Figure 8.18. Flow field at a bridge opening with guide bank.

8.5.3 Spurs and Bendway Weirs

Spurs and bendway weirs are used to protect channel banks from erosion and can be used to better align flow at bridges where channel migration has occurred. Figure 8.19 (a) and (b) show two dimensional modeling results of bankfull flow at an unprotected bend and the same bend protected with spurs. As discussed in HEC-23 (FHWA 2009a), the hydraulic function and design of these two countermeasures are significantly different in that bendway weirs are designed to overtop during in-channel flows and spurs are not. The primary similarities of these structures are that they extend from the bankline and usually have unprotected channel bank between structures. In the models shown in Figure 8.19, high velocities reach the toe of bank in the unprotected bank but low velocity circulation occurs along the bank when protected by spurs. Shear stress (or any other hydraulic parameter) can also be computed, contoured, and compared between models using the two-dimensional model interface or CADD/GIS software.

These models also illustrate that the upstream spur is subjected to the highest flow velocity and that the spurs are likely to shift the thalweg and may erode and shift the opposite bankline because of the increased flow velocities away from the spurs. One-dimensional modeling could also be used to simulate these conditions but the results would be more indicative of average conditions at a cross section rather than the detailed distributed results of the two-dimensional flow field. It should be recognized, however, that even a very refined two-dimensional model network is not a complete representation of the flow characteristics, especially when structure overtopping occurs.

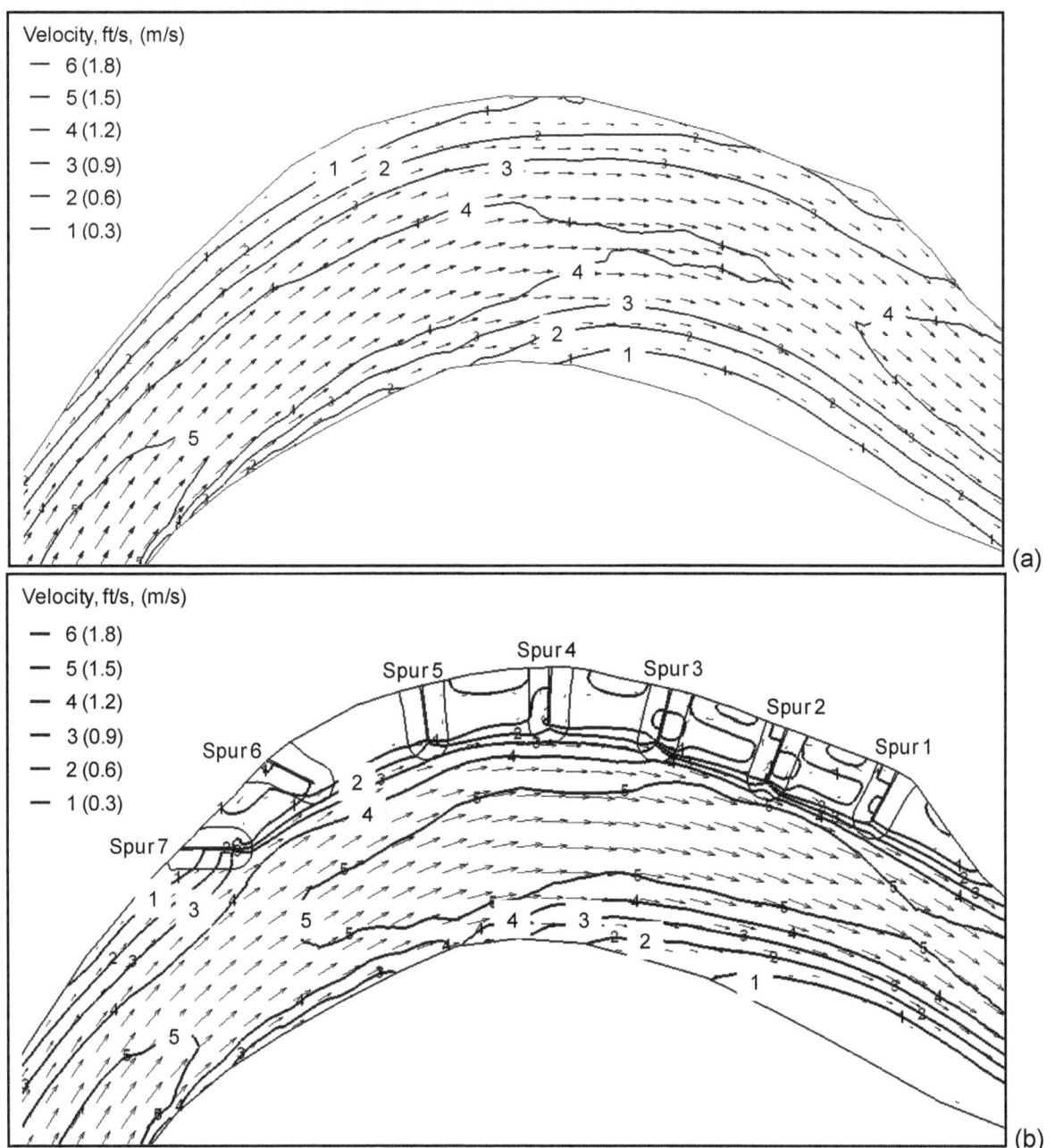

Figure 8.19. Two-dimensional analysis of flow along spurs, (a) flow field without spurs, and (b) flow field with spurs.

(page intentionally left blank)

CHAPTER 9

SEDIMENT TRANSPORT AND ALLUVIAL CHANNEL CONCEPTS

9.1 INTRODUCTION

Safe bridge design includes the recognition that channels are not stationary, but that they may adjust their bed and banks during the life of the bridge. The HEC-20 (FHWA 2012a) and HDS 6 manuals (FHWA 2001) are the primary FHWA manuals related to stream instability and sediment transport topics. Another reference that provides broad coverage of this topic is Sedimentation Engineering (ASCE 2008). HDS 6 states that "The moveable boundary of the alluvial river adds another dimension to the design problem and can compound environmental concerns. Therefore, the design of highway crossing and encroachments in the river environment requires knowledge of the mechanics of alluvial channel flow." This chapter provides an overview of these topics in the context of bridge design.

Most channels and floodplains that roads cross are alluvial. Alluvial channels are formed by materials that have been transported and deposited by flowing water and can be transported by the channel in the future. Channel adjustments include aggradation, degradation, width adjustment, and lateral shifting. Aggradation and degradation are the overall raising or lowering of a channel bed over time from sediment accumulation or erosion. Channel widening and shifting are the result of bank erosion due to hydraulic forces or by mass failure of the bank.

Sediment transport analyses can play a role in several aspects of safe bridge design. Of primary concern is whether the channel will experience long-term aggradation or degradation. Aggradation decreases flow conveyance and has the potential of increased frequency and magnitude of flooding, road overtopping, and loss of service. Degradation threatens bridge foundations by removing support and making the bridge more vulnerable to scour during floods. A related concern is that the bridge could alter the prevailing flow conditions and cause aggradation or degradation. Departments of Transportation may also conduct channel restoration as part of a bridge replacement. Sediment transport analysis is needed to determine the potential impacts of the restoration to avoid creating a channel that does not adequately convey sediment supplied from upstream. Another role that sediment transport can play in bridge design is that contraction scour can be computed from a sediment transport model rather than from the standard contraction scour equation. This would be done if there was significant uncertainty in the use of the standard contraction scour equation or if there was a significant potential benefit from applying a more detailed analysis. In summary, sediment transport analyses should be considered as part of a bridge design for the following reasons.

- Evaluation of long-term aggradation or degradation potential
- Concerns over a bridge replacement impacting channel vertical stability
- Evaluation of channel restoration project impacts on sediment transport and channel vertical stability
- More detailed evaluation of potential contraction scour, especially for short duration flows

ASCE (2008) indicates that one-dimensional sediment transport models are most often applied to simulations involving extended river reaches and extended time periods, typically to determine the long-term response of a river to natural or man-made changes. This is because of the computational efficiency of one-dimensional models as compared to two-dimensional models. This makes one-dimensional models well-suited to address the topics listed above. As indicated by ASCE (2008), one-dimensional models cannot resolve local details of flow and mobile bed dynamics, which two- or three-dimensional models provide the possibility of resolving, though currently for relatively small scale problems over relatively short time periods.

Channel stability and sediment transport are complex processes that interact to produce the existing channel form and future channel adjustments. This is why HEC-20 (FHWA 2012a) emphasizes that qualitative evaluation (Level 1), and standard engineering analyses (Level 2) should be conducted even when advanced numerical sediment transport modeling (Level 3) is performed. Factors that influence sediment transport include sediment properties, hydrology, watershed and land-use conditions, channel geometry, and vegetation.

Sediment properties include size, gradation, cohesion, density, shape, porosity of the sediment mixture, angle of repose, and sediment layer depths. Many, if not all, aspects of hydrology also play a role in sediment transport analyses. These include not only peak flow rates, but also individual flood hydrographs, and the durations of all flows. The entire range of flow may be significant because even though the highest flows have the highest rates of sediment transport, lower flows may have significantly longer durations and produce the greatest cumulative sediment transport. Channels respond and adjust to changes in flow and sediment supply. Therefore, changing watershed conditions often result in adjustments in channel geometry. Channel geometry, bed material, and vegetation determine hydraulic variables (velocity, depth, etc.), which in turn control sediment transport capacity. Consequently, sediment transport and channel stability depend not only on the specific physical processes, but also the history of natural and human-induced factors in the watershed.

The following sections provide a general overview of sediment transport concepts and processes. Other resources are available to provide the in-depth information required to perform these analyses. These resources include HDS 6 (FHWA 2001), Sedimentation Engineering (ASCE 2008), textbooks (Simons and Senturk 1992, Yang 2003, Julien 2010), and the manuals for specific numerical models that incorporate sediment transport.

9.2 SEDIMENT CONTINUITY

The amount of material transported, eroded, or deposited in an alluvial channel is a function of sediment supply and channel transport capacity. Sediment supply is provided from the tributary watershed and from erosion occurring in the upstream channel bed and banks. Sediment transport capacity is primarily a function of sediment size and the hydraulic properties of the channel. When the transport capacity of the flow equals sediment supply from upstream, a state of equilibrium exists.

Application of the sediment continuity concept to a channel reach illustrates the relationship between sediment supply and transport capacity. The sediment continuity concept states that the sediment inflow minus the sediment outflow equals the rate of change of sediment volume in a given reach. More simply stated, during a given time period the amount of sediment coming into the reach minus the amount leaving the downstream end of the reach equals the change in the amount of sediment stored in that reach (Figure 9.1). The sediment inflow to a given reach is defined by the sediment supply from the watershed and channel (upstream of the study reach plus lateral input directly to the study reach). The transport capacity of the channel within the given reach defines the sediment outflow. Changes in the sediment volume within the reach occur when the total input to the reach (sediment supply) is not equal to the downstream output (sediment transport capacity). When the sediment supply is less than the transport capacity, erosion (degradation) will occur in the reach so that the transport capacity at the outlet is satisfied, unless controls exist that limit erosion.

Conversely, when the sediment supply is greater than the transport capacity, deposition (aggradation) will occur in the reach.

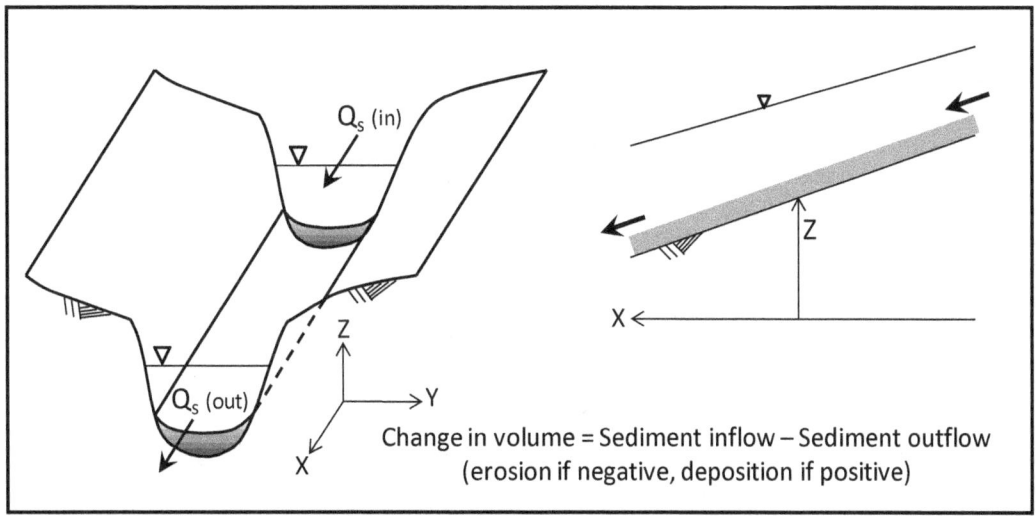

Figure 9.1 Definition sketch of the sediment continuity concept.

Controls that limit erosion may either be human induced or natural. Human-induced controls included bank protection works, grade control structures, and stabilized bridge or culvert crossings. Natural controls can be geologic, such as outcrops, or the presence of significant coarse sediment material in the channel. The presence of coarse material can result in the formation of a surface armor layer of larger sediments.

The Exner equation describes the sediment continuity equation mathematically. The one-dimensional differential form of the equation it is:

$$(1-\eta)W\frac{\partial Z}{\partial t} = -\frac{\partial Q_s}{\partial X} \tag{9.1}$$

where:

- W = Active width of the channel, ft (m)
- Z = Channel bed elevation, ft (m)
- t = Time (s)
- η = Bed material porosity (volume of voids/total volume)
- Q_s = Sediment transport rate, ft³/s (m³/s)
- X = Distance along channel, ft (m)

Applied to a channel reach, the sediment continuity equation is:

$$\Delta Z = \frac{\Delta \forall}{WL(1-\eta)} = \frac{\Delta t(Q_{s(in)} - Q_{s(out)})}{WL(1-\eta)} \tag{9.2}$$

where:

- $\Delta \forall$ = Change in volume of sediment particles stored or eroded in the reach, ft³ (m³)
- L = Reach length, ft (m)

9.3 SEDIMENT TRANSPORT CONCEPTS

9.3.1 Initiation of Motion

Initiation of motion of bed material particles exposed to flowing water is difficult to define precisely. The particles are subjected to drag and lift forces by the flowing water. Flow field near the boundary, turbulence fluctuations, and particle size, shape and relative position with respect to other particles all contribute to these forces. Particle size, shape, and relative position to other particles also contribute to forces that resist motion, including gravitational and external support forces acting on the particle (friction and other point contacts between grains). This problem has been simplified and studied empirically by many scientists for laboratory conditions, dating back to Shields (1935). Detailed discussions are available from many sources including HDS 6 (FHWA 2001). Shields related the beginning of motion to particle size, particle submerged unit weight, and flow shear stress to predict the initiation of motion.

Standing water exerts hydrostatic pressure on the channel bed. For uniform flow with small slopes, the flowing water exerts a time-average shear stress in the direction of flow equal to the hydrostatic pressure times the channel slope:

$$\tau_0 = \gamma y S_o \tag{9.3}$$

where:

- τ_0 = Shear stress, lb/ft² (Pa)
- γ = Specific weight of water, lb/ft³ (N/m³)
- y = Flow depth (hydraulic radius, hydraulic depth for wide channels, or local depth) ft (m)
- S_o = Bed slope (or energy slope for gradually varied flow)

Another useful formula for estimating average shear stress for gradually varied flow conditions is:

$$\tau_0 = \frac{\gamma}{y^{(1/3)}} \left(\frac{nV}{K_u}\right)^2 \tag{9.4}$$

where:

- n = Manning roughness coefficient
- V = Flow velocity, ft/s (m/s)
- K_u = 1.486 U.S. Customary units
- K_u = 1.0 SI

Equation 9.4 shows the relationship between velocity and shear stress; shear stress is proportional to velocity squared. The Shields parameter relates critical shear stress to particle size and specific weight by.

$$\tau_c = k_s D_s (\gamma_s - \gamma) \tag{9.5}$$

where:

- τ_c = Critical shear stress for beginning of motion, lb/ft² (Pa)
- k_s = Shields parameter
- D_s = Particle size, ft (m)
- γ_s = Specific weight of the particle, lb/ft³ (N/m³)

Shields parameter ranges from 0.03 to 0.10 for natural sediments and depends on particle shape, angularity, gradation and imbrication. The use of 0.047 is common for sand sizes. When the shear stress of the flow exceeds the critical shear stress of the particle, the channel bed begins to mobilize and bed material is transported downstream. Particle motion begins as sliding and rolling of individual particles along the bed. It is important to recognize that the Shields equation is not a sediment transport equation because it does not provide any estimate of the amount of sediment in motion. It is also important to note that only the shear stress acting on the particles, or grain friction, should be used in applying this relationship.

9.3.2 Modes of Sediment Transport

Once the critical shear stress is exceeded, bed material begins to move (roll, slide, and saltate) along the bed surface. This material is referred to as bed load or contact load because it is in almost continuous contact with the bed. For small amounts of positive excess shear stress (defined as $\tau_o - \tau_c$), this is the only mode of bed material transport. As excess shear stress increases, turbulence begins to suspend some of the particles. The turbulence acts to mix the particles in the water column and gravity causes the particles to settle. Therefore, bed material can also transported downstream as suspended bed material load. The two types of bed material load are illustrated in Figure 9.2.

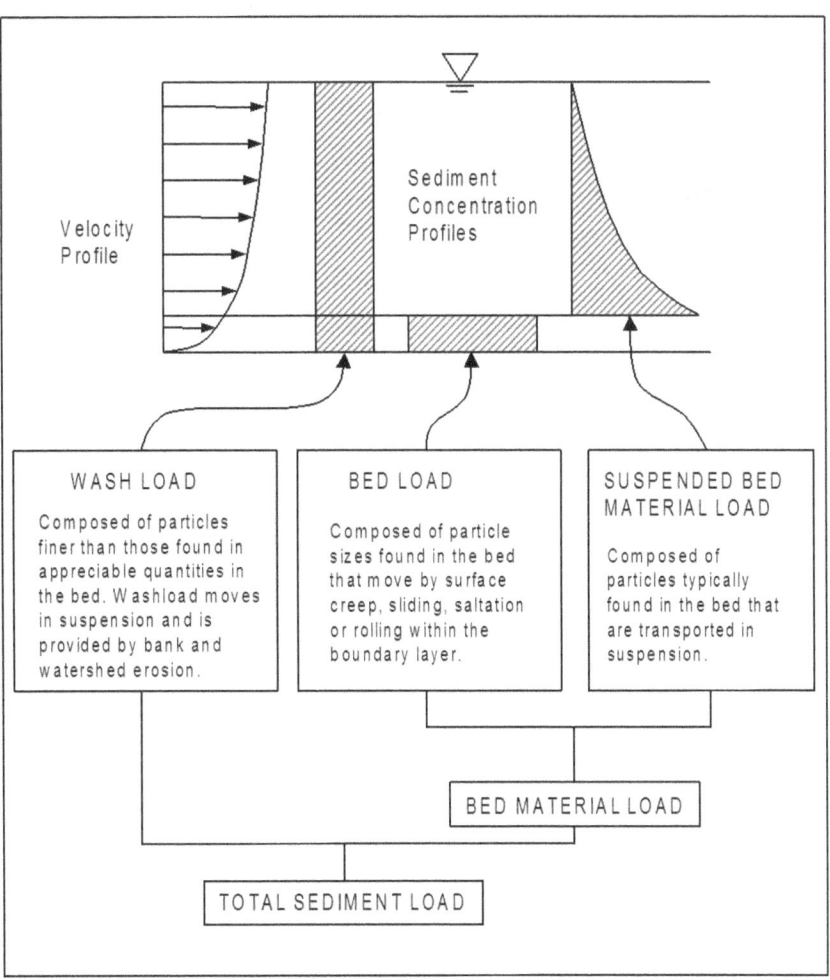

Figure 9.2. Definitions of sediment load components (FHWA 2012a).

The suspended bed material load shown in Figure 9.2 is a result of the interaction between gravity and turbulence. Because gravity is causing particles to settle, they are concentrated near the bed. Turbulence mixes the particles in the water column and, depending on the size and density of the particles, relatively few particles may reach the surface. The suspension of particles is illustrated in Figure 9.3, which shows the concentration profile for various particle sizes in a turbulent flow field. The equation that describes the concentration profiles is:

$$\frac{c}{c_a} = \left[\left(\frac{y_o - y}{y}\right)\left(\frac{a}{y_o - a}\right)\right]^z \quad (9.6)$$

where:

- c = Sediment concentration at height y from the bed
- c_a = Sediment concentration at height a above the bed
- y_o = Total flow depth (from surface to bed) ft (m)
- z = Rouse number = $\omega/(\beta \kappa v_*)$
- ω = Fall velocity of the particle in quiescent water, ft/s (m/s)
- β = Parameter relating particle and momentum transfer due to turbulence, approximately equal to 1.0 for fine particles
- κ = Von Karman's constant of 0.4
- v_* = Shear velocity = $\sqrt{\tau_o/\rho} = \sqrt{gRS}$
- ρ = Water density, slugs/ft³ (kg/m³)
- g = Acceleration due to gravity, ft/s² (m/s²)
- R = Hydraulic radius, ft (m)

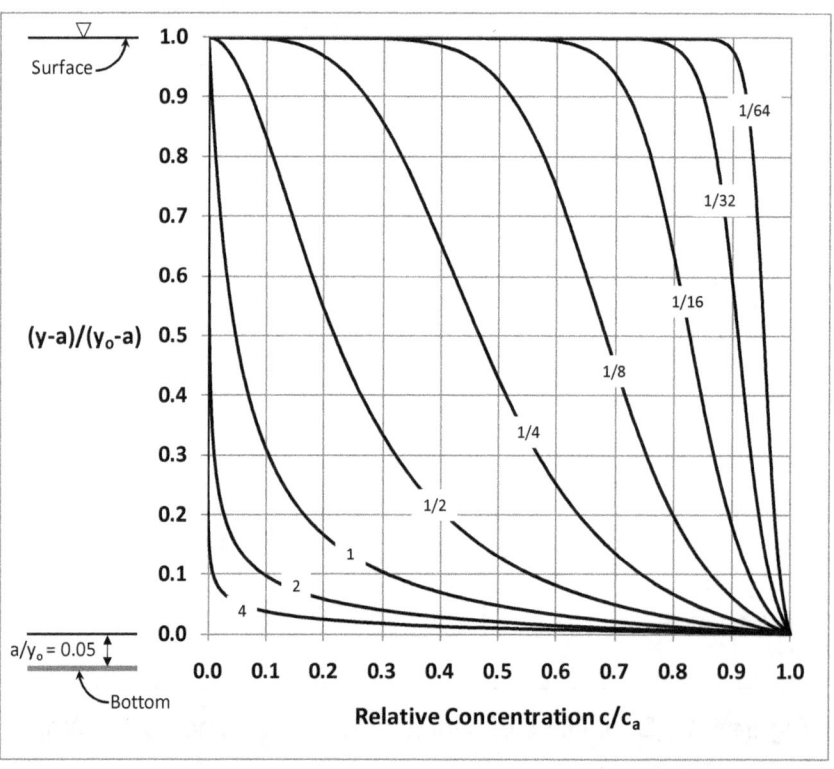

Figure 9.3. Suspended sediment concentration profiles (Rouse 1937).

Larger particles have greater fall velocities and larger Rouse numbers. Therefore, Figure 9.3 shows that for a given level of turbulence (as represented by the shear velocity), large particles will remain close to the bed. Finer particles have smaller Rouse numbers, are mixed higher into the flow and have higher concentrations. Julien (2010) indicates that particle sizes with Rouse numbers less than 0.025 (1/40) will have essentially uniform concentration profiles. These particles are extremely fine, primarily silts and clays, and have very small fall velocities. They are defined as wash load, which are derived primarily from upland erosion and bank erosion of floodplain materials. Wash load material is not found in appreciable quantities in the channel bed.

In summary, bed material is transported in contact with the bed (bed load) and in suspension (suspended bed material load). The total sediment load transported by the channel also includes wash load, which is supplied to the channel rather than derived from the bed. Wash load is also transported in suspension. In coarse bed channels, such as cobble-bed and boulder-bed streams, sand may act as wash load because it is not found in appreciable quantities in the bed and because the supply is far less than the channel capacity to transport this size.

9.3.3 Bed-Forms

In sand-bed streams, sand material is easily eroded and is continually being moved and shaped by the flow. The interaction between the flow of the water-sediment mixture and the sand-bed creates different bed configurations which change the resistance to flow, velocity, water surface elevation and sediment transport. Consequently, an understanding of the different types of bed forms that may occur, bed form geometry, resistance to flow, and sediment transport associated with each bed form can help in analyzing flow in an alluvial channel.

Flow Regime. Flow in alluvial sand-bed channels is divided into two regimes separated by a transition zone. Forms of bed roughness in sand-bed channels are shown in Figure 9.4. There is no direct relationship between the classification of upper and lower flow regime and Froude Number (supercritical/subcritical flow). The flow regimes are:

- The lower flow regime, where resistance to flow is large and sediment transport is small. The bed form is either ripples or dunes or some combination of the two. Water surface undulations are out of phase with the bed surface, and there is a relatively large separation zone downstream from the crest of each ripple or dune. The velocity of the downstream movement of the ripples or dunes depends on their height and the velocity of the grains moving up their backs.

- The transition zone, where the bed configuration may range from that typical of the lower flow regime to that typical of the upper flow regime, depending mainly on antecedent conditions. If the antecedent bed configuration is dunes, the depth or slope can be increased to values more consistent with those of the upper flow regime without changing the bed form; or, conversely, if the antecedent bed is plane, depth and slope can be decreased to values more consistent with those of the lower flow regime without changing the bed form. Resistance to flow and sediment transport also have the same variability as the bed configuration in the transition zone. This phenomenon can be explained by the changes in resistance to flow and, consequently, the changes in depth and slope as the bed form changes.

- The upper flow regime, in which resistance to flow is small and sediment transport is large. The usual bed forms are plane bed or antidunes. The water surface is in phase with the bed surface and normally the fluid does not separate from the boundary, except when an antidune breaks.

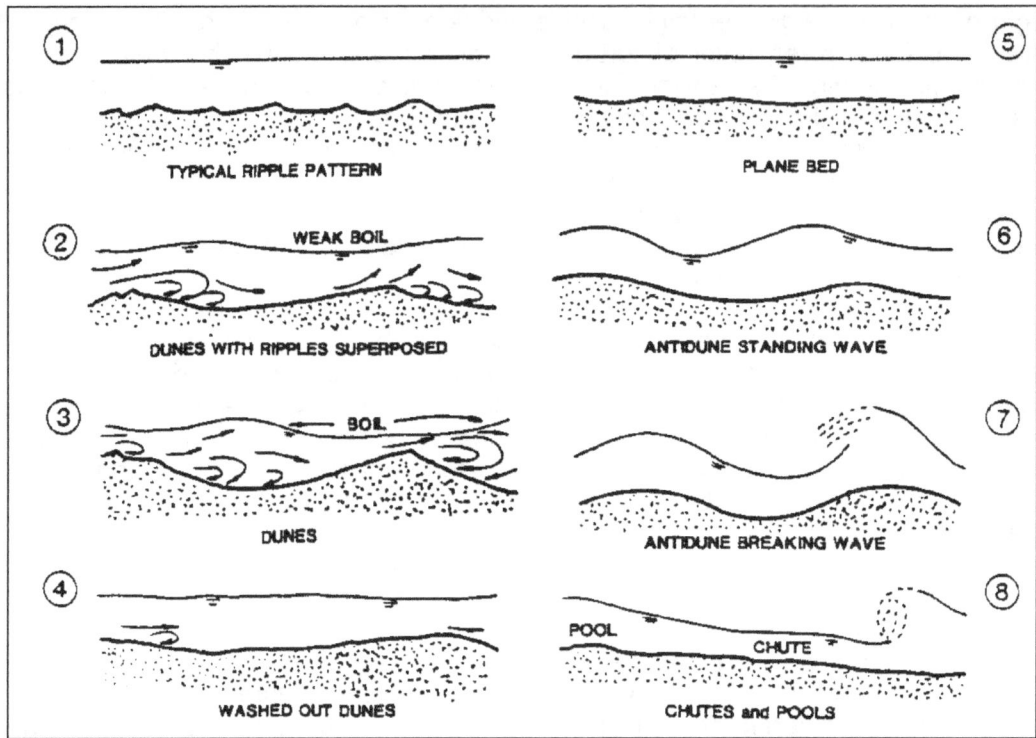

Figure 9.4. Bed forms in sand channels (after HDS 6 - FHWA 2001).

<u>Effects of Bed Forms at Stream Crossings</u>. At high flows, most sand-bed stream channels shift from a dune bed to a transition or a plane bed configuration. The resistance to flow is then decreased by one-half to one-third of that preceding the shift in bed form. The increase in velocity and corresponding decrease in depth may increase scour around bridge piers, abutments, spur dikes or banks and may increase the required size of riprap.

Another effect of bed forms on highway crossings is that with dunes on the bed, there is a fluctuating pattern of scour on the bed. Methods for computing bed-form geometry can be found in Julien and Klaassen (1995) and Karim (1999). Karim included laboratory and field data where the crest-to-trough height, Δ, for dunes ranged from less than 0.1y to up to 0.5y. Karim also showed a range of antidune heights between 0.1y and 0.4y. Bennet (USGS 1997) indicated an approximate upper limit as $\Delta < 0.4y$. The average dune height equation by Julien and Klaassen is:

$$\frac{\Delta}{y} = 2.5 \left(\frac{D_{50}}{y} \right)^{0.3} \tag{9.7}$$

The lower and upper bounds on dune heights (95 percent) range from 0.3 to 3.2 times this average height. Dune lengths can be approximated as 6.25 times the flow depth. Care must be used in analyzing crossings of sand-bed streams in order to anticipate changes that may occur in bed forms and the impact of these changes on the resistance to flow, sediment transport, and the stability of the reach and highway structures. With a dune bed, the Manning n could be more than twice as large as a plane bed (see Figure 9.5). A change from a dune bed to a plane bed, or the reverse, can have an appreciable effect on depth and velocity. In the design of a bridge or a stream stability or scour countermeasure, it is good engineering practice to assume a dune bed (large n value) when establishing the water surface elevations, and a plane bed (low n value) for calculations involving velocity.

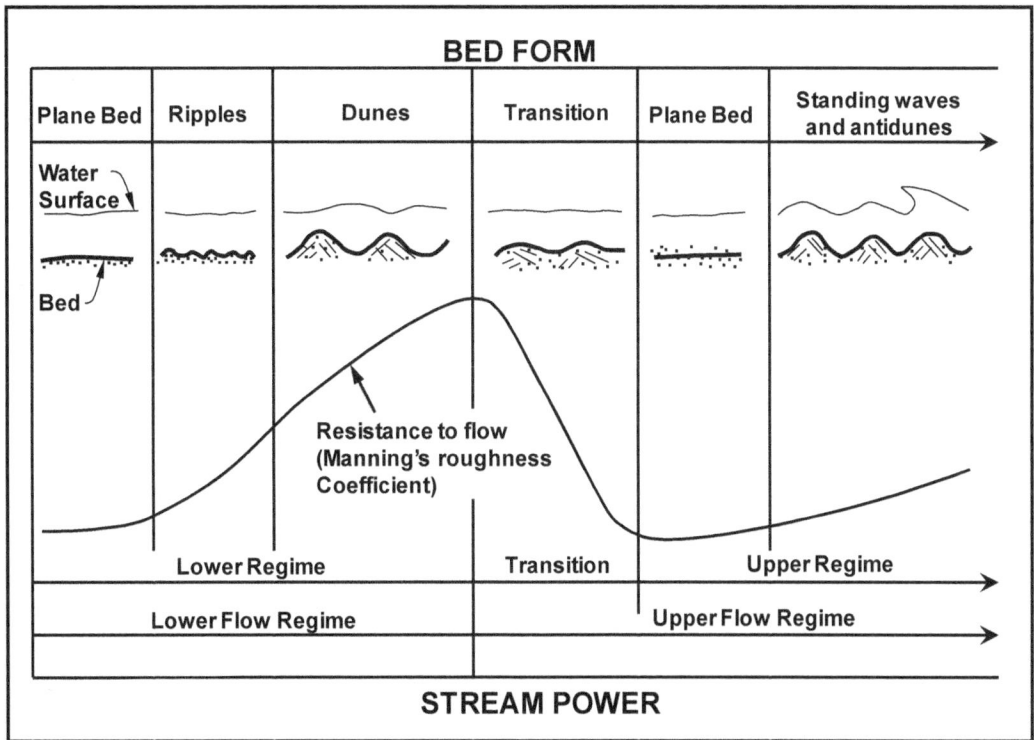

Figure 9.5. Relative resistance to flow in sand-bed channels (after USGS 1989).

9.4 OVERVIEW OF SEDIMENT TRANSPORT EQUATIONS AND SELECTION

Equations for predicting bed material sediment transport differ depending on the mode of sediment transport. ASCE (2008) includes 16 bed load equations. The Meyer-Peter and Müller (1948) equation is considered to be a classic bed load equation. The equation has the basic form of:

$$q_b = a(\tau_0 - \tau_c)^{3/2} \tag{9.8}$$

where:

q_b = Bed load discharge per unit width of channel, ft²/s (m²/s)
a = Empirical coefficient

As with the analysis of incipient motion, only grain friction should be included in the bed shear (τ_0) variable. Many of the equations presented in ASCE (2008) include excess shear stress ($\tau_0 - \tau_c$) to the 1.5 power. Because bed shear is proportional to velocity squared (see Equation 9.4), bed-load dominated sediment transport, such as in gravel-bed rivers, is generally proportional to velocity cubed.

Another classic method for predicting sediment transport is the Colby (1964) graphical method for bed material load in sand-bed rivers. The Colby method is discussed in detail in HDS 6 (FHWA 2001). Sand-bed channels are dominated by suspended sediment transport for most flow conditions. The first step in the Colby method is to determine an uncorrected sediment discharge based on flow velocity. The Colby curves follow a trend of sediment discharge proportional to velocity to the power of between 3.5 and 6. These large powers indicate that suspension is more effective in transporting sediment in sand-bed channels. They also indicate that uncertainty in velocity generates extreme uncertainty in sediment transport calculations. Based on these observations, a 10 percent change in velocity can result in a 40 to 80 percent change in sediment transport rate. The Colby method also includes correction factors for water temperature and wash load concentration because these factors affect the fluid viscosity and particle fall velocity.

As flow velocity and shear stress increase bed load increases, but suspended load increases rapidly and can easily dominate the sediment transport process. This is because bed load is transported in a small fraction of the flow depth (often considered as twice the median (D_{50}) sediment diameter and because the flow velocity (and bed load velocity) is low near the bed. Suspended load is carried through more, and potentially all of the flow depth (see Figure 9.3). Velocity quickly increases with distance above the bed so suspended load is carried downstream at a much higher velocity than bed load.

Although the Colby method provides insight into the sediment transport process, suspended load was more rigorously investigated by Einstein (1950). The Einstein suspended load equation is described in HDS 6 (2001), and is a solution to the suspended load equation:

$$q_s = \int_a^{y_0} vc\,dy \tag{9.9}$$

Where the variables are defined as in Equation 9.6 and:

q_s = Suspended load discharge per unit width, ft²/s (m²/s)
v = Velocity at height y above the bed, ft/s (m/s)

The solution of the integral uses Equation 9.6 for sediment concentration and a logarithmic velocity profile equation (vertical velocity distribution is discussed in Chapter 6). The concentration and velocity profiles are illustrated in Figure 9.6. This integration depends on a reference concentration that is determined from the bed load. ASCE (2008) presents nine equations for determining the reference concentration and an easily applied equation (Abad and Garcia 2006) to solve the integration of Equation 9.9. Because the rate of bed load transport and the concentration profile depend on grain size, the integration is performed for the range of grain sizes in the bed material and the total bed material load is the sum of the proportionate transport rates computed for each size class. Julien (2010) used Equation 9.9 to show that bed load comprises 80 percent or more of the total load when shear velocity divided by fall velocity ($v*/\omega$) is less than 0.5, and that suspended load comprises 80 percent or more of the total load when $v*/\omega > 2$. For $0.5 < v*/\omega < 2$ the sediment transport is considered to be mixed load.

ASCE (2008) also presents six empirically based equations for determining total sediment load. These equations have the advantage of being more easily applied, but should only be used within the limits of the data used in their development. This concept applies to the use of any sediment transport equation. The HDS 6 manual (FHWA 2001) includes 20 sediment transport equations and the applicability to various grain sizes. The HEC-RAS Reference Manual (USACE 2010c) and SAM reference manual (USACE 2002) include information on the range of data (particle size, specific gravity, velocity, depth, slope, channel width and temperature) used to develop many of the sediment transport equations used for sand and gravel sizes. Any equation that is considered for use should be evaluated for applicability to the specific conditions.

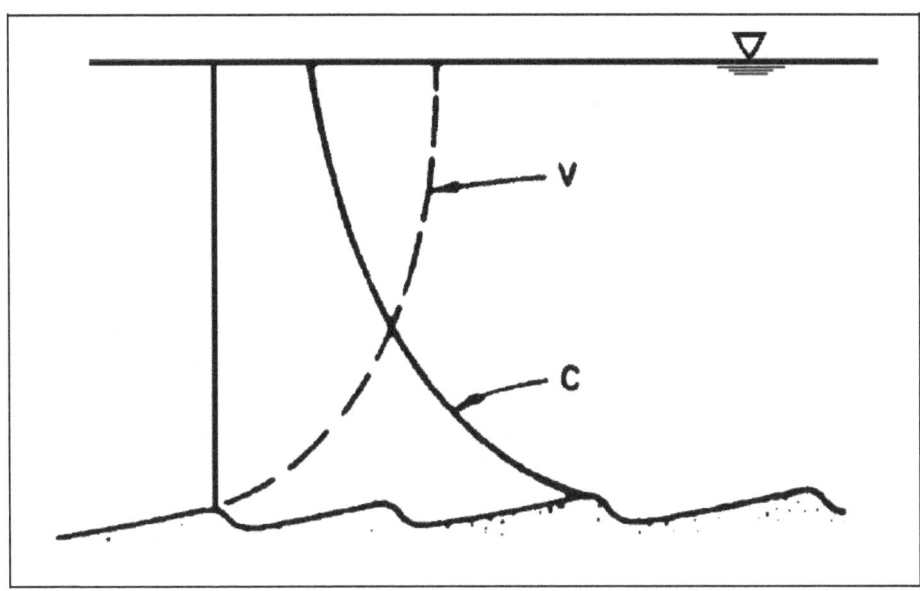

Figure 9.6. Velocity and sediment concentration profiles.

It is important to note that there are several ways of expressing and calculating rates sediment transport. These include volumetric (ft^3/s, m^3/s), mass and weight (tons/day, metric-tons/day), and concentration (ppm, mg/l), sediment volume/total volume, and sediment weight/total weight). HDS 6 provides exact and approximate equations for converting between these expressions.

9.5 OVERVIEW OF SEDIMENT TRANSPORT MODELING

The HEC-RAS Reference Manual (USACE 2010c) states that "Sediment transport modeling is notoriously difficult." This is because the high degree of variability and uncertainty in much of the data and because the equations are highly sensitive to the input variables. As indicated in the previous section, a small change in velocity can significantly change the sediment transport capacity. Changes in sediment size also dramatically impact transport capacity. Another reason for the difficulty in modeling is that the sediment transport process is extremely complex.

The HEC-RAS Users Manual (USACE 2010b) describes three types of sediment transport analysis capabilities within HEC-RAS. In order of increasing complexity, they are (1) Sediment Transport Capacity, (2) Sediment Impact Analysis Methods (SIAM), and (3) Sediment Routing. For each of these types of analysis, six sediment transport equations are available.

The sediment transport capacity function is simply a sediment transport calculator. The potential transport capacity is determined for each cross section in a user-defined reach and bed material grain size distribution. These calculations can be reviewed to identify imbalances between individual cross sections or reaches. Cross sections that have significantly different transport capacity should be reviewed to determine if there are errors on inconsistencies in the cross section, or if there are other conditions that limit sediment transport. Bridge constrictions often have very different sediment transport capacity for flood conditions due to the flow constriction that causes contraction scour. Bridge constrictions should have little impact on transport capacity for in-bank flows.

The next level of complexity is SIAM. SIAM is a sediment budget tool. It combines channel reach-weighted hydraulics, annual flow duration information, bed material gradation, other sediment properties, and information on sediment sources to compute the annual sediment transport capacity for each reach. The engineer must identify channel reaches with similar hydraulic and sediment properties. The results can be used to locate potential instabilities and sediment imbalances (surpluses and deficits) between reaches.

The third, and most challenging capability is sediment transport routing. In sediment routing, the sediment transport capacity is used to update cross section geometry, which is then used to update the hydraulic calculations. The geometry is updated for individual cross sections, though the hydraulic variables can be weighted with up- and downstream cross sections. A flood hydrograph or long-term flow hydrograph is entered as a series of constant flows. Within each flow time step, many sediment transport and cross section updating time steps are often required. The model does not assume that transport capacity is reached at every cross section, but limits erosion based on potential entrainment rates and limits deposition based on fall velocity, flow velocity and water depth. Sediment layer depths, as well as lateral limits for erosion and deposition are also input. Sediment transport modeling also requires greater model upstream and downstream extent, as well as careful consideration of all boundary conditions (hydraulic and sediment).

Figure 9.7 shows channel profiles for Las Vegas Wash, a channel with a history of degradation. The channel has experienced increased flow over time and sediment supply is limited by upstream channel stabilization. The bridge crossing location degraded over 30 ft (9 m) between 1970 and 1999. Equilibrium slope calculations indicated that an additional 40 feet (12 m) of degradation could occur based on expected discharge and sediment supply rates. Equilibrium slope is defined as the slope a channel will seek based on an expected combination of sediment supply and water discharge, and is described in detail in the HEC-20 manual (FHWA 2012a). Because equilibrium slope calculations do not provide the amount of time it will take to reach equilibrium, a sediment transport model was developed to provide an independent estimate of channel degradation at the bridge and the time it would take to reach various amounts of degradation. The final profile from the sediment transport model is at equilibrium with the expected flow and sediment and was achieved approximately 10 years into the simulation. The bridge was protected with grade control structures and the predicted final degradation has occurred downstream of the bridge. Note that the final profile shows aggradation downstream of station 6000 ft (1830 m). This aggradation is due to sediment accumulating in the pool of Lake Mead, which is a downstream, though highly variable, control.

Because of the sensitivity to the hydraulic conditions, a sediment transport routing model will often highlight deficiencies in a hydraulic model. When velocity or conveyance change significantly between cross sections, the change in sediment transport capacity may result in unrealistic amounts of aggradation or degradation, or create unrecoverable numerical instabilities during the model run. Sediment transport routing is inherently non-uniform and unsteady. It is non-uniform because the cross section geometry will change as erosion and deposition occur. It is unsteady because the rate of sediment transport imbalance determines the amount of cross section change (Equations 9.1 and 9.2).

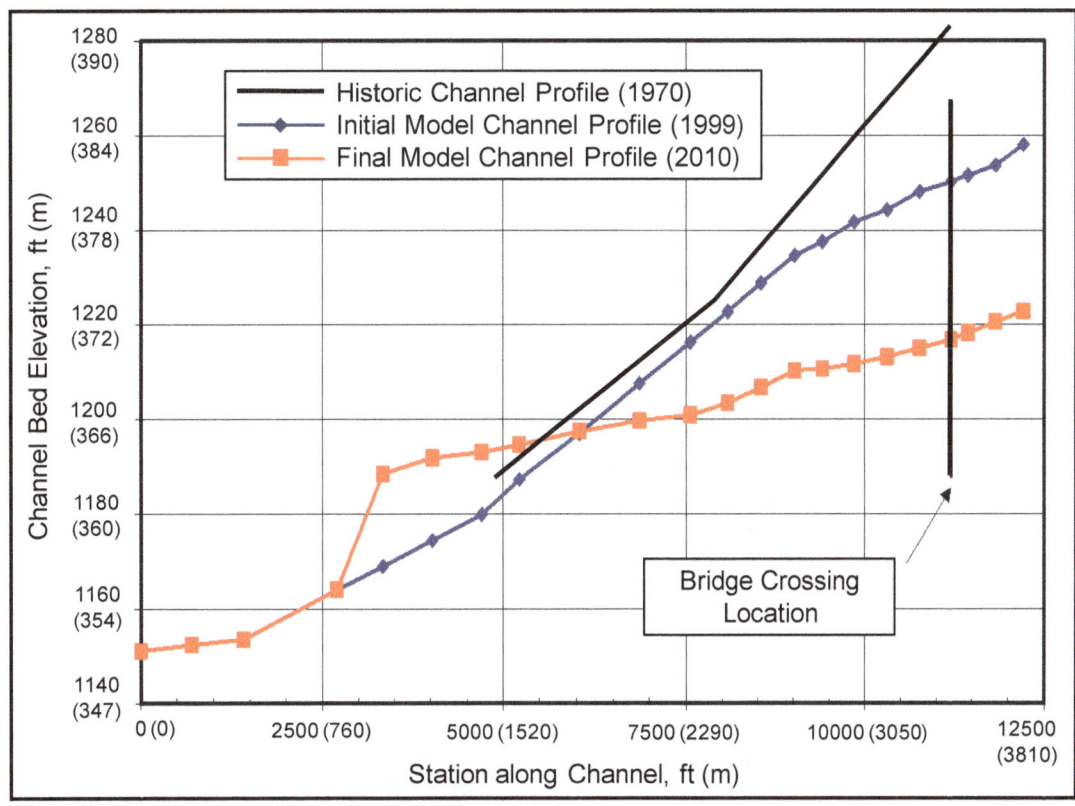

Figure 9.7. Channel profiles from sediment routing model.

The following is a partial list of the information needed for a typical sediment routing analysis:

- Channel and floodplain geometry
- Channel and floodplain roughness
- Structure geometry
- Geologic or structural vertical controls
- Hydraulic boundary conditions
- Inflow hydrographs (including tributary)
- Sediment supply boundary conditions (including tributary)
- Bed material gradations
- Depth of alluvium and sediment layers
- Sediment transport relationships

There are many decisions that impact the results of a bridge hydraulic analysis. Selecting high Manning n values results in conservative water surface elevations. Selecting low roughness values results in more conservative velocity estimates. Using a fixed bed model for bridge hydraulics may also produce conservative estimates of backwater. This is because contraction scour enlarges the bridge opening and reduces the velocity in the bridge. Therefore, in many cases, backwater actually caused by bridges is less than a fixed bed model predicts. In some cases, use of a mobile-bed model, or incorporating contraction scour in the bridge opening of a fixed-bed model, better represents actual flow conditions at the bridge. This is illustrated in Figure 9.8, which shows the water structure for a fixed-bed model run for natural (no bridge) and bridge conditions and a mobile-bed model. The bed profile shows the construction scour caused by the bridge constriction. In this case, the mobile-bed model computed approximately 40% less backwater due to the 2.6 ft (0.79 m) of contraction scour that resulted based on sediment transport.

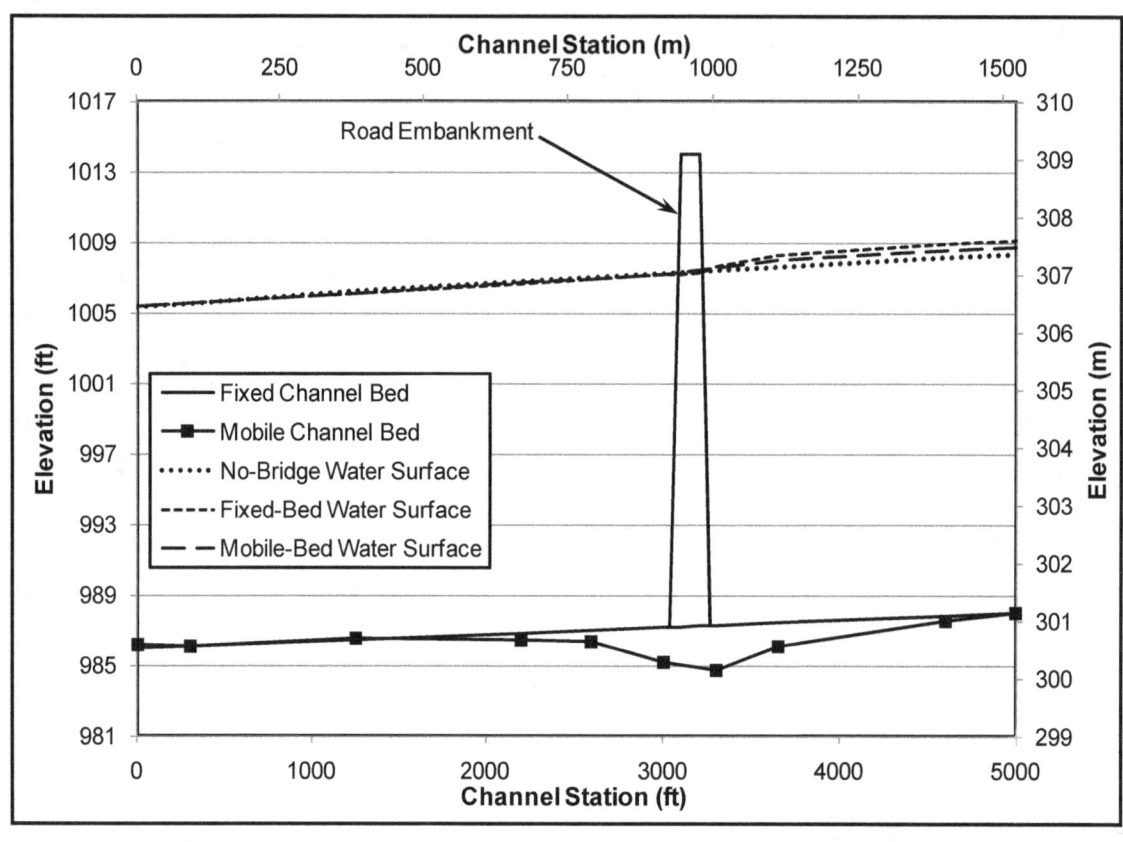

Figure 9.8. Contraction scour and water surface for fixed-bed and mobile-bed models.

9.6 ALLUVIAL FANS

Alluvial fans are very dynamic sedimentary landforms that can create significant hazards to highways as a result of floods, debris flows, deposition, channel incision, and avulsion (Schumm and Lagasse 1998). They can occur where there is a change from a steep to a flat gradient, especially in mountainous regions. The National Research Council Committee on Alluvial Fan Flooding (NRC 1996) defined alluvial fans as sedimentary deposits that are convex in cross-profile and located at a topographic break, such as the base a mountain, escarpment, or valley side, that is composed of stream flow and/or debris flow sediments and that has the shape of a fan either fully or partially extended. As the bed material and water reaches the flatter section of the stream, the coarser bed materials can no longer be transported because of the sudden reduction in both slope and velocity. Consequently, a cone or fan builds out as the material is dropped. Alluvial fans are often characterized by unstable channel geometries and rapid lateral movement. The steep channel tends to drop part of its sediment load in the main channel building out into the main stream. In some instances, the main stream can make drastic changes, or avulsions, during major floods. The NRC committee determined that alluvial fan hazards can include (1) flow path uncertainty below the fan apex, (2) abrupt deposition and ensuing erosion of sediment as a stream or debris flow loses competence to carry material eroded from the steeper, upstream source area, and (3) the combination of sediment availability, slope, and topography creates ultra hazardous conditions that elevation or fill will not reliably mitigate risk.

The potential for avulsion, deposition, channel blockage, and channel incision are important for highway design. To minimize these impacts on highways, a reconnaissance of the fan and its drainage should be undertaken so that potential changes can be identified and countermeasures taken. Any study of alluvial fans should include a geomorphic map delineating active and inactive portions of the fan and the identification of problem sites within the active portions of the fan. For example, local aggradation in a channel can lead to avulsion because avulsion is likely to occur in places where deposition has raised the floor of the channel to a level that is nearly as high as the surrounding fan surface. This condition can be identified in the field by observation or by the surveying of cross-fan profiles (Schumm and Lagasse 1998).

French (1987) cautions that alluvial fan hydraulics are highly unsteady and two-dimensional. Analyzing hydraulic and sediment transport conditions on alluvial fans should not be conducted without in-depth geomorphic evaluation. ASCE (2008) indicates that there are two-dimensional models available for modeling flow and sediment transport on alluvial fans, specifically FLO-2D (Obrien 2009). FLO-2D is a grid-based finite difference model that is well-suited for simulating unconfined flow and sediment conditions that occur on alluvial fans, including mud- and debris-flow conditions. Although the grid-based approach is less suited for determining the detailed hydraulic results often desired for bridge applications, the highly unsteady, unconfined flow conditions on alluvial fans are the dominant processes and make this approach necessary.

(page intentionally left blank)

CHAPTER 10

OTHER CONSIDERATIONS

10.1 HYDRAULIC FORCES ON BRIDGE ELEMENTS

10.1.1 General

Bridge design engineers must analyze the stability of the bridge as a whole and elements of the bridge under various loading conditions. Rivers, streams and coastal water bodies exert significant forces on bridge structures especially during times of flood or storm surge. The hydraulic forces potentially acting on a bridge include hydrostatic, buoyancy, drag and wave forces. Impact by vessels and forces exerted by debris or ice are also closely tied to hydraulics. Bridge designers require information from the results of the hydraulic analysis to evaluate the hydraulic forces on bridge elements.

Bridge designers typically follow the AASHTO LRFD Bridge Design Specifications (herein referred to as the LRFD Specifications) (AASHTO 2010), with some state-specific modifications, in evaluating forces and loads on bridges. The guidelines of LRFD Specifications, along with information and insights from other references, are briefly summarized in this section.

10.1.2 Hydrostatic Force

The weight of water exerts hydrostatic pressure in all directions. It is calculated as the product of the height of the water surface above the point of interest and the unit weight of water. Thus the pressure is greatest at the lowest point of a submerged element and is zero at the water surface elevation.

The hydrostatic force acting on a bridge element in a particular direction is the summation, or integral, of the product of the pressure and the surface area of the bridge element projected in the plane perpendicular to the direction of the force. Hydrostatic forces on one side of a bridge are at least partly balanced by opposing hydrostatic forces acting on the other side. Any imbalance in the hydrostatic force is due to variation in the water surface elevation. Bridge designers must be informed of the water surface elevation upstream and downstream of the bridge for the design flood in order to evaluate the hydrostatic forces.

10.1.3 Buoyancy Force

Buoyancy is an uplift force equivalent to the weight of water displaced by the submerged element. It can be a threat to a submerged bridge superstructure if the superstructure design incorporates large enclosed voids as with a box-girder or if air pockets develop between girders beneath the deck. Buoyancy is also a factor in evaluating wave-related forces on bridge decks, discussed later in this chapter. If a pier is constructed with a large empty void, the buoyant uplift force acting on the pier may be significant. Bridge designers must be informed of the water surface elevation upstream and downstream of the bridge for the design flood in order to evaluate the buoyancy forces.

10.1.4 Stream Pressure and Lift

Stream pressure is the name in the LRFD Specifications for the pressure associated with the drag exerted on the structure by flowing water. By the LRFD Specifications, the stream pressure on a bridge element is computed as a simple function of the square of the impinging flow velocity multiplied by a drag coefficient.

<u>Stream Pressure on Piers</u>

The drag coefficient for piers is a function of the shape of the pier nose (upstream end), the plan view shape of the pier, the skew (if any) of the pier axis versus the flow direction, and the presence or absence of debris on the pier. The hydraulic engineer must inform the bridge designer of the magnitude and direction of the local impinging flow velocity for the design event, as well as the flow depth and debris collection potential, in order to evaluate the stream pressure on a pier.

<u>Stream Pressure and Lift on Bridge Superstructures</u>

Recent research provides refined guidance on evaluating the stream pressure and forces acting on submerged bridge superstructures. The FHWA (2009c) used physical modeling and three-dimensional computational fluid dynamics (CFD) modeling to investigate the hydrodynamic forces on inundated bridge decks, specifically the drag force acting parallel to the flow direction and tending to push the superstructure off of the piers and the abutments; lift force acting vertically and tending to lift the superstructure; and the overturning moment resulting from unevenly distributed forces and tending to rotate the superstructure about its center of gravity. The physical modeling and CFD modeling both focused on three different superstructure design types: one with six flanged girders, one with three larger rectangular girders, and a third with a highly streamlined cross sectional shape. Figure 10.1 shows a CFD results plot for a six-girder bridge model.

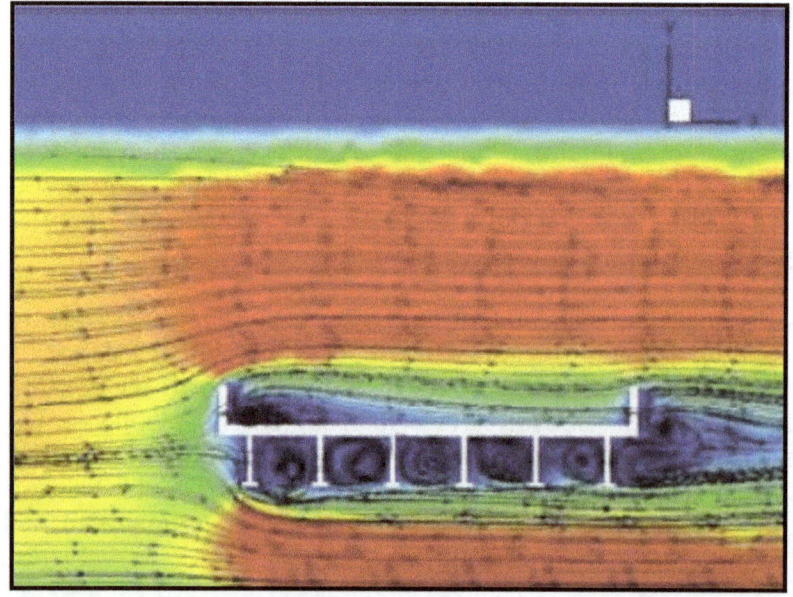

Figure 10.1. CFD results plot showing velocity direction and magnitude from a model of a six-girder bridge (from FHWA 2009c).

The resulting report, entitled "Hydrodynamic Forces on Bridge Decks" (FHWA 2009c) provides equations for use in determining the drag coefficient, lift coefficient and moment coefficient as functions of the inundation ratio, for each of the three superstructure types investigated. The inundation ratio is a measure of the degree of submergence of the superstructure. It is defined as the vertical distance measured down from the water surface to the bridge low chord divided by the depth of the superstructure measured vertically from the top of the parapet to the low chord. For the six-girder superstructure, the equations yield drag coefficients roughly ranging from 0.7 to 2.2.

For inundated bridge decks, lift is another force component that should be considered in bridge design. FHWA (2009c) provides equations for lift, as well as the resulting turning moment that the combined drag and lift forces create. The deck may not react as a single unit depending on the interconnection of the girders, so lift and drag may be more severe for individual deck elements. Figure 10.2 is a definition sketch for drag, lift, and turning moment variables.

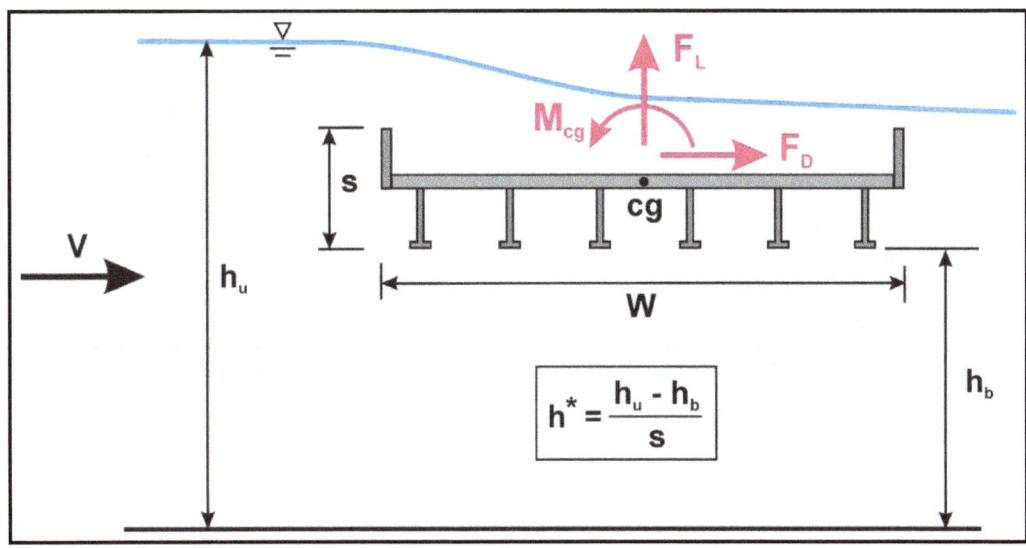

Figure 10.2. Definition sketch for deck force variables (FHWA 2009c).

Equations for computing drag, lift, and moment per unit length of bridge (FHWA 2009c) are:

$$h^* = \frac{h_u - h_b}{s} \tag{10.1}$$

$$F_D = \frac{\rho C_D V^2 s}{2}; \text{ for } h^* > 1 \tag{10.2}$$

$$F_D = \frac{\rho C_D V^2 s h^*}{2}; \text{ for } h^* < 1 \tag{10.3}$$

$$F_L = \frac{\rho C_L V^2 W}{2} \tag{10.4}$$

$$M_{cg} = \frac{\rho C_M V^2 W^2}{2} \tag{10.5}$$

The values of the drag, lift and moment coefficients for a six-girder bridge are shown in Figures 10.3 through 10.5. FHWA (2009c) also provides charts of these coefficients for three-girder bridges.

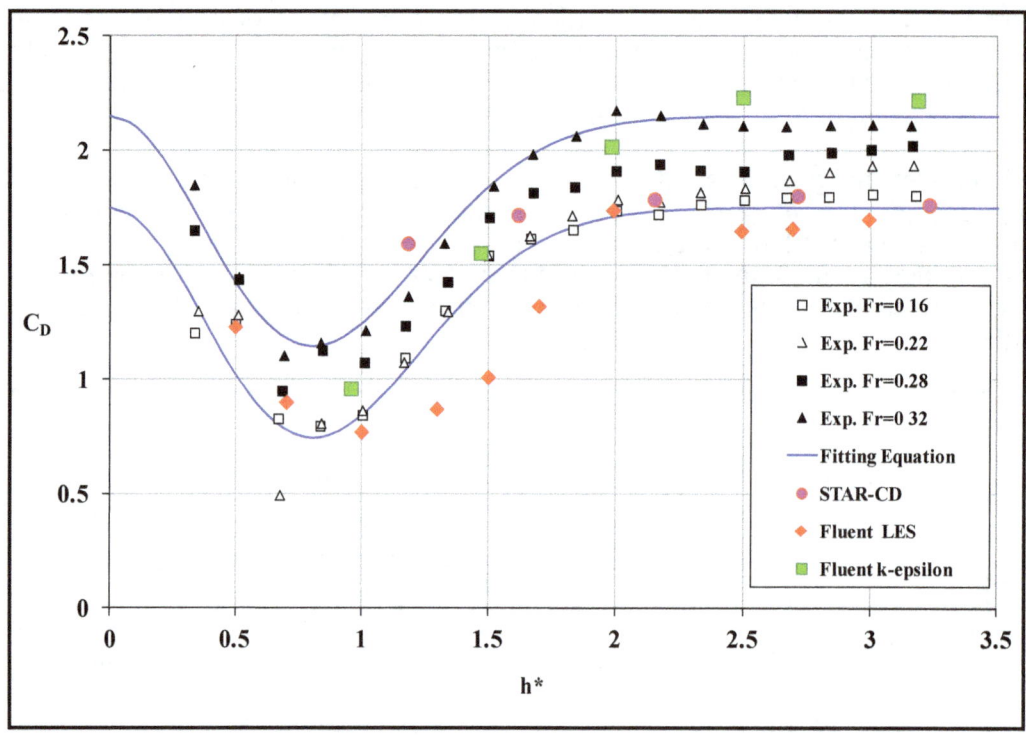

Figure 10.3. Drag coefficient for 6-girder bridge (FHWA 2009c).

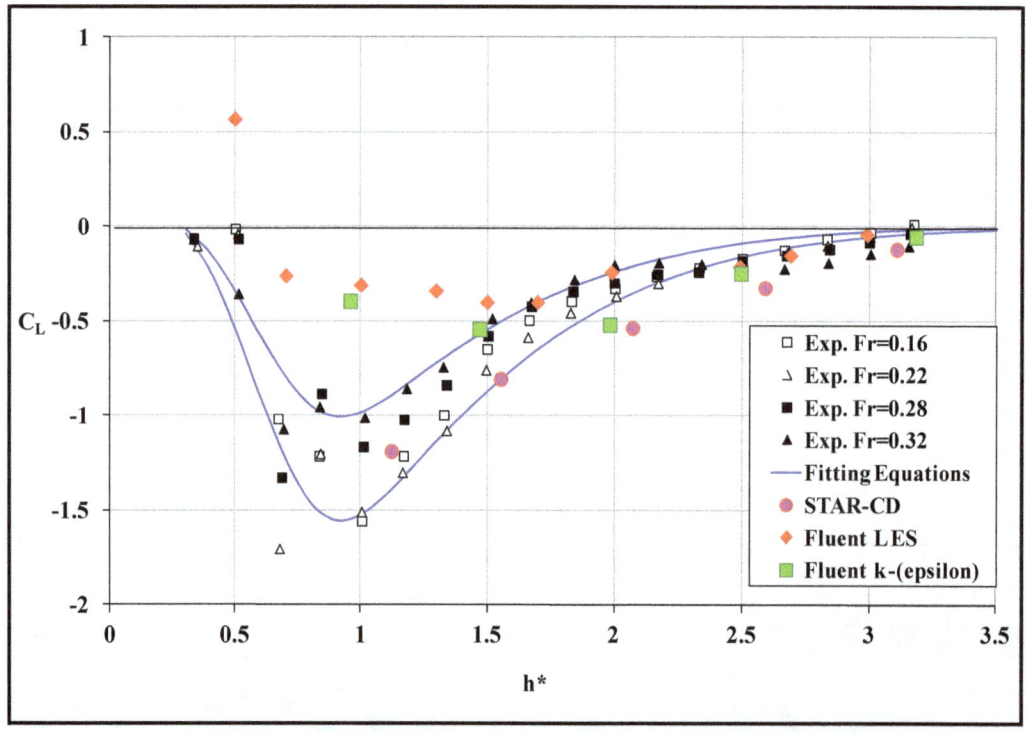

Figure 10.4. Lift coefficient for 6-girder bridge (FHWA 2009c).

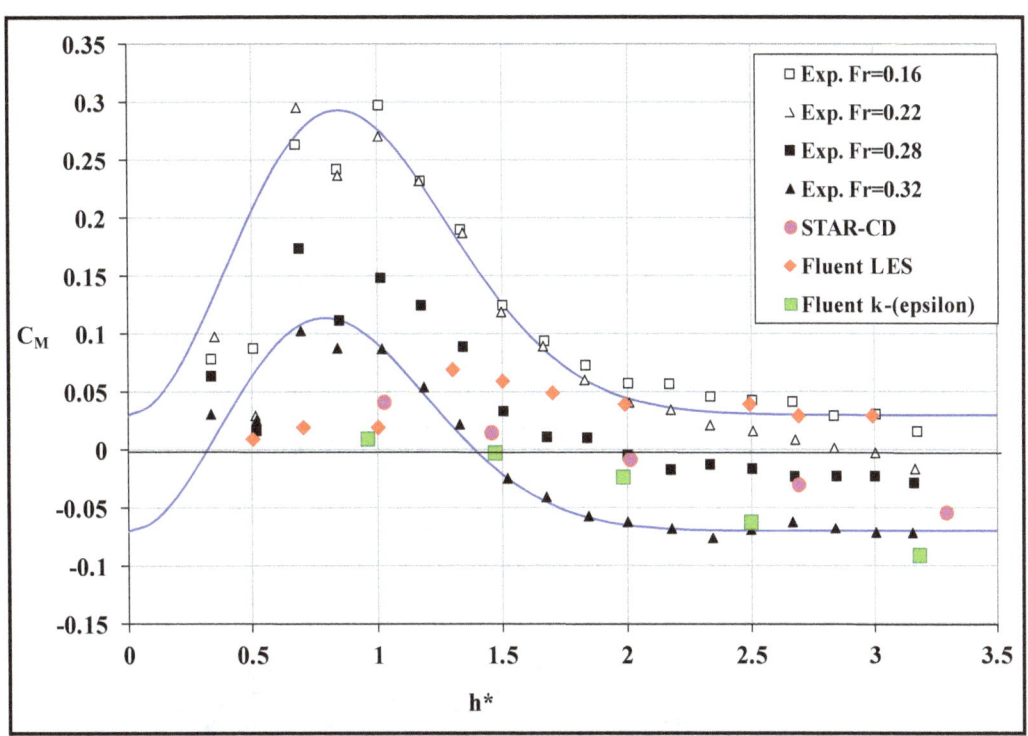

Figure 10.5. Moment coefficient for 6-girder bridge (FHWA 2009c).

The hydraulic engineer must inform the bridge designer of certain information from the hydraulics analysis to determine these forces. The required information includes the water surface elevation, depth and velocity upstream of the bridge, along with the elevation of the bridge low chord. It will be necessary to qualitatively adjust the drag coefficient to accommodate a bridge superstructure type other than a six-girder or three-girder bridge design.

10.1.5 Wave Forces

The design of bridges in coastal settings must consider the potential for significant wave forces. The FHWA document HEC-25 "Highways in the Coastal Environment" (FHWA 2008), documents extensive damage incurred by bridges along U.S. gulf coast during Hurricane Ivan in 2004 and Hurricane Katrina in 2005. The landfall of each hurricane caused a high surge in the water level, allowing the large waves generated by the storm to affect the superstructures of bridges (FHWA 2008). Figure 10.6 is a photograph of superstructure damage to a U.S. Highway 90 in Mississippi from Hurricane Katrina.

Waves striking a bridge superstructure impart forces acting both horizontally and vertically. The magnitudes of the forces depend upon several factors including the tide level, storm surge, and properties of the anticipated waves. The FHWA conducted a pooled-fund study to develop guidelines and specifications for the design of bridges subject to wave forces in coastal settings.

The resulting recommendations were published by AASHTO in the document "Guide Specifications for Bridges Vulnerable to Coastal Storms," (AASHTO 2008). A bridge designer following this AASHTO document requires certain information about the tidal hydraulics and the wave setting. The hydraulic engineer should be prepared to provide the following information, with input from a coastal engineer:

- Maximum probable wave height for the design event
- Wave length
- Wave period
- Upwind fetch over which wave can be generated
- Storm tide water surface elevation at the bridge for the design event, including local wind setup where appropriate
- Stream bed elevation at the bridge
- Current velocity from tidal hydraulic modeling for the design event

Figure 10.6. Photograph of a bridge damaged by Hurricane Katrina from HEC-25 (FHWA 2008).

The wave height and other wave properties can be computed using equations from the Shore Protection Manual (USACE 1984) or from the Coastal Engineering Manual (USACE 2008) or determined through the application of numerical wave modeling software. The wave properties are generally dependent upon the wind speed, duration and direction, the upwind fetch length, the water depth at the bridge and the water depth over the fetch.

10.1.6 Effects of Debris

Debris accumulations on bridges can dramatically increase the hydraulic forces exerted on both piers and superstructures. The LRFD Specifications provide guidance on incorporating debris potential into the stream pressure calculations, with respect to assigning a drag coefficient and estimating the cross sectional area of the debris blockage.

A research project by the NCHRP used physical modeling to examine debris forces on bridges. The report, titled "Design Specifications for Debris Forces on Highway Bridges," (NCHRP 2000, Report 445) recommends separate evaluation of the drag force and hydrostatic force from debris accumulations. For evaluating the drag force, the report provides envelope curves and tables to aid in assigning the drag coefficient for debris on piers and superstructures as a function of the amount of blockage caused by the debris and the Froude number in the contracted section. The report also provides useful guidance on the selection of the reference velocity for use in the drag force, or stream pressure, calculations. The hydrostatic force is calculated based on the difference in water surface from the upstream side of the debris accumulation to the downstream side of the bridge.

Another research project by the NCHRP used field observations, a photographic database and extensive physical modeling to investigate the affects of debris on bridge pier scour. The resulting report, titled "Effects of Debris on Bridge Pier Scour," (NCHRP 2010b, Report 653) provides refined guidance on estimating the potential dimensions of a debris flow blockage, on incorporating debris into one- and two-dimensional hydraulic models, and on computing an effective pier width for pier scour calculations based on the estimated debris dimensions. When the potential for debris accumulation on the bridge is significant, the hydraulic engineer should be prepared to provide the bridge designer with the estimated dimensions and reference elevation of the potential debris blockage. The hydraulic engineer should also recommend an appropriate drag coefficient for the debris, based on NCHRP Report 445.

10.1.7 Effects of Ice

When ice accumulates at a bridge and forms a substantial ice jam, significant problems can develop. Some of the negative consequences include bridge scour and bank erosion, even during times of low streamflow. Ice jams also impart significant lateral forces on the bridge. Similar to debris blockages, ice jams magnify the stream pressure forces by increasing the surface area to which the stream pressure is applied. The upstream water surface elevation (and consequently the hydrostatic force) is affected by the inordinate amount of backwater that often accompanies ice jams. The elevation at which ice is expected to accumulate has a significant influence on the bridge stability calculations. Extensive discussion on evaluation of ice forces is provided in the LRFD Specifications.

The design team should perform site-specific research to assess whether ice jamming is a relevant concern. If it is a concern, the hydraulic engineer may be required to develop hydrologic and hydraulic information to assist the bridge designer in evaluating ice forces. It may be beneficial, for instance, to determine the months of the year when ice jamming is most likely to occur. Streamflow records would then be studied to assess the potential for flooding during the most likely ice jamming months, and to identify a streamflow rate that represents a reasonable yet conservative flow rate for assessing the potential elevation of an ice jam on the bridge. Field reconnaissance may reveal evidence of the elevation range within which ice jams typically form. The Transportation Association of Canada has published the "Guide to Bridge Hydraulics" (TAC 2004), which includes information on estimating the stage and thickness of ice jams. If needed, the hydraulic engineer can develop hydraulic model simulations of ice jam situations. The HEC-RAS program includes the capability to incorporate ice cover into its simulations.

Ice can exert other forces on a bridge besides the increase in stream pressure and hydrostatic force mentioned above. Large ice floes striking bridge piers can generate significant impact forces. Large sheets of ice can experience thermal expansion, generating lateral pressure on the bridge. Ice adhering to the bridge structure during water level increases can impart uplift forces. The hydraulic engineer should be prepared to assist the bridge designer in assessing the potential range of water levels associated with these forces.

10.1.8 Vessel Collision

When a bridge is to cross a navigable waterway, the design should consider the potential for impact forces from vessel collisions. Bridges should be designed, wherever practicable, to minimize the probability of a vessel impact. Advisable practices include providing appropriate vertical clearance above the water surface, keeping piers as far away from navigation channels as practicable, and avoiding the placement of piers near a bend in a navigation channel. Navigating large ships and barges can be very difficult, especially at bends, and especially in high-velocity waterways. Locating one or more bridge piers near a bend in a high-velocity waterway with barge or ship traffic dramatically increases the risk of a vessel impact.

After taking appropriate precautions in locating the bridge and bridge piers, it is still necessary to allow for some probability of vessel collision. The type of vessel to be considered depends upon the waterway being crossed and the typical boat traffic. The LRFD Specifications provide significant guidance on selecting an appropriate design vessel and assessing the probability of a vessel collision. The bridge designer typically evaluates more than one vessel collision scenario.

One potential scenario spelled out by the LRFD Specifications is a case of an empty barge breaking free from its mooring and hitting a bridge pier under peak 100-year flood conditions. The flood conditions include the presence of half of the long-term scour and half of the flood-specific scour at the time the vessel strikes the pier. The hydraulic engineer should inform the bridge designer about the peak 100-year flood velocity, flow direction, depth, water surface elevation and total scour to enable evaluation of the impact force. The required velocity is the local velocity impinging on the pier in question. It is usually appropriate to report the same velocity, flow direction and depth that were used in the scour calculations when providing information for vessel impact forces under flood conditions.

Another commonly considered case is a fully loaded vessel motoring along the navigation channel and errantly striking the bridge during typical waterway conditions. The LFRD Specifications state that the appropriate velocity and water surface for such a scenario are those associated with yearly mean conditions, combined with half of the estimated long-term scour depth. If streamflow records are available for the stream reach being crossed, the annual mean of the daily mean flow rates can be used to represent yearly mean conditions. In a tidal waterway it is more meaningful to select one or more specific tidal levels, such as mean high water, to represent typical waterway conditions.

Some bridge pier locations, for instance in the vicinity of seaports or major shipping channels, may be exposed to very large vessel impact forces that cannot readily be accommodated in the bridge structure design. In such cases it is common to incorporate separate structural dolphins, with or without fender racks, to prevent a bridge impact. Care should be taken in the design of dolphin installations to avoid aggravating the scour potential at the bridge piers they are protecting.

10.2 BACKWATER EFFECTS OF BRIDGE PIERS

Hydraulic drag at bridge piers is experienced by the bridge as a force that must be resisted by structural stability. It is experienced by the stream flow as resistance to flow that must be overcome by an increase in energy driving flow through the bridge waterway. The increase in energy takes the form of backwater. The total backwater upstream of a bridge is often dominated by the constriction associated with the road embankments and bridge abutments, with the piers making only a small contribution. The relatively small backwater contribution of piers, however, can be a significant factor in bridge design, especially in the context of highly restrictive, no-rise floodway regulations (see Chapter 2).

Most bridges crossing regulatory floodways require piers to be located within the floodway to keep the span lengths within a cost effective range and to avoid unreasonable superstructure depths. The placement of piers in the floodway, however, can lead to regulatory challenges since the piers are likely to cause some small amount of backwater. In such cases the hydraulic engineer and the design team should work together to develop a design for the spans and piers that satisfies the regulatory constraints without unacceptable cost impacts. Aligning the piers with the flow direction, using hydraulically streamlined pier geometries and elevating the low chord to a reasonable freeboard height above the 100-year flood elevation are best practices that should be incorporated to the extent feasible. In such cases the bridge hydraulics should be analyzed using a methodology that accounts for the hydraulic benefits of streamlined pier geometries. The momentum method within HEC-RAS, for instance (see Chapter 5) uses a pier drag coefficient that is a function of the pier geometry. Yarnell's equation also incorporates a pier shape factor.

The Texas Department of Transportation commissioned a research study to aid in the evaluation of the magnitude and nature of backwater associated with bridge piers. The researchers conducted physical modeling to correlate the pier drag coefficient and the relative backwater (backwater depth divided by flow depth) to the Froude number downstream of the pier, for a range of pier sizes and flow contraction ratios. The results of the study (Charbeneau and Holly 2001) led to a recommended equation for calculating the backwater effects of pier. The recommended equation follows the form of Yarnell's equation but incorporates modifications for improved correlation to the physical modeling results. In general it was found that the observed relative backwater depth was consistently less than Yarnell's equation would predict.

Another analysis strategy that can be useful in dealing with no-rise floodway regulations is to perform a simulation that includes only the bridge elements that are actually located within the floodway, excluding elements of the crossing that are outside the floodway. This simulation allows the hydraulic engineer to isolate the impacts caused by work in the floodway. Only work in the floodway is regulated to the no-rise standard per FEMA regulations, though local ordinances may regulate to a no-rise standard outside the FEMA regulatory floodway.

10.3 COINCIDENT FLOWS AT CONFLUENCES

10.3.1 Significance of Coincident Flows at Confluences

When a bridge over a stream is located near a confluence with another stream, the engineer must consider the potential influence of the other stream on the hydraulics at the crossing. Questions to consider include:

- If the bridge is upstream of the confluence: How will the other stream affect the water surface profile through the bridge waterway for various flood recurrence intervals?
- If the bridge is within or very near the floodplain confluence zone: How will the interaction between the flows from the two streams affect the distribution and direction of flow throughout the confluence area?

In order to appropriately consider the effects of the confluence, it is necessary to estimate the coincident flow probabilities of the two streams. Consider a bridge crossing a minor tributary stream a short distance upstream from a major river, as illustrated in Figure 10.7. The tributary would likely have a much smaller contributing drainage area than the river. Major flooding on the river may be driven by different factors than those that cause major flooding on the tributary. Floods on large rivers, for instance, are often driven by spring runoff supplemented by long-duration spring rainfall, and may last for weeks. Floods on smaller tributaries are often driven by intense thunderstorms at other times of year, and typically last a few hours.

Figure 10.7. Illustration of a confluence situation.

The engineer will intuitively recognize that the 100-year flood on a small tributary is not likely to coincide with 100-year flood levels on a large receiving river. It is necessary, however, to determine the possible combinations of flow in the tributary and receiving river that have a 100-year recurrence interval. Overestimating the coincident-probability discharge rates will lead to the engineer adopting a design condition with a greater recurrence interval than intended. Underestimating the values could lead to negative consequences, such as a bridge low chord profile that provides less freeboard than intended for the design event.

10.3.2 Available Guidance

Available guidance on coincident flow frequencies at confluences has been scarce. FHWA document HEC-22 "Urban Drainage Design Manual," (FHWA 2009b) provides guidance on coincident flow frequencies for the design of storm drain outfalls into rivers and streams. The guidance takes the form of a table indicating the relative flood frequencies of the tributary and main stream as a function of the ratio of the contributing drainage areas. For example, if the drainage area of the main stream is 100 times the drainage area of the tributary, the engineer would determine the outfall tailwater based on a 25-year flood profile on the main stream for a 100-year storm drain design. The table in HEC-22 provides a convenient format. The basis of the table, however, is not well documented but was apparently based on data for a limited number of watersheds in a specific geographic location in coastal Virginia.

NCHRP conducted a research project (Project 15-36) NCHRP (2010c) with the objective of developing practical, reliable procedures for estimating coincident flow probabilities at confluences. The work has been completed and at present publication is pending. The resulting report, once published, is expected to provide significantly improved guidance on handling the issue of coincident flows at confluences.

10.4 ADVANCED BRIDGE HYDRAULICS MODELING

10.4.1 Background

One-dimensional and two-dimensional hydraulic analysis techniques are sufficiently rigorous for the needs of most bridge design projects. Occasionally, however, a project calls for more advanced hydraulic modeling techniques, such as physical modeling or computational fluid dynamics (CFD) modeling, which do not require the simplifications inherent in one- and two-dimensional modeling.

Physical hydraulic modeling refers to simulations conducted in a geometrically-scaled physical representation of the bridge or, more often, an element or section of the bridge along with the surrounding channel or waterway. CFD modeling refers to a highly detailed three-dimensional mathematical representation of the bridge element and waterway. Both techniques allow for the investigation of flow patterns and hydrodynamic phenomena at a degree of resolution, detail and rigor that is not readily obtainable with one-dimensional or two-dimensional analysis.

10.4.2 Applications of Advanced Modeling

Physical Modeling. Numerous physical model studies have been conducted to assess scour potential in situations involving large piers with complex geometry. Physical modeling has also been used to evaluate vertical contraction scour as illustrated in Figure 10.8. HEC-18 (FHWA 2012b) provides recommended equations for estimating the scour potential at complex piers, but their range of reliable application does not cover the full range of possible complex configurations. Physical modeling, therefore, is sometimes used to enhance the reliability of the scour estimates. Physical modeling provides the benefit of demonstrating, resolving and displaying the complex flow behavior without reliance on numerical formulations that are, of necessity, only approximate representations of the real physical conditions. Physical modeling also allows a more detailed understanding of the geometric configuration of scour around the pier.

Physical modeling for scour investigations is conducted in moveable-bed flumes. The flumes are constructed as geometrically scaled models of the prototypes they represent. Most of the limitations of physical modeling stem from the challenge of scaling the hydraulic conditions from the prototype to the model. Hydraulic scaling for bridge hydraulics applications is usually based on the Froude number, meaning that the Froude number in the model is set to equal the Froude number in the prototype under design conditions. Even with Froude number scaling, challenges can arise which are described later in this section. To support physical modeling, two-dimensional computer modeling is often conducted to determine the velocity magnitude and direction and the depth of flow at each pier in the prototype for the design flow conditions.

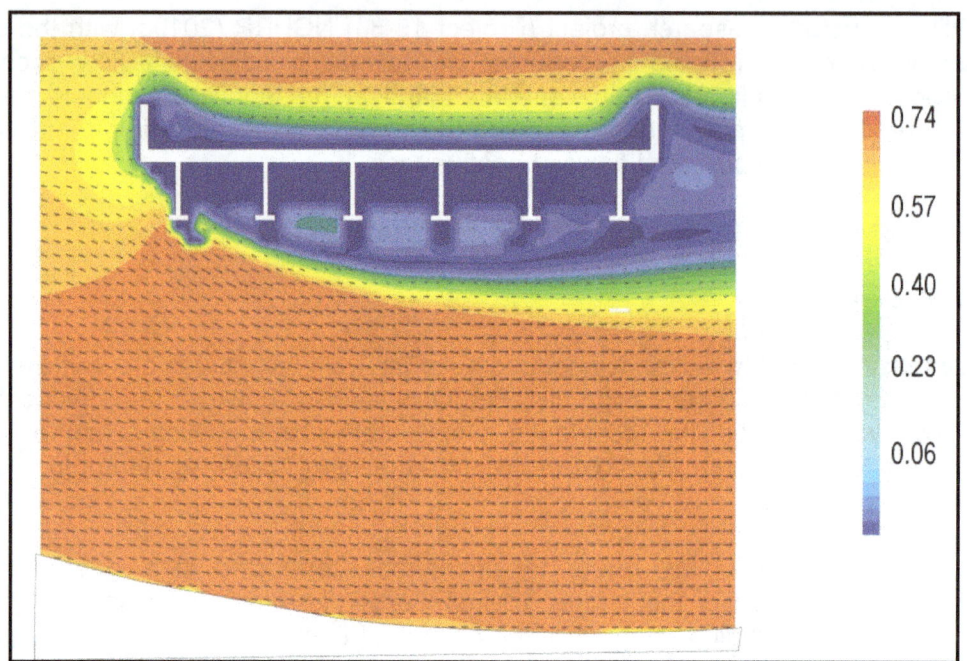

Figure 10.8. Velocity from physical modeling using Particle Image Velocimetry (PIV).

CFD Modeling. CFD modeling employs numerical methods to solve the Navier-Stokes equation in analyzing detailed three-dimensional fluid flow patterns for a wide range of applications, from stream hydraulics to aircraft design to medical studies of flow through blood vessels. CFD modeling is able to resolve complex near-field flow patterns, such as the vortices in the vicinity of flow obstructions, provided the grid cells of the model are properly sized and configured. Figure 10.9 is an example of CFD modeling of a submerged bridge deck. The general applicability of CFD to bridge hydraulics is, to date, somewhat limited. When CFD is applied to bridge hydraulics, it is usually directed toward local scour prediction or the analysis of hydraulic forces on bridge piers and superstructures.

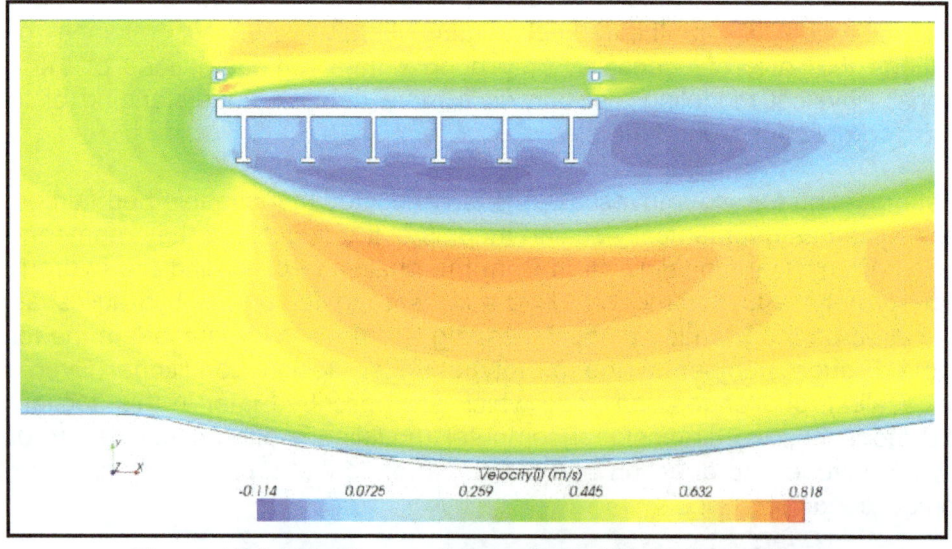

Figure 10.9 Velocity result from CFD (RANS) modeling.

<u>Hybrid Approach</u>. Compared to the convenience offered by modern computational facility, physical modeling has become a much more expensive, time-consuming, and inflexible option for investigating a large variety of bridge hydraulics subjects. However, there are specific aspects in CFD that requires input from physical modeling. Examples of such needs include turbulence models, scaling effect, roughness simulation, and sediment transport. It is, in many occasions, most efficient to have a relatively small number of high-precision physical experiments that spans through the range of important parameters and their critical values, and use the result to set up or calibrate corresponding CFD models. Subsequently, these CFD models can be used to conduct more detailed parametric study at a modest cost.

Figure 10.10 shows an example of such process. In an investigation on rectangular pier scour, a physical model of the pier was set up in a hydraulic flume that simulates specified flow velocity, flow depth, and bed material (a). The shape of the equilibrium scour hole was surveyed using a laser distance scanning system (b). A CFD model was established based on the bathymetry from the survey (c). This model was used in a series of LES simulation to study the horseshoe vortex systems (d). This process eliminated the need of a sophisticated sediment entrainment model in CFD by using the scour data from physical modeling, and therefore allowed computational effort to be focused on the behavior of the horseshoe vortices.

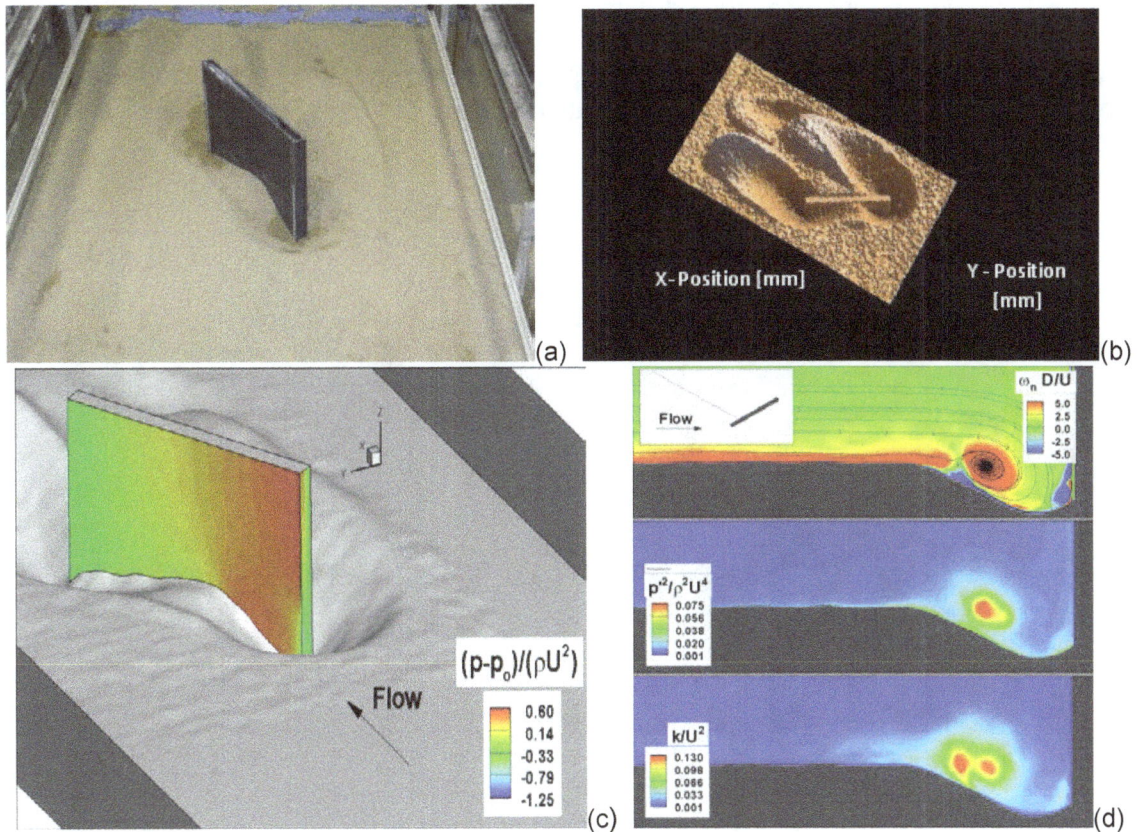

Figure 10.10. Hybrid modeling of pier scour.

10.4.3 Example of Advanced Modeling Applied to Bridge Design Projects

The Woodrow Wilson Bridge carries I-95 over the Potomac River between Washington, D.C. and Alexandria, Virginia. The original structure was opened to traffic in 1961. Dramatic growth in the traffic volume required a replacement of the original bridge. Construction of the new bridge was completed in 2008 at a cost exceeding $2 billion. The engineering effort for the design of the replacement bridge included extensive investigation of the scour potential using state-of-the-art techniques. The large investment in substructure foundations for the new bridge justified significant effort to refine the scour estimates. The advanced techniques employed in the scour evaluation included:

- Two-dimensional hydraulic modeling of the river and floodplain encompassing a reach of roughly 6 miles, to provide local velocities and angles of attack at pier locations
- Analysis of scour in cohesive materials
- Three-dimensional CFD modeling of flow and scour at complex piers with large dolphins
- Large-scale and small-scale physical modeling of scour at complex piers with large dolphins and fenders

The CFD modeling was conducted using an enhanced version of the CCHE3-D software. The modeling combined 3-dimensional flow dynamics and fully coupled sediment transport (Dou et al. 2001). The work focused on scour simulations at a limited number of channel piers, and the model domain for the simulation of each pier was limited to the near vicinity of the pier, as shown in Figure 10.11.

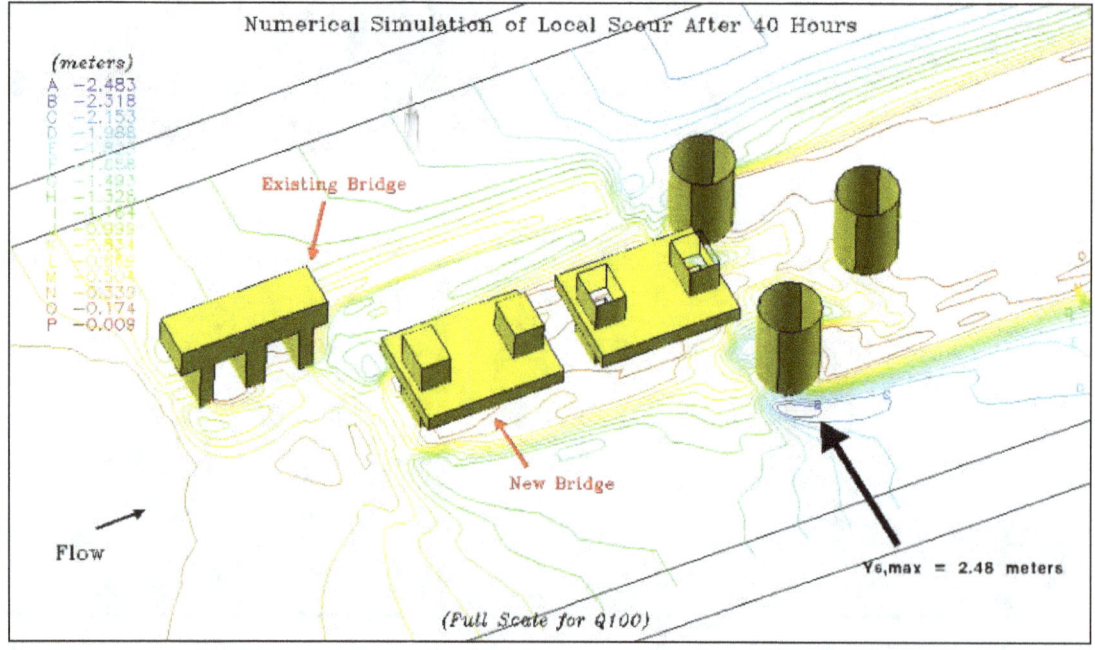

Figure 10.11. Illustration of CFD modeling of Woodrow Wilson Bridge pier and dolphins (from Dou et al. 2001).

The large-scale tests at the Turner Fairbank Highway Research Center were much more costly to conduct. Consequently only four large-scale experiments were performed. The large-scale tests had model-to-prototype scale ratios of 1 to 28 for one test and 1 to 50 for the other three. The purpose of the large-scale models was to investigate the scale effects by comparison with small-scale models of the same conditions. The comparison between large- and small-scale models showed enough similarity to provide confidence in the use of the small-scale tests to predict the scour at the piers (Jones 2000). The physical models provided significant value in comparing the effects of different design options. They showed, for example, that the use of three 45-foot diameter dolphins to protect the bascule piers from vessel collision could double the scour potential at the piers, while the use of an alternate fender ring could actually reduce the scour potential. Figure 10.12 is a photograph of one of the experiments at the J. Sterling Jones Hydraulics Research Laboratory.

Figure 10.12. Photograph of the post-scour condition of a small-scale physical model test of a Woodrow Wilson Bridge pier.

10.4.4 Limitations of Advanced Modeling

Physical modeling and CFD provide the benefit of detailed hydraulic analysis without the simplifications that are required to analyze the problem with one- or two-dimensional models. Certain practical and technical challenges exist, however, that have thus far prevented widespread application of advanced modeling techniques to bridge design projects.

Limitations of Physical Models. The practical limitations of physical models arise from the limited number of suitable facilities available for model testing and the relatively high cost of constructing and running the experiments. The testing should be done at a facility with appropriate flumes, measurement equipment, water supply, and, most importantly, expert personnel. Such facilities are typically associated with universities and/or government agencies, although some are owned by private interests. Hydraulic laboratory facilities typically maintain a significant backlog with respect to both flume space and personnel. If physical modeling is to be employed for a bridge design project, therefore, the arrangements must be made with significant lead time before the bridge design must be complete. Physical model studies can be highly labor intensive, which corresponds to significant cost. Personnel are needed to fabricate model elements, install them in the flumes, install and calibrate measurement devices, run the experiments, record the results, analyze the results, and refine the experiments as necessary.

Technical challenges in physical modeling are typically related to scaling between the model and the prototype. The most common type of scaling for open-channel hydraulic studies is Froude-based scaling, which generally means that the geometric configuration is scaled down by some uniform scale ratio, but hydraulic properties have different scale ratios such that the Froude number in the model is the same as in the prototype. Froude scaling is a reasonably straightforward way to establish similitude in models of open-channel flow. Unfortunately, the scaling of sediment sizes and sediment transport in a physical model are not straightforward for Froude scaling. Recognizing this limitation, physical model studies for scour evaluation often use a bed material size that has a critical velocity that is just less than the model velocity, rather than attempting to scale the sediment size from the prototype.

The scaling requirements of physical modeling lead to another limitation in bridge hydraulic modeling. Unless the depth to width ratio is distorted in the model, it is usually not feasible to physically model the entire bridge waterway and floodplain unless the available flume facility is uncommonly wide. Most physical model studies applied to bridge hydraulics, therefore, are designed to represent the flow around and adjacent to a specific bridge element, such as a pier or abutment. Supplemental two-dimensional computer modeling is often employed in order to apply the correct local velocity and flow direction in the physical model.

Limitations of CFD Models. CFD modeling is not yet in widespread use for bridge hydraulics. Practical limitations of CFD are associated with the required amount of computational resources and the limited availability of personnel qualified to develop and apply CFD models. The examples to date of CFD being applied to bridge hydraulics problems required the use of very powerful computers, which are not widely available. As a result of the computational intensity, most CFD studies applied to bridge hydraulics have focused on a specific local flow field, for instance at a pier, rather than attempting to model the entire bridge waterway. Therefore CFD is usually a supplement to, rather than a substitute for, one- or two-dimensional modeling. As with physical modeling, the number of personnel with expertise in CFD modeling, especially as applied to bridges, is relatively small.

The current technical limitations of CFD in bridge hydraulics, as with physical modeling, relate to sediment transport and scour processes. The Woodrow Wilson Bridge example cited above required the simplifying assumption of uniform sand, where the actual bed material was varied and included cohesive soil. Significant refinement is required to the computational algorithms of CFD models if they are to be validated for direct use in predicting scour depths.

10.5 BRIDGE DECK DRAINAGE DESIGN

10.5.1 Objectives of Bridge Deck Drainage Design

The design of a bridge should include consideration of bridge deck drainage in order to protect public safety, support efficient traffic flow and prevent or minimize water related damage to the bridge. Relevant design measures include the use of appropriate cross slopes and longitudinal slopes on the bridge deck, along with hardware such as inlets, scuppers, and drainage pipes. While the concerns and design approaches are comparable to roadway pavement drainage design, significant differences exist because of the physical and geometric constraints of installing a drainage system on a bridge.

FHWA document HEC-21 "Design of Bridge Deck Drainage," (FHWA 1993) provides extensive guidance on the design of deck drainage systems. This section briefly summarizes the design considerations for bridge deck drainage, drawing heavily from HEC-21.

10.5.2 Bridge Deck Drainage Considerations

<u>Minimizing Spread Width and Flow Depth</u>. Runoff flow spreading into traffic lanes on a bridge deck causes safety risks and reduces levels of traffic service. The flow encroaching into traffic lanes, if deep enough, can cause hydroplaning, an extremely hazardous condition in which a film of water separates vehicle tires from the road surface. The spread width and depth of flow are both functions of the runoff discharge, the shoulder width, the cross slope of the deck and the longitudinal grade of the bridge. Figure 10.13 is a cross-section sketch illustrating the concept of spread width.

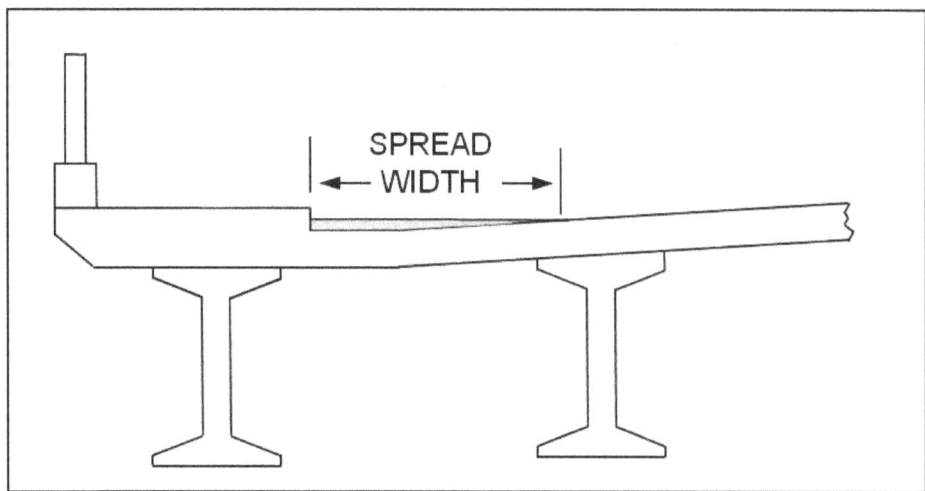

Figure 10.13. Sketch illustrating spread width of bridge deck drainage.

Excessive spread width and depth can be mitigated by removing all or a portion of the runoff from the deck surface. Various types of inlets or scuppers can be used to remove runoff. Multiple inlets may be needed to keep the spread width and depth below objectionable amounts. Shorter bridges or those on a steeper slope may not require inlets anywhere along the deck to achieve acceptable spread width. HEC-21 (FHWA 1993) describes a methodology and provides equations to assist in determining the inlet spacing requirements for a bridge.

Most transportation agencies have established design criteria regarding the acceptable spread width associated with the design rainfall event. The route classification, traffic flow and design speed are factors in setting the spread width criteria. A typical requirement for high-speed, high-volume routes is that the spread width caused by the design rainfall event may not encroach beyond the shoulder into a traffic lane.

<u>Superelevation Transitions</u>. Superelevation transitions on a bridge deck can be problematic for drainage design. Potential issues that can arise include:

- Flow switching from the gutter on one side of the road to the other side
- A sag in the gutter profile, causing water to pond
- A locally flattened cross slope allowing excessively wide flow spread

If a superelevation transition cannot be avoided, the engineer should consider mitigating potential problems by placing one or more inlets just upslope of the beginning of the transition.

Protecting Road Embankments at Bridge Ends. Erosion damage commonly occurs on the road embankment slopes adjacent to bridge ends because of inadequate control of bridge deck drainage. The problem can be minimized by designing the drainage system to deliver the flow safely to the bottom of the embankment without erosion. HEC-21 advises intercepting gutter flow with roadway drainage inlets on the approaches at both ends of the bridge. The intercepted flow is typically either conveyed by pipes to the bottom of the embankment or is delivered through pipes to an existing storm drain system. Some flow can be allowed to bypass the specified roadway inlet if there is curbing of sufficient height and length to convey the bypassed flow to the next drainage inlet or erosion-protected outfall location, thus protecting the embankment from erosion damage.

Minimizing Drainage-Related Damage at Bridge Joints. Water seeping from the deck through bridge joints can cause corrosion damage to the girders, bearings and substructure. For that reason, some transportation agencies require inlets on the bridge deck to capture flow before it runs across the joint at the downslope end of the bridge, even if there are no other inlets on the deck. The interception capacity of a single bridge deck inlet is quite limited. Therefore the beneficial function of bridge deck inlets placed at the downslope end of a bridge deck is primarily to intercept nuisance flows such as runoff from minor rainfall events, snowmelt, and landscape watering, rather than to keep joints dry during high-intensity rainfall events.

10.5.3 Design Rainfall Intensity

The runoff flow rate that must be accommodated in bridge deck drainage design is directly related to the short-duration rainfall intensity, which is the expected temporal rate of rainfall over a brief period of time (usually 10 minutes or less). Transportation agencies typically link the criteria for acceptable spread width and protection of the embankments at the bridge ends to a standard design recurrence interval (frequency) for rainfall intensity. The 10-year rainfall intensity is commonly used as the design standard for moderate- to high-volume roads. HEC-22 (FHWA 2009b) provides guidance on selecting design rainfall frequency for deck drainage.

10.5.4 Practical Considerations in Design of Bridge Deck Inlets and Drainage Systems

Dimensional Limitations of Bridge Deck Inlets. The pavement drainage inlets used in roadway pavement drainage applications are generally unsuitable for bridge deck applications because they cannot be easily integrated into the structural dimensions of a bridge deck. Roadway pavement drainage typically drops through a long curb opening or gutter grate into a large concrete catch basin, from which it is discharged through a pipe into an outfall or a storm drain system.

Bridge deck inlets, by necessity, usually have a smaller footprint on the bridge deck surface. Large openings may cause extensive complications in the design and construction of deck reinforcement. Bridge deck inlets are typically rectangular or round cast iron grates that allow runoff to drop into shallow inlet chambers constructed of formed concrete, ductile iron or welded steel. HEC-21 provides illustrations of several common inlet configurations, and also explains the factors that affect the interception capacity of bridge deck inlets. Grates with bars parallel to the traffic direction are the most hydraulically efficient. Many new bridges and bridge widenings, however, are being designed to accommodate bicycle traffic. Such bridges require bicycle-safe grates, which have bars perpendicular to the traffic direction. Perpendicular-bar grates can be made more efficient with vane grates, which are tilted or curved with the top edges inclined in the upstream direction.

Handling Intercepted Runoff. From the shallow inlet chambers, the drainage is discharged either into a vertical scupper or an underdeck drainage system. A vertical scupper may discharge drainage water directly into the air under the bridge or may extend down to the ground along the height of a pier. In many situations the runoff cannot be discharged directly into receiving waters beneath the bridge due to storm water quality concerns or regulations. In such cases any water intercepted from the bridge deck must be conveyed in an underdeck drainage system to a point on the stream bank or shore, where the drainage can then be discharged to an underground storm drain system or to an appropriate storm water quality feature. Direct discharge of drainage into the air below the bridge can also be restricted for other reasons. Roads, railroads, and residential, commercial or industrial development beneath the bridge are examples of settings in which direct discharge from the bridge deck is unacceptable.

As a general rule, underdeck drainage systems are problematic to the construction, maintenance and aesthetics of bridges, and should be avoided unless they are required by the setting or by regulations. If they cannot be avoided, they should be kept as short as possible. Underdeck drainage pipe is usually ductile iron, PVC or fiberglass and is typically of smaller diameter than the conduit used in underground storm drains. Figure 10.14 is a photograph of an installed underdeck drainage system, constructed of fiberglass pipe.

Figure 10.14. Installed underdeck bridge drainage system.

Water Quality Impacts on Receiving Waters. When a bridge is to cross a wide waterway, the design of bridge deck drainage can be a significant challenge due to the need to avoid negatively impacting the water quality of the waterway being crossed. When environmentally sensitive waters are present, the state's water quality regulations will prohibit direct discharge of the bridge deck runoff into the stream beneath the bridge. In such cases the runoff must be captured and conveyed off of the bridge to an acceptable stormwater quality mitigation feature (termed a stormwater best-management practice, or BMP). The NCHRP report titled "Assessing the Impacts of Bridge Deck Runoff Contaminants in Receiving Waters" is a resource to aid in identifying, assessing and managing the water quality aspects of bridge deck runoff (NCHRP 2002).

Maintenance Considerations. Even under the best conditions, bridge deck inlets tend to become plugged by debris. To minimize the required maintenance effort and promote the efficiency of the bridge deck drainage system, inlets and under deck bridge drainage systems should be designed to keep debris at or above the bridge deck surface, and should be located in areas that are easy to reach and safe for maintenance crews to service. Inlets should be placed at the outer edge of the shoulder, and the shoulder should be as wide as feasible. Inlets should not be located within traffic lanes, unless current and projected traffic volumes are very low.

CHAPTER 11

LITERATURE CITED

Abad, J.D. and Garcia, M.H., 2006, Discussion of "Efficient Algorithm for Computing Einstein Integrals by Junke Guo and Pierre Y. Julien," Journal of Hydraulic Engineering, Vol. 130, No. 12, pp. 1198-1201, 2004), ASCE, 132(3), 332-334.

American Association of State Highway and Transportation Officials (AASHTO), 2005, "AASHTO Standard Specifications for Highway Bridges," 17th Edition.

American Association of State Highway and Transportation Officials (AASHTO), 2007, "AASHTO Hydraulic Drainage Guidelines," 4th Edition.

American Association of State Highway and Transportation Officials (AASHTO), 2008, "Guide Specifications for Bridges Vulnerable to Coastal Storms, Single User Digital Publication."

American Association of State Highway and Transportation Officials (AASHTO), 2010, "LRFD Bridge Design Specifications," Fifth Edition.

American Society of Civil Engineers (ASCE), 2008, "Sedimentation Engineering, Processes, Measurements, Modeling, and Practice," ASCE Manuals and Reports on Engineering Practice No. 110 (edited by Garcia, M.H.).

Aquaveo, 2011, "Surface-water Modeling System (SMS, Version 11.0), Aquaveo LLC, www.aquaveo.com.

Byrne, A.T., 1893, "A Treatise on Highway Construction," Textbook and Work of Reference on Location, Construction, or Maintenance of Roads, Streets, and Pavements, Second Edition, John Wiley & Sons, NY.

Charbeneau, R.J. and Holley, E.R., "Backwater Effects of Bridge Piers in Subcritical Flow," TxDOT, Research Report No. 0-1805-1.

Charbeneau, R.J., Klenzendorf, B., and Barrett, M.E., 2008, "Hydraulic Performance of Bridge Rails," Texas Department of Transportation, Research and Technology Implementation Office, (TX-09/0-5492-1).

Chaudhry, M.H., 2008, "Open-Channel Flow," Second Edition, Springer Science and Business Media, New York, NY.

Chow, V.T., 1959, "Open-Channel Hydraulics," McGraw-Hill Book Company, NY.

Colby, B.R., 1964, "Practical Computations of Bed-Material Discharge," ASCE Hydr. Div., Jour, Vol. 90, No. HY2.

Dou, X, S. Stein, and J.S. Jones, 2001, "Woodrow Wilson Bridge Replacement Scour Study: Using a 3-D Numerical Model, ASCE Conference Proceedings "World Water and Environmental Resources Congress 2001."

Einstein, H.A., 1950, "The Bed Load Function for Sediment Transportation in Open Channels," Technical Bulletin 1026, U.S. Department of Agriculture, Soil Conservation Service, Washington, D.C.

Federal Emergency Management Agency (FEMA), 2009, "Guidelines and Specifications for Flood Hazard Mapping Partners," Appendix C: Guidance for Riverine Flooding Analyses and Mapping.

Federal Emergency Management Agency (FEMA), 2010, "Executive Order 11988: Floodplain Management," http://www.fema.gov/plan/ehp/ehplaws/eo11988.shtm, Last Modified: Wednesday, 11-Aug-2010 14:15:30 EDT.

Federal Highway Administration (FHWA), 1978, "Hydraulics of Bridge Waterways," Hydraulic Design Series Number 1, Second Edition, U.S. Department of Transportation, Washington, D.C. (Bradley, J.N.)

Federal Highway Administration (FHWA), 1984a, "Navigation Channel Widths and Bends," FHWA Technical Advisory 5140.2.

Federal Highway Administration (FHWA), 1984b, "Guide for Selecting Manning's Roughness Coefficients for Channels and Flood Plains," FHWA-TS-84-204 (Arcement, G.J. and V.R. Schneider).

Federal Highway Administration (FHWA), 1986, "Bridge Waterways Analysis Model: Research Report," FHWA/RD-86/108, July.

Federal Highway Administration (FHWA), 1993, "Design of Bridge Deck Drainage," Hydraulic Engineering Circular 21 (HEC-21), Report No. FHWA-SA-92-010 (Young, G.K., S.E. Walker, F. Chang).

Federal Highway Administration (FHWA), 1998, "User's Manual for WSPRO--A Computer Model for Water Surface Profile Computers," Report No. FHWA-SA-98-080 (Arneson, L.A. and J.O. Shearman).

Federal Highway Administration (FHWA), 2001, "River Engineering For Highway Encroachments – Highways in the River Environment," FHWA Hydraulic Design Series Number 6, Federal Highway Administration, Washington, D.C. (Richardson, E.V., D.B. Simons, P.F. Lagasse).

Federal Highway Administration (FHWA), 2002, "Highway Hydrology," Hydraulic Design Series Number 2, Second Edition (HDS 2), Report No. FHWA-NHI-02-001 (McCuen, R.H., Johnson, P.A., Ragan, R.M.)

Federal Highway Administration (FHWA), 2003, "User's Manual for FESWMS FST2DH, Two-dimensional Depth-averaged Flow and Sediment Transport Model," Release 3, Report No. FHWA-RD-03-053, University of Kentucky Research Foundation, Lexington, KY (Froehlich, D.C.), 209 p.

Federal Highway Administration (FHWA), 2004, "Tidal Hydrology, Hydraulics and Scour at Bridges," HEC-25, 1st Edition (Zevenbergen, L.W., P.F. Lagasse, and B.L. Edge).

Federal Highway Administration (FHWA), 2005, "Debris Control Structures Evaluation and Countermeasures," Hydraulic Engineering Circular No. 9 (Bradley, J.B., D.L. Richards, C.D. Bahner).

Federal Highway Administration (FHWA), 2008, "Highways in the Coastal Environment," Hydraulic Engineering Circular 25, Second Edition (Douglass, S.L. and J. Krolak).

Federal Highway Administration (FHWA), 2009a, "Bridge Scour and Stream Instability Countermeasures Experience, Selection, and Design Guidance, Volumes 1 and 2," HEC-23, Third Edition, Publication No. FHWA-NHI-09-111 (Lagasse, P.F., P.E. Clopper, J.E. Pagán-Ortiz, L.W. Zevenbergen, L.A. Arneson, J.D. Schall, L.G. Girard).

Federal Highway Administration (FHWA), 2009b, "Urban Drainage Design Manual," Hydraulic Engineering Circular 22 (HEC-22), Third Edition, Report No. FHWA-NHI-10-009 (Brown, S.A., J.D. Schall, J.L. Morris, C.L. Doherty, S.M. Stein, J.C. Warner), 478 p.

Federal Highway Administration (FHWA), 2009c, "Hydrodynamic Forces on Inundated Bridge Decks," Report No. FHWA-HRT-09-028 (Kerenyi, K., T. Sofu, and J. Guo), 44 p.

Federal Highway Administration (FHWA), 2012a, "Stream Stability at Highway Structures," HEC-20 Fourth Edition, Report No. HIF-FHWA-12-004 (Lagasse, P.F., L.W. Zevenbergen, W.J. Spitz, L.A. Arneson).

Federal Highway Administration (FHWA), 2012b, "Evaluating Scour at Bridges," HEC-18, Fifth Edition, Report No. HIF-FHWA-12-003 (Arneson, L.A, L.W. Zevenbergen, P.F. Lagasse, P.E. Clopper).

Federal Highway Administration (FHWA), 2012c, "Submerged-Flow Bridge Scour under Clear-Water Condition," Federal Highway Administration, Report No. FHWA-HRT-12-034 (Suaznabar, O., H. Shan, Z. Xie, J. Shen, and K. Kerenyi).

Fread, D.L., 1974, "Numerical Properties of Implicit Four-Point Finite Difference Equations of Unsteady Flow, HRL-45, NOAA Technical Memorandum NWS HYDRO-18, Hydrologic Research Laboratory, National Weather Service, Silver Spring, MD.

Fread, D.L., 1981, Some Limitations of Contemporary Dam-Breach Flood Routing Models, Preprint 81-525: Annual Meeting of American Society of Civil Engineers, St. Louis, MO, 15 pp.

French, R.H., 1987, "Hydraulic Processes on Alluvial Fans," Elsevier Science Publishers B.V.

Hager, W.H., 1987, "Lateral Outflow over Side Weirs," Journal of Hydraulic Engineering, ASCE, Vol. 113, No. 4.

Henderson, F.M., 1966, "Open Channel Flow," Prentice-Hall Inc., Library of Congress card number: 66-10695.

Hicks, D.M., and Mason, P.D., 1991, "Roughness Characteristics of New Zealand Rivers, New Zealand Water Resources Survey.

Jarrett, Robert D., 1984, "Hydraulics of High-Gradient Streams," Journal of Hydraulic Engineering, ASCE, Vol. 110, No. 11, November.

Jin, M. and D.L. Fread, 1997, "Dynamic Flood Routing with Explicit and Implicit Numerical Solution Schemes," Journal of Hydraulic Engineering.

Jones, J.S., 2000, "Hydraulic Testing of Wilson Bridge Designs," Public Roads, Volume 63 No. 5, FHWA.

Julien, P.Y., 2010, "Erosion and Sedimentation," Second Edition, Cambridge University Press. Cambridge, UK.

Julien, P.Y. and G.J. Klaassen, 1995, "Sand-Dune Geometry of Large Rivers During Floods," American Society of Civil Engineers, Journal of Hydraulic Engineering, Vol. 121, No. 9.

Karim, F., 1999, "Bed-Form Geometry in Sand-Bed Flows," Journal of Hydraulic Engineering, Vol. 125, No. 12, December.

King, H.W., 1918, "Handbook of Hydraulics for the Solution of Hydraulic Problems," First Edition, McGraw-Hill, NY.

King, H.W. and Brater, E.F., 1963, "Handbook of Hydraulics," Fifth Edition, McGraw Hill Book Company, New York.

Lindsey, W.F., 1938, "Drag of Cylinders of Simple Shapes," National Advisory Committee on Aeronautics, 8 pp.

Liu, H.K., J.N. Bradley, and E.J. Plate, 1957, "Backwater Effects of Piers and Abutments," Colorado State University, Civil Engineering Section, Report CER57HKLI0, 364 pp.

McEnroe, B.M., 2006, "Downstream Effects of Culvert and Bridge Replacement," Kansas Department of Transportation, Final Report No. K-TRAN: KU-04-9, University of Kansas, Lawrence, KS.

McEnroe, B.M., 2007, "Sizing of Highway Culverts and Bridges: A Historical Review of Methods and Criteria," Kansas Department of Transportation, Report No. K-TRAN: KU-05-4, University of Kansas, Lawrence, KS.

Manning, R., 1889, "On the Flow of Water in Open Channels and Pipes," Transactions of the Institution of Civil Engineers of Ireland.

Meyer-Peter, E. and Muller, R., 1948, "Formulas for Bed-Load Transport," Proc. 3d Meeting IAHR, Stockholm, pp. 39-64.

Munson, B.R., Young, D.F., Okiishi, T.H., and Huebsch, W.W., 2010, "Fundamentals of Fluid Mechanics," John Wiley and Sons, Inc.

National Cooperative Highway Research Program (NCHRP), 2000, "Debris Forces on Highway Bridges," NCHRP Report 445 (Parola, A.C., J.A. Colin, M.A. Jempson).

National Cooperative Highway Research Program (NCHRP), 2002, "Assessing the Impacts of Bridge Deck Runoff Contaminants in Receiving Waters, Volume 1," NCHRP Report 474, Transportation Research Board, Washington, D.C. (Dupuis, T.V.).

National Cooperative Highway Research Program (NCHRP), 2006, "Criteria for Selecting Hydraulic Models,) Web-Only Document 106, Transportation Research Board (Gosselin, M., D.M. Sheppard, S. McLemore).

National Cooperative Highway Research Program (NCHRP), 2010a, "Estimation of Scour Depth at Bridge Abutments," Draft Final Report, NCHRP 24-20, Transportation Research Board (Ettema, R., T. Nakato, and M. Muste).

National Cooperative Highway Research Program (NCHRP), 2010b, "Effects of Debris on Bridge Pier Scour," NCHRP Report 653, Transportation Research Board, Washington, D.C. (Lagasse, P.F., P.E. Clopper, L.W. Zevenbergen, W.J. Spitz, L.G. Girard).

National Cooperative Highway Research Program (NCHRP), 2010c, "Estimating Joint Probabilities of Design Coincident Flows at Stream Confluences," NCHRP Project 15-36 (Kilgore, R.T., D.B. Thompson, and D. Ford).

National Cooperative Highway Research Program (NCHRP), 2011a, "Evaluation of Bridge Pier Scour Research: Scour Processes and Prediction," NCHRP Project 24-27(01), Transportation Research Board, National Academy of Science, Washington, D.C. (Ettema, R., Constantinescu, G., and B.W. Melville).

National Cooperative Highway Research Program (NCHRP), 2011b, "Evaluation of Bridge-Scour Research: Abutment and Contraction Scour Processes and Prediction," NCHRP 24-27(02), Transportation Research Board (Sturm, T., R. Ettema, B. Melville).

National Research Council (NRC), 1996, "Alluvial Fan Flooding," National Academy Press, Washington, D.C.

National Research Council (NRC), 2009, "Mapping the Zone, Improving Flood Map Accuracy," The National Academies Press, Washington, D.C.

NRCS, 1963, "Guide for Selecting Roughness Coefficient," Values for Channels, Soil Conservation Service (currently the Natural Resource Conservation Service) (Fasken, G.B.).

O'Brien, J.S., 2009, "FLO-2D Reference Manual," Version 2009.

Reed, J.R. and A.J. Wolfkill, 1976, "Evaluation of Friction Slope Models," River 76, Symposium on Inland Waterways for Navigation Flood Control and Water Diversions, Colorado State University, CO.

Rouse, H., 1937, "Modern Conceptions of the Mechanics of Fluid Turbulence," ASCE Trans., Vol. 102, Reston, VA.

Schumm, S.A. and P.F. Lagasse, 1998, "Alluvial Fan Dynamics - Hazards to Highways," ASCE Water Resources Engineering 98, Proceedings of the International Water Resource Engineering Conference, Memphis, TN.

Schwartz, M.L., 2005, "Numerical Modeling," Encyclopedia of Coastal Science, (Schwartz, M.L., (Ed.)).

Shields, I.A., 1935, "Anwendung der Aenlichkeitsmechanik und der Turbulenzforschung auf die Geschiebebewegung," Berlin, Germany, translated to English by W.P. Ott and J.C. van Uchelen, California Institute of Technology, Pasadena, CA.

Simons, D.B. and F. Senturk, 1992, "Sediment Transport Technology," Water Resources Publications, Littleton, CO.

Transportation Association of Canada (TAC), 2004, "Guide to Bridge Hydraulics," Second Edition (Neill, C.R., (Ed.)).

U.S. Army Corps of Engineers (USACE), 1984, "Shore Protection Manual," Second Printing, Coastal Engineering Research Center.

U.S. Army Corps of Engineers (USACE), 1986a, "Accuracy of Computed Water Surface Profiles," Research Document No. 26, (Burnham, M. and D.W. Davis), 198 p.

U.S. Army Corps of Engineers (USACE), 1986, "Storm Surge Analysis," Engineer Manual, EM 1110-2-1412.

U.S. Army Corps of Engineers (USACE), 1990, "Hydraulic Design of Spillways, EM 1110-2-1603, US Army Corps of Engineers, Washington, D.C.

U.S. Army Corps of Engineers (USACE), 1992, "HEC-2 Water Surface Profiles Program," Technical Paper presented at the ASCE Water Forum 1992, Baltimore, MD.

U.S. Army Corps of Engineers (USACE), 1995, "Flow Transitions in Bridge Backwater Analysis," Research Document 42, Hydrologic Engineering Center, Davis, CA, September.

U.S. Army Corps of Engineers (USACE), 2001, "UNET One-Dimensional Unsteady Flow Through a Full Network of Open Channels, User's Manual," Report No. CPD-66, (Barkau, R.L.), 360 p.

U.S. Army Corps of Engineers (USACE), 2002, "SAM Hydraulic Design Package for Channels," Coastal and Hydraulics Laboratory, U.S. Army Engineer Research and Development Center, Vicksburg, MS (Thomas, M.A., E.R. Copeland, and D.N. McComas).

U.S. Army Corps of Engineers (USACE), 2008, "Coastal Engineering Manual," Publication Number EM 1110-2-1100.

U.S. Army Corps of Engineers (USACE), 2009, "Users Guide to RMA2 WES Version 4.5," (Donnell, B.P., J.V. Letter, Jr., Dr. W.H. McAnally, Jr., and W.A. Thomas).

U.S. Army Corps of Engineers (USACE), 2010a, "HEC-RAS, River Analysis System Applications Guide," Report No. CPD-70 (Warner, J.C., G.W. Brunner, B.C. Wolfe, and S.S. Piper).

U.S. Army Corps of Engineers (USACE), 2010b, "HEC-RAS, River Analysis System User's Manual," Version 4.1, Report No. CPD-68 (Brunner, G.W., CEIWR-HEC).

U.S. Army Corps of Engineers (USACE), 2010c, "HEC-RAS, River Analysis System Hydraulic Reference Manual," Report No. CPD-69 (Brunner, G.W.).

U.S. Bureau of Reclamation (USBR), 1987, "Design of Small Dams," United State Department of the Interior, Bureau of Reclamation, A Water Resources Technical Publication, Third Edition.

U.S. Geological Survey (USGS), 1956, "Estimating Hydraulic Roughness Coefficients," Agricultural Engineering, V. 37, No. 7, p. 473-475 (Cowan, W.L.).

U.S. Geological Survey (USGS), 1967, "Roughness Characteristics of Natural Channels," USGS Water Supply Paper 1849 (Barnes, H.H., Jr.).

U.S. Geological Survey (USGS), 1976, "Computer Applications for Step-backwater and Floodway Analysis," U.S. Geological Survey Open-File Report 76-499, 103 p.

U.S. Geological Survey (USGS), 1989, "Guide for Selecting Manning's Roughness Coefficients for Natural Channels and Flood Plains," Water Supply Paper 2339 (Arcement, Jr., G.J. and V.R. Schneider).

U.S. Geological Survey (USGS), 1997, "Resistance, Sediment Transport, and Bedform Geometry Relationships in Sand-Bed Channels," in: Proceedings of the U.S. Geological Survey (USGS) Sediment Workshop, February 4-7 (Bennett, J.P.).

U.S. Geological Survey (USGS), 2008, "StreamStats: A Water Resources Web Application" (Ries, III, K.G., J.D. Guthrie, A.H. Rea, P.A. Steeves, and D.W. Stewart).

Yang, C.T., 2003, "Sediment Transport: Theory and Practice," Krieger Publishing Company.

(page intentionally left blank)

APPENDIX A

Metric System, Conversion Factors, and Water Properties

APPENDIX A

Metric System, Conversion Factors, and Water Properties

The following information is summarized from the Federal Highway Administration, National Highway Institute (NHI) Course No. 12301, "Metric (SI) Training for Highway Agencies." For additional information, refer to the Participant Notebook for NHI Course No. 12301.

In SI there are seven base units, many derived units and two supplemental units (Table A.1). Base units uniquely describe a property requiring measurement. One of the most common units in civil engineering is length, with a base unit of meters in SI. Decimal multiples of meter include the kilometer (1000m), the centimeter (1m/100) and the millimeter (1m/1000). The second base unit relevant to highway applications is the kilogram, a measure of mass which is the inertial of an object. There is a subtle difference between mass and weight. In SI, mass is a base unit, while weight is a derived quantity related to mass and the acceleration of gravity, sometimes referred to as the force of gravity. In SI the unit of mass is the kilogram and the unit of weight/force is the newton. Table A.2 illustrates the relationship of mass and weight. The unit of time is the same in SI as in the U.S. Customary system (seconds). The measurement of temperature is Centigrade. The following equation converts Fahrenheit temperatures to Centigrade, $°C = 5/9 (°F - 32)$.

Derived units are formed by combining base units to express other characteristics. Common derived units in highway drainage engineering include area, volume, velocity, and density. Some derived units have special names (Table A.3).

Table A.4 provides useful conversion factors from U.S. Customary to SI units. The symbols used in this table for metric units, including the use of upper and lower case (e.g., kilometer is "km" and a newton is "N") are the standards that should be followed. Table A.5 provides the standard SI prefixes and their definitions.

Table A.6 provides physical properties of water at atmospheric pressure in SI system of units. Table A.7 gives the sediment grade scale and Table A.8 gives some common equivalent hydraulic units.

Table A.1. Overview of SI Units.

	Base Units	Units	Symbol
Base units	length	meter	m
	mass	kilogram	kg
	time	second	s
	temperature*	kelvin	K
	electrical current	ampere	A
	luminous intensity	candela	cd
	amount of material	mole	mol
Supplementary units	angles in the plane	radian	rad
	solid angles	steradian	sr

*Use degrees Celsius (°C), which has a more common usage than kelvin.

Table A.2. Relationship of Mass and Weight.

System	Mass	Weight or Force of Gravity	Force
U.S. Customary	slug, pound-mass	pound, pound-force	pound, pound-force
SI	kilogram	newton	newton

Table A.3. Derived Units With Special Names.			
Quantity	Name	Symbol	Expression
Frequency	hertz	Hz	s^{-1}
Force	newton	N	$kg \cdot m/s^2$
Pressure, stress	pascal	Pa	N/m^2
Energy, work, quantity of heat	joule	J	$N \cdot m$
Power, radiant flux	watt	W	J/s
Electric charge, quantity	coulomb	C	$A \cdot s$
Electric potential	volt	V	W/A
Capacitance	farad	F	C/V
Electric resistance	ohm	Ω	V/A
Electric conductance	siemens	S	A/V
Magnetic flux	weber	Wb	$V \cdot s$
Magnetic flux density	tesla	T	Wb/m^2
Inductance	henry	H	Wb/A
Luminous flux	lumen	lm	$cd \cdot sr$
Illuminance	lux	lx	lm/m^2

Table A.4. Useful Conversion Factors.

Quantity	From U.S. Customary Units	To Metric Units	Multiply by *
Length	mile	km	1.609
	yard	m	0.9144
	foot	m	<u>0.3048</u>
	inch	mm	<u>25.40</u>
Area	square mile	km^2	2.590
	acre	m^2	4047
	acre	hectare	0.4047
	square yard	m^2	0.8361
	square foot	m^2	0.09290
	square inch	mm^2	645.2
Volume	acre foot	m^3	1233
	cubic yard	m^3	0.7646
	cubic foot	m^3	0.02832
	cubic foot	L (1000 cm^3)	28.32
	100 board feet	m^3	0.2360
	gallon	L (1000 cm^3)	3.785
	cubic inch	cm^3	16.39
Mass	lb	kg	0.4536
	kip (1000 lb)	metric ton (1000 kg)	0.4536
Mass/unit length	plf	kg/m	1.488
Mass/unit area	psf	kg/m^2	4.882
Mass density	pcf	kg/m^3	16.02
Force	lb	N	4.448
	kip	kN	4.448
Force/unit length	plf	N/m	14.59
	klf	kN/m	14.59
Pressure, stress, modulus of elasticity	psf	Pa	47.88
	ksf	kPa	47.88
	psi	kPa	6.895
	ksi	MPa	6.895
Bending moment, torque	ft-lb	N · m	1.356
	ft-kip	kN · m	1.356
Moment of mass	lb · ft	m	0.1383
Moment of inertia	lb · ft^2	kg · m^2	0.04214
Second moment of area	in^4	mm^4	416200
Section modulus	in^3	mm^3	16390
Power	ton (refrig)	kW	3.517
	Btu/s	kW	1.054
	hp (electric)	W	745.7
	Btu/h	W	0.2931

*4 significant figures; underline denotes exact conversion

Table A.4. Useful Conversion Factors (continued).			
Quantity	From U.S. Customary Units	To Metric Units	Multiply by *
Volume rate of flow	ft^3/s	m^3/s	0.02832
	cfm	m^3/s	0.0004719
	cfm	L/s	0.4719
	mgd	m^3/s	0.0438
Velocity, speed	ft/s	m/s	<u>0.3048</u>
Acceleration	f/s^2	m/s^2	<u>0.3048</u>
Momentum	lb · ft/sec	kg · m/s	0.1383
Angular momentum	lb · ft^2/s	kg · m^2/s	0.04214
Plane angle	degree	rad	0.01745
	degree	mrad	17.45

*4 significant figures; underline denotes exact conversion

Table A.5. Prefixes.					
Submultiple Name	Submultiple Factor	Submultiple Symbol	Multiple Name	Multiple Factor	Multiple Symbol
deci	10^{-1}	d	deka	10^1	da
centi	10^{-2}	c	hecto	10^2	h
milli	10^{-3}	m	kilo	10^3	k
micro	10^{-6}	µ	mega	10^6	M
nano	10^{-9}	n	giga	10^9	G
pica	10^{-12}	p	tera	10^{12}	T
femto	10^{-15}	f	peta	10^{15}	P
atto	10^{-18}	a	exa	10^{18}	E
zepto	10^{-21}	z	zetta	10^{21}	Z
yocto	10^{-24}	y	yotto	10^{24}	Y

Table A.6. Physical Properties of Water at Atmospheric Pressure in SI Units.

Temperature		Density	Specific weight	Dynamic Viscosity	Kinematic Viscosity	Vapor Pressure	Surface Tension[1]	Bulk Modulus
Centigrade	Fahrenheit	kg/m^3	N/m^3	N·s/m^2	m^2/s	N/m^2 abs.	N/m	GN/m^2
0	32	1,000	9,810	1.79×10^{-3}	1.79×10^{-6}	611	0.0756	1.99
5	41	1,000	9,810	1.51×10^{-3}	1.51×10^{-6}	872	0.0749	2.05
10	50	1,000	9,810	1.31×10^{-3}	1.31×10^{-6}	1,230	0.0742	2.11
15	59	999	9,800	1.14×10^{-3}	1.14×10^{-6}	1,700	0.0735	2.16
20	68	996	9,790	1.00×10^{-3}	1.00×10^{-6}	2,340	0.0728	2.20
25	77	997	9,781	8.91×10^{-4}	8.94×10^{-7}	3,170	0.0720	2.23
30	86	996	9,771	7.97×10^{-4}	8.00×10^{-7}	4,250	0.0712	2.25
35	95	994	9,751	7.20×10^{-4}	7.24×10^{-7}	5,630	0.0704	2.27
40	104	992	9,732	8.53×10^{-4}	6.58×10^{-7}	7,380	0.0696	2.28
50	122	988	9,693	5.47×10^{-4}	5.53×10^{-7}	12,300	0.0679	
60	140	983	9,843	4.68×10^{-4}	4.74×10^{-7}	20,000	0.0662	
70	158	978	9,694	4.04×10^{-4}	4.13×10^{-7}	31,200	0.0644	
80	176	972	9,535	3.54×10^{-4}	3.64×10^{-7}	47,400	0.0626	
90	194	965	9,467	3.15×10^{-4}	3.26×10^{-7}	70,100	0.0607	
100	212	958	9,398	2.82×10^{-4}	2.94×10^{-7}	101,300	0.0589	

[1] Surface tension of water in contact with air

Table A.7. Physical Properties of Water at Atmospheric Pressure in U.S. Customary Units.

Temperature		Density	Specific Weight	Dynamic Viscosity lb-sec/ft^2 x 10^{-4}	Kinematic Viscosity ft^2/sec x 10^{-5}	Vapor Pressure	Surface Tension[1]	Bulk Modulus
Fahrenheit	Centigrade	Slugs/ft^3	lb/ft^3			lb/in^2	lb/ft	lb/in^2
32	0	1.940	62.416	0.374	1.93	0.09	0.00518	287,000
39.2	4.0	1.940	62.424					
40	4.4	1.940	62.423	0.323	1.67	0.12	0.00514	296,000
50	10.0	1.940	62.408	0.273	1.41	0.18	0.00508	305,000
60	15.6	1.939	62.366	0.235	1.21	0.26	0.00504	313,000
70	21.1	1.936	62.300	0.205	1.06	0.36	0.00497	319,000
80	26.7	1.934	62.217	0.180	0.929	0.51	0.00492	325,000
90	32.2	1.931	62.118	0.160	0.828	0.70	0.00486	329,000
100	37.8	1.927	61.998	0.143	0.741	0.95	0.00479	331,000
120	48.9	1.918	61.719	0.117	0.610	1.69	0.00466	332,000
140	60.0	1.908	61.386	0.0979	0.513	2.89		
160	71.1	1.896	61.006	0.0835	0.440	4.74		
180	82.2	1.883	60.586	0.0726	0.385	7.51		
200	93.3	1.869	60.135	0.0637	0.341	11.52		
212	100	1.847	59.843	0.0593	0.319	14.70		

[1] Surface tension of water in contact with air

Table A.8. Sediment Particles Grade Scale.

Size			Approximate Sieve Mesh Openings Per Inch		Class
Millimeters	Microns	Inches	Tyler	U.S. Standard	Name
4000-2000		160-80			Very large boulders
2000-1000		80-40			Large boulders
1000-500		40-20			Medium boulders
500-250		20-10			Small boulders
250-130		10-5			Large cobbles
130-64		5-2.5			Small cobbles
64-32		2.5-1.3			Very coarse gravel
32-16		1.3-0.6			Coarse gravel
16-8		0.6-0.3	2.5		Medium gravel
8-4		0.3-0.16	5	5	Fine gravel
4-2		0.16-0.08	9	10	Very fine gravel
2-1	2.00-1.00	2000-1000	16	18	Very coarse sand
1-1/2	1.00-0.50	1000-500	32	35	Coarse sand
1/2-1/4	0.50-0.25	500-250	60	60	Medium sand
1/4-1/8	0.25-0.125	250-125	115	120	Fine sand
1/8-1/16	0.125-0.062	125-62	250	230	Very fine sand
1/16-1/32	0062-0031	62-31			Coarse silt
1/32-1/64	0.031-0.016	31-16			Medium silt
1/64-1/128	0.016-0.008	16-8			Fine silt
1/128-1/256	0.008-0.004	8-4			Very fine silt
1/256-1/512	0.004-0.0020	4-2			Coarse clay
1/512-1/1024	0.0020-0.0010	2-1			Medium clay
1/1024-1/2048	0.0010-0.0005	1-0.5			Fine clay
1/2048-1/4096	0.0005-0.0002	0.5-0.24			Very fine clay

Table A.9. Common Equivalent Hydraulic Units.

Volume

Unit	cubic inch	liter	U.S. gallon	cubic foot	cubic yard	cubic meter	acre-foot	sec-foot-day
liter	61.02	1	0.264 2	0.035 31	0.001 31	0.001	810.6 E-9	408.7 E-9
U.S. gallon	231	3.785	1	0.133 7	0.004 95	0.003 79	3.068 E-6	1.547 E-6
cubic foot	1,728	28.32	7.481	1	0.037 04	0.028 32	22.96 E-6	11.57 E-6
cubic yard	46,660	764.6	202	27	1	0.746 60	619.8 E-6	312.5 E-6
meter^3	61,020	1,000	264.2	35.31	1.308	1	810.6 E-6	408.7 E-6
acre-foot	75.27 E+6	1,233,000	325,900	43,560	1,613	1,233	1	0.5042
sec-foot-day	149.3 E+6	2,447,000	646,400	86,400	3,200	2,447	1.983	1

Discharge (Flow Rate, Volume/Time)

Unit	gallon / minute	liter / second	acre-foot / day	foot^3 / second	million gallon / day	meter^3 / second
gallon / minute	1	0.063 09	0.004 419	0.002 228	0.001 440	63.09 E-06
liter / second	15.85	1	0.070 05	0.035 31	0.022 82	0.001
acre-foot / day	226.3	14.28	1	0.504 2	0.325 9	0.014 28
feet^3 / second	448.8	28.32	1.983	1	0.646 3	0.028 32
million gallon / day	694.4	43.81	3.068	1.547	1	0.043 82
meter^3 / second	15,850	1,000	70.04	35.31	22.82	1